COASTAL DESERTS
Their Natural and Human Environments

UNIVERSITY OF ARIZONA PRESS

Related Books on Arid Lands

Arid Lands in Perspective $18.00
William G. McGinnies and Bram J. Goldman, eds.

**Deserts of the World: An Appraisal of Research
Into Their Physical and Biological Environments** $18.00
William G. McGinnies, Bram J. Goldman, and Patricia Paylore, eds.

Food, Fiber, and the Arid Lands $24.00
William G. McGinnies, Bram J. Goldman, and Patricia Paylore, eds.

Polar Deserts and Modern Man (1974)
Terah L. Smiley and James H. Zumberge, eds.

Arid-Lands Research Institutions (1967): A World Directory (paperback) $5.00
Patricia Paylore, ed.

COASTAL DESERTS
Their Natural and Human Environments

Editors:
DAVID H. K. AMIRAN
ANDREW W. WILSON

THE UNIVERSITY OF ARIZONA PRESS
Tucson, Arizona

THE UNIVERSITY OF ARIZONA PRESS

Copyright © 1973
The Arizona Board of Regents
All Rights Reserved
Manufactured in the U.S.A.

I. S. B. N.-0-8165-0312-5
L. C. No. 73-76305

CONTENTS

Preface xi
Acknowledgments xiii

PART ONE
GENERAL CONSIDERATIONS

1. World Distribution of Coastal Deserts — 3
   — *Peveril Meigs*
2. Remote Sensing of the World's Arid Lands — 13
   — *Leonard W. Bowden*
3. Problems and Implications in the Development of Arid Lands — 25
   — *David H. K. Amiran*
4. The Larger Urban Centers of the Coastal Deserts — 33
   — *Andrew W. Wilson*
5. The Changing Role of Water in Arid Lands — 37
   — *Gilbert F. White*
6. Desalted Seawater for Agriculture: Is It Economical? — 45
   — *Marion Clawson, Hans H. Landsberg, and Lyle T. Alexander*
7. Plastic Oases for Arid Seashores — 55
   — *Carl N. Hodges, Merle H. Jensen, and Carle O. Hodge*
8. Offshore Desert Islands as Centers of Development — 63
   — *Homer Aschmann*
9. On the Causes of Aridity Along a Selected Group of Coasts — 67
   — *Paul E. Lydolph*

PART TWO
LATIN AMERICAN DESERTS

10. On the Origin of the Dry Climate in Northern South America and the Southern Caribbean — 75
    — *James F. Lahey*
11. New Evidence on the Climatic Controls Along the Peruvian Coast — 91
    — *Federico J. Prohaska*
12. Distinctive Hydrogeological Characteristics of Some Pampas of the Peruvian Coastal Region — 109
    — *Pierre Taltasse*
13. The Coastal Desert of Chile — 111
    — *Reynaldo Börgel O.*
14. A Climatic Profile of the North Chilean Desert at Latitude 20° South — 115
    — *César Caviedes L.*
15. New Directions in the Chilean North — 123
    — *Donald D. MacPhail and Harold E. Jackson*
16. The Fish-Meal Industry of Iquique — 137
    — *Rolando Salinas M.*

17. The Plio-Quaternary Climatic Changes Along the Semiarid Seaboard of Chile 147
    — Roland Paskoff
18. The Semiarid Coastal Region of Northeastern Brazil 153
    — Manuel Correia de Andrade

PART THREE
THE OLD WORLD DESERTS

19. Coastal Deserts of the Old World and Their Reclamation 157
    — Leonid E. Rodin
20. Climatic-Geomorphological Zones and Land Utilization in the Coastal Deserts of the North Sahara 159
    — Wolfgang Meckelein
21. The Crisis of the Saharan Oases 167
    — Jean Despois
22. Eilat: Seaside Town in the Desert of Israel 171
    — David H. K. Amiran
23. The Utilization of the Namib Desert, South West Africa 177
    — Richard F. Logan

PART FOUR
AUSTRALIAN DESERTS

24. Economic Development of the Australian Coastal Deserts 189
    — Joseph Gentilli

Index 205

ILLUSTRATIONS

WORLD DISTRIBUTION
1- 1. Climatic types in Arabia and Africa 4
1- 2. Climatic types in Baja California, Mexico, South Africa, and Australia 5
1- 3. Kathiawar-Makran coastal deserts 7
1- 4. Coastal deserts of the Persian Gulf area 7
1- 5. Coastal deserts of southeastern Arabia 7
1- 6. Coastal deserts around the Red Sea 8
1- 7. Coastal deserts of the Somali Republic 8
1- 8. Coastal deserts along the Mediterranean Sea 8
1- 9. Coastal deserts of the Nile Delta 9
1-10. Atlantic-Sahara coastal desert 9
1-11. Coastal deserts of southwestern Africa 9
1-12. Coastal deserts of western Australia 10
1-13. Coastal deserts of southern Australia 10
1-14. Coastal deserts of western Mexico and Baja California 11
1-15. Peruvian coastal deserts 11
1-16. The Atacama desert 12
1-17. Coastal deserts of Patagonia 12

REMOTE SENSING
2- 1. Characteristics of the electromagnetic spectrum 14
2- 2. Indian subcontinent photographed from Gemini XI 15
2- 3. Cloud-enshrouded Earth imaged from 23,000 miles 15
2- 4. The Nile Delta photographed from Gemini V 16

2– 5.	Gemini IX photo of the Peruvian coast	16
2– 6.	High-altitude photograph of Indio, California, area	17
2– 7.	Gemini V imagery of Salton Sea region, California	17
2– 8.	The Salton Sea region, from Apollo 9 color photograph	18
2– 9.	Land-use map prepared as a model from spacecraft imagery	18
2–10.	Photo of the Alamo Delta, Salton Sea	20
2–11.	Quarry in the boxed area in figure 2–10	20
2–12.	Multicamera images show small-scale photography	20
2–13.	Reading gathered at night over the Imperial Valley	21
2–14.	Thermal infrared image of the Alamo Delta	21
2–15.	Multi-sensor remote-sensing photo	22
2–16.	Visible-range photo showing same field patterns as figure 2–15	22
2–17.	Infrared and ultraviolet scans in the Coachella Valley	23
2–18.	Correlation of radar scatterometer return with land use and vegetation	24
2–19.	Microwave image of Salton Sea at 37,000 feet	24
2–20.	Microwave imagery between Imperial and Mexicali Valleys	24

PLASTIC OASES

7– 1.	"Package" approach at Puerto Peñasco	56
7– 2.	Air-inflated horticultural enclosures	57
7– 3.	Vegetables harvested from controlled environments	58

CAUSES OF ARIDITY

9– 1.	Precipitation along American, African, and Australian coasts	68
9– 2.	Precipitation along American, African, and Australian coasts	70

ORIGIN OF DRY CLIMATE

10– 1.	Northern coastal margin of South America and southern Caribbean Sea	75
10– 2.	Dry area along the northern coast of South America	76
10– 3.	Frequency of rain	76
10– 4.	Topography and rainfall along northern coast of South America	77
10– 5.	January rainfall: northern coast of South America	77
10– 6.	April rainfall: northern coast of South America	77
10– 7.	July rainfall: northern coast of South America	77
10– 8.	October rainfall: northern coast of South America	77
10– 9.	Rainfall frequencies for Netherlands West Indies	77
10–10.	Spring windflow over northern South America	78
10–11.	Summer windflow over northern South America	78
10–12.	Autumn windflow over northern South America	79
10–13.	Winter windflow over northern South America	79
10–14.	Spring windflow over northern South America	80
10–15.	Summer windflow over northern South America	80
10–16.	Autumn windflow over northern South America	81
10–17.	Winter windflow over northern South America	81
10–18.	Bellamy triangle network	81
10–19.	Windflow over northern coast of South America	82
10–20.	Spring windflow divergence over northern South America	82
10–21.	Summer windflow divergence over northern South America	82
10–22.	Autumn windflow divergence over northern South America	82
10–23.	Winter windflow divergence over northern South America	83
10–24.	January temperature and surface winds over southern Caribbean Sea	84
10–25.	April temperature and surface winds over southern Caribbean Sea	84

10–26.	July temperature and surface winds over southern Caribbean Sea	84
10–27.	October temperature and surface winds over southern Caribbean Sea	84
10–28.	Air-mass divergence over Caribbean Sea	84
10–29.	Direction of replacement of air masses moving out over the Caribbean Sea	85
10–30.	Clouds evaporating and dissipating during July	85
10–31.	Three-dimensional structure of waves	86
10–32.	Mean 500-millibar heights for Netherlands West Indies rainfall types	86
10–33.	Mean 500-millibar heights for Netherlands West Indies rainfall types	87
10–34.	Resultant broadscale windflow over Caribbean	88
10–35.	Mean 500-millibar heights for Netherlands West Indies rainfall types	89
10–36.	Mean 500-millibar heights for Netherlands West Indies rainfall types	89

CLIMATIC CONTROLS

11– 1.	Atmospheric pressure and temperature along west coast of the Americas	93
11– 2.	Annual wind roses at Lima	94
11– 3.	Temperature, dew point, and relative humidity over Lima	96
11– 4.	Winter inversion characteristics and temperatures at Lima	97
11– 5.	West coast of South America as seen by satellite	100
11– 6.	Summer inversion characteristics and temperatures at Lima	101
11– 7.	Relative humidity over Lima	102
11– 8.	West coast of South America as seen by satellite	103
11– 9.	Inversion, surface temperature, and cloud cover at Lima	105

CLIMATIC PROFILE

14– 1.	Climatic regions of the north Chilean desert	116
14– 2.	Isopleths of insolation for the north Chilean desert	117
14– 3.	Average temperatures in the north Chilean desert	118

THE CHILEAN NORTH

15– 1.	The Norte Grande population distribution	124
15– 2.	Interior basin and range topography of the Atacama	125
15– 3.	Terraces at the oasis of Aiquina on the Rio Salado, Antofagasta Province	125
15– 4.	The industrial district of Iquique	126
15– 5.	Vertical view of Arica and entrenched Azapa Valley	127
15– 6.	Port of Arica with famous landmark, El Morro	127
15– 7.	Great open pit mine of Chuquicamata	128
15– 8.	Geological exploration team on the Chilean Altiplano	129
15– 9.	Fishing fleet of the Port of Antofagasta	131
15–10.	Aerial view of coastal fog, the camanchaca	133
15–11.	Fog trap (captaniebla) near Antofagasta	133
15–12.	The llareta, a plant used as fuel	134

THE FISH-MEAL INDUSTRY

16– 1.	Fish-meal factories and fishing fleets along the Chilean coast	138
16– 2.	Hinterland of Iquique	139
16– 3.	Industrial section of Iquique	141

CONTENTS

16- 4. Fish-meal factory in the port of Iquique — 141
16- 5. Port of Iquique, showing business and fishing sections — 141

PLIO-QUATERNARY CLIMATIC CHANGES
17- 1. The semiarid part of Chile — 147
17- 2. Synoptic map of semiarid Chile — 150

DESERTS OF THE NORTH SAHARA
20- 1. Climatic-geomorphological areas along the Libyan desert — 160

EILAT: SEASIDE TOWN
22- 1. Port cities of Eilat (Israel) and Aqaba (Jordan) — 171
22- 2. Population growth of Eilat — 172
22- 3. Number of passengers on Tel Aviv–Eilat flights — 174

THE NAMIB DESERT
23- 1. The Namib Desert — 177
23- 2. Landforms of the Namib — 178
23- 3. Namib hydrography — 179
23- 4. Namib land allocations — 180
23- 5. Namib transportation routes — 184

AUSTRALIAN COASTAL DESERTS
24- 1. Arid areas of Australia in relation to rainfall — 189
24- 2. Rainfall subdivisions of northwestern Australia — 191
24- 3. Subregions of Australia, according to rainfall — 192
24- 4. Shark Bay salt ponds at Useless Loop — 193
24- 5. Subregions IV (Mardie to Cape Keraudren) and V (Cape Keraudren to Anna Plains) — 199
24- 6. The Port Hedland area — 202

TABLES

WORLD DISTRIBUTION
1- 1. Principal types of coastal deserts — 3

PROBLEMS AND IMPLICATIONS
3- 1. Increasing world population — 26
3- 2. Irrigation statistics and cultivated lands — 26

LARGER URBAN CENTERS
4- 1. Coastal desert cities over 100,000 — 33
4- 2. Comparison of population growth rates — 34
4- 3. Non-agricultural employment in San Diego and in United States — 35
4- 4. Population, employment, and income data for arid-land cities — 35

PLASTIC OASES
7- 1. Temperatures and condensation in a greenhouse at Puerto Peñasco — 58
7- 2. Growing and harvest periods at Puerto Peñasco — 59
7- 3. Yields from controlled environments in Puerto Peñasco — 59
7- 4. Yields for greenhouse cucumbers at Puerto Peñasco — 60
7- 5. Climatic data for Puerto Peñasco, Tucson, and Abu Dhabi — 60

ORIGIN OF DRY CLIMATE
10- 1. Windfields over northern South America — 80

CLIMATIC CONTROLS

11– 1.	Pressure differences and winds along Peruvian coast	95
11– 2.	Wind direction and speed over Lima	95
11– 3.	Air and sea temperatures at Lima	99
11– 4.	Precipitation profile of the western Andean slope	104

CLIMATIC PROFILE

14– 1.	Extreme temperatures at several Chilean stations	116
14– 2.	Relative humidity at several Chilean stations	117
14– 3.	Cloudiness in northern Chile	117
14– 4.	Rainfall in the piedmont oases	119
14– 5.	Rainfall at stations in the high Andes	119

THE FISH-MEAL INDUSTRY

16– 1.	Fish-meal factories and productive capacity in Chile	140
16– 2.	Fish-meal factories according to capacity in 1966	140
16– 3.	Fish-meal production in ports of Chile	140
16– 4.	Evolution of fish-meal exports	143
16– 5.	Evolution of fishing, according to tons caught	144

COASTAL DESERTS AND RECLAMATION

19– 1.	Climatic factors of coastal desert regions	157

DESERTS OF THE NORTH SAHARA

20– 1.	Natural factors in the North Sahara coastal desert area	164

AUSTRALIAN COASTAL DESERTS

24– 1.	Subregions of the coastal desert area	190
24– 2.	Agricultural activities in the Carnarvon area	195

PREFACE

Coastal deserts are commanding increasing interest after centuries of neglect for them and for all arid lands. In this continuously changing world, four changes presently are contributing to a new awareness of the potential for development of coastal deserts.

First, the world's population is increasing at a more rapid rate than ever before. Increasing population pressures in congested areas have led many persons to look more closely at the sparsely populated arid lands. Low population densities alone would not be sufficient to encourage the increasing interest in coastal deserts if it were not for advances in water technology that may change greatly the capacity of coastal deserts to support adequately more people.

The second change involves increasing availability of water. The main reason for the low-density occupance of deserts has been the scarcity of water, without which no settlement is possible, certainly not under the conditions of an arid climate. Although desert man has to his credit astonishing achievements in water-engineering over long periods of cultural development, it is our advanced, modern technology that makes water supply in arid lands today more of an economic than a technical problem. At least in theory, these presently developed technologies are applicable to all arid lands.

On the other hand, perfection still has to be achieved on the hoped-for technological breakthrough that would provide inexpensive desalinated seawater. It is awaited eagerly by people in countries with coastal deserts, in particular, as well as by mankind at large. This advance will revolutionize the utilization of coastal deserts. True, seawater already has been desalted, and a number of coastal stations and towns have been obtaining their water supply from the sea for some years. However, the cost of this water has not yet been brought down to a level for universal application. It has been possible only for an oil capital like Kuwait, for example. A city operating at normal levels of income, not to mention agricultural consumers, cannot afford this cost. Desalinated seawater at something like conventional water prices — which in the future are more likely to increase than remain the same — would revolutionize the whole development potential of coastal deserts, giving them a singular advantage over other areas. Unfortunately, even after the technology and economics of water desalination have been solved, prospects are not clear as to what may develop.

Progress in transportation technology is the third factor. Including air transportation and refrigerator ships, it makes coastal deserts easily accessible and their products exportable. Thus, new developments have been facilitated. The increasing demand in the world for resorts can be admirably met by those coastal deserts that have clear skies throughout most of the year. However, recreational uses are less attractive in those parts of the west coast deserts that have a high incidence of fog or low cloud, the result of cold currents or upwelling along these coasts. Furthermore, the cold water is another disadvantage, discouraging bathing off these desert coasts. For example, off the coast of Peru at 14.5° south latitude, surface water temperatures are 16° to 19°C (60° to 66°F) in summer and 14° to 15°C (57° to 59°F) in winter. The relatively few east coast deserts, marginal west coast deserts, and coastal deserts removed from the open ocean — such as the desert parts of the coastline of the Mediterranean Sea — have a definite advantage for recreational purposes.

While the cool waters of west coast deserts are poor for recreation, they are good for fishing. Furthermore, modern transportation has made it possible to utilize these rich fishing grounds, and the progress made in this field in recent years is impressive. Food and Agriculture Organization statistics for 1969 show that Peru, with a catch of 9.2 million tons of fish, is the world's first-ranking fishing nation once more. South Africa, with 1.3 million tons of fish, is another country with a prominent west coast desert. It ranked sixth among the world's fishing nations in 1969.

Although, for the present at least, the utilization of coastal deserts as the world's future "food factories" must be considered a doubtful and certainly not an immediate prospect, a great change in the form of urbanization and industrialization already is affecting certain of the coastal deserts. The major incentives stimulating this trend are the availability of ample unoccupied and, therefore, generally rather cheap land; the dry and often sunny climate reducing investment in plant structures, even permitting some stages of production or assemblage and of storage to be carried out in the open; easy accessibility due to the coastal location; and the attraction of a warm dry climate to the working population. The experience of California in the second third of the twentieth century well illustrates this last point.

The fourth world change pertains to mineral discoveries, which have been a major cause of development in some areas of the coastal deserts. Around the Persian Gulf (also known as the Arabian Gulf), and also in Libya, petroleum is of major importance. In Western Australia iron ore principally accounts for mining development, although other minerals also are important.

Some definite limitations are inherent in an arid environment, however, and more specifically in an arid climate. The aridity causes certain corrective self-regulating processes in nature to develop at a slow pace, as compared to such activity in humid areas. Arid nature, therefore, does not act correctively with the speed necessary to keep pace with industrial and urban developments, when these are developed under the same procedures as practiced in humid zone conditions. Air pollution, despoliation by urban and industrial waste, and water pollution and salinization may occur in arid lands to a degree affecting the environment not less than in humid areas, at comparable levels of human activity.

On the whole, natural resources of arid lands are rather limited in variety and amount, or, at least, not yet identified. Such limitations become particularly stringent with increases in demand, so that the process of decision-making among alternative uses becomes exceedingly difficult, for there is but little leeway for error. In coastal deserts in particular, where intensive development tends to be compressed into limited areas, the nature of the arid zone requires great care in decisions on multiple use. Industry, resort utilization, specialty agriculture, and fisheries easily can become conflicting and even mutually exclusive uses. One corrective against unsound or excessive development often is the high price of arid-zone products, as a result of both high production costs and high costs of transportation. This directs industrial and agricultural production to products for the local or regional market, or to those products commanding high unit prices.

To achieve a positive and balanced development under the confining conditions characteristic of coastal deserts requires expert planning and control. Damage to the environment can be done quickly, yet long periods of time are required for redress, if this is possible or practical at all.

Most coastal deserts, until quite recently, have been isolated from the mainstream of the world's technical and scientific development; thus few of them can muster locally the qualified manpower required to cope successfully with the problems inherent in their development. Educational, economic, and social problems at all levels, therefore, are an integral part of the arid-zone problem.

These problems carry particularly heavy impact in west coast deserts. Here, as a result of general climatic conditions, areas of full or even extreme aridity occupy a comparatively great part of the coastal deserts. An interesting exception is the case of Western Australia.

A Case Study in Peru

Field studies in Peru have presented some clear-cut examples of the problems involved in balancing present-day technological development with the environment of a coastal desert. These same studies, nevertheless, included an illustration of enlightened and progressive environmental management that brought about a well-adjusted and interrelated agricultural use of a coastal desert. The Mallares farm, managed for generations by the Romero family, provided this illustration on a 1967 basis.

Mallares is located close to the Rio Chira, a sizable river descending from the Andes and traversing the coastal desert of Peru at about 5° south latitude, south of the Talara district, Peru's major oilfield. In April 1967 the flow of the Chira amounted to some 150 cubic meters per second; however, as with most mountain and all desert rivers, the flow can be much larger, as in 1965 when the Chira discharged 5,000 cubic meters per second. Even at normal rates of discharge, most of the water goes to waste in the sea. The Chira water, flowing down a deep Andean valley, carries a high load of suspended matter, valuable as natural top dressing in cultivation.

The Mallares farm comprised some 6,000 hectares (14,800 acres), of which 2,500 hectares were cultivated. Over the years, adjoining or nearby properties were acquired for the sole purpose of obtaining water rights but without any intention of cultivating the land. This practice has numerous parallels in other arid countries.

The Mallares farm cultivated, in descending order of crop area, cotton, rice, oranges, maize, coconuts, avocados, mandarines, grapefruit, and mango; that is, most of its cultivated area was devoted to three field crops, the rest to fruit trees.

Cotton and rice were irrigated in basin irrigation, as this kills insect eggs and restricts to a minimum the need for spraying with insecticides. The application of insecticides was further restricted by biological controls. Dense rows of trees, mainly algarobbo, were planted to line blocs of fields. The trees housed birds, which assisted in controlling insect pests. Farm labor trained in distinguishing between harmful and useful insects checked the fields daily, and only in the rare cases when harmful insects proliferate was recourse taken to spraying. Obviously, the restriction in applying insecticides was desirable both from environmental and economical points of view. As a result, the rice fields — of the high yielding Surinam variety — made an excellent impression at Mallares.

In this area of the Peruvian coastal desert, all stages of cultivation are strictly regulated on a regional basis, at least in the case of cotton. No farmer may spray his fields unless authorization has been obtained from the district inspector of agriculture. Equally, farmers may not begin harvesting before, or delay beyond, the dates fixed by the inspector. A farmer is fined for leaving cotton in the fields after the last harvesting date. All these are measures to prevent the spread of cotton pests.

Beneath the fruit orchards no ground crop is grown, because at the tropical temperature levels prevailing here, pests on surface crops and those on tree crops would mutually interact, making an efficient pest-control program exceedingly difficult.

Cotton and rice both require heavy applications of water. Cotton gets 10,000 cubic meters per hectare in six applications of water; rice gets as much as 30,000 cubic meters per hectare, being kept permanently under water of course. The area cultivated in rice is limited according to the availability of water. Only by applying water at such high rates, together with meticulous drainage and strict crop rotation, is alkalinization of soils prevented, something much in evidence elsewhere in the area. In particular, drainage ditches at Mallares were dug to at least two meters below field level, thus draining off all the water and preventing salinization.

The Mallares farm illustrates an achievement in the severe circumstances of an extreme coastal desert. This achievement by highly intelligent and careful local management has resulted in a prosperous farm enterprise, integrating advanced agricultural techniques, biological pest control and minimum use of pesticides and artificial fertilizer, and first and foremost a strict irrigation-cum-drainage regime. Comprehensive drainage, starting simultaneously with irrigation, has prevented deterioration of the soil, a premium resource in arid areas.

Mallares farm does, however, use large amounts of water, an actuality that the farm manager considers regrettable. The need to do so is the result of local food habits, which require rice as the main staple. If, over the years, education could bring about the substitution for rice of a cereal less demanding in water — for example, maize — the water economy could be significantly improved. Such an achievement would increase the efficiency of utilization of a second premium resource of arid lands — water.

Unfortunately Mallares is rather the exception than the rule. Its particular significance is in demonstrating what can be achieved by balanced and intelligent management of the environment. The Mallares experience is rather singular, too, in comparison with urban and industrial environments in the South American coastal deserts, which are far from achieving the harmonious integration observable at Mallares.

Great efforts are being devoted to making desalination of seawater a practical proposition, applicable even to agricultural production. The achievement of this goal will require the solution, alongside the technological problems involved, of a host of problems in environmental management in order to achieve the same degree of integration with the stark nature of coastal deserts that is illustrated by the Mallares farm in Peru.

The complex of problems is specific to coastal deserts. The solution of technological problems, and the satisfactory achievement of management of the environment, will determine whether coastal deserts can be developed or whether they will deteriorate.

ACKNOWLEDGMENTS

The pressure of man and his industries on the environment of a coastal desert, the difficulties in properly assessing priorities with a concern for the limitations posed by the arid environment, and consequently the problems of making decisions regarding development in probably the world's most extreme coastal desert, were vividly brought home to a group of arid-zone scientists during a symposium on coastal deserts held in Peru in March and April 1967, in which the editors of the present volume participated.

The initiative for this symposium originated with the Commission on the Arid Zone of the International Geographical Union, following the publication of the *Geography of Coastal Deserts* (UNESCO, Arid Zone Research, vol. 28, 1966) by Peveril Meigs, Chairman of the Commission for many years. Peru was an obvious choice as a locale for a symposium on this topic. The symposium was sponsored jointly by the International Geographical Union, UNESCO, and the government of Peru, which acted as host through its Ministry of Agriculture. The three sponsors provided the necessary funds for the symposium. The program included a week of discussions at Lima, preceded by field study of the northern coastal desert between Piura and Tumbes, and followed by a field study of the southern coastal desert between Arequipa and Tacna. The approximately one hundred fifty participants in the symposium owe special thanks to the government of Peru, the then Minister of Agriculture, Dr. Javier Silva Ruete, and to Inginiero Raul Vallés Escardo, the energetic and inspiring chairman of the local symposium committee, and to all his colleagues. Thanks for assistance rendered are due furthermore to UNESCO, and especially to M. Batisse, and to the International Geographical Union and to Professor Hans Boesch, its secretary-general at the time.

The animated discussions during the symposium, and the work resulting from it, made the preparation and publication of this present volume desirable. Some of the chapters in the book evolved from papers presented at Lima; other chapters were prepared especially for it.

Thanks are extended to the translation office of the Hebrew University in Jerusalem, Israel, for English renditions of manuscripts written originally in French and Spanish. The Division of Economic and Business Research of the College of Business and Public Administration, University of Arizona, contributed the final typing of the manuscripts, which service is much appreciated.

Special thanks are due William G. McGinnies, Director Emeritus, Office of Arid Lands Studies, The University of Arizona, for his initiative in the publication of the volume. Thanks also are due Patricia Paylore and Kathryn Gloyd, Office of Arid Lands Studies, for their work in bibliographic verification and standardization of form, and for preparation of the Index.

Finally, appreciative thanks are extended to Marshall Townsend, Director, University of Arizona Press, for his assistance in the publication of the volume; to Elaine Nantkes, associate editor, who did so much to improve the readability of the book; to Douglas Peck, production manager, who is responsible for the high quality printing and design; and to their willing assistants who aided in bringing about the professional quality of the book.

D.H.K.A.
A.W.W.

PART ONE

GENERAL CONSIDERATIONS

CHAPTER 1

WORLD DISTRIBUTION OF COASTAL DESERTS

Peveril Meigs

U.S. Quartermaster Corps. (ret.)

The coastal deserts have received world interest since the cost of extracting fresh water from the sea has been reduced in the past few years. Today, in the early 1970s, there is no coastal desert that cannot have water if the economy justifies it (Meigs, 1967). The water may be desalted by solar radiation, by electrodialysis, by steam methods, or perhaps by freezing, depending on the climate, the size of the installation, or the availability of cheap fuel.

The future of the economy of the coastal deserts consists, above all, in minerals exploitation (particularly oil), fishing, harvesting of seaweed, sea trade, irrigation from streams that originate in a moister climate (desalination is not yet cheap enough to provide ordinary irrigation water), and the enjoyment of nature that can be summarized under the heading of tourism. Unused vast tracts of great beauty still remain along the coasts, and they should be preserved, perhaps as national parks, before industrialism has left its mark upon them. Sometimes a city that was founded for one of the preceding reasons could develop manufacturing if it had a supply of fresh water.

Profound environmental differences exist among the coastal deserts of the world — differences that significantly affect the potentiality for forms of economic development. On a broad worldwide basis, climate is the best criterion for subdivision. Climatically, the coastal deserts fall into four principal types, which may be called hot, warm, cool, and cold. The approximate limits of these types are given in table 1-1, based upon the temperatures of the warmest and coldest months. The types are mapped in figures 1-1 and 1-2, in part digested from maps previously published (Meigs, 1953).

The hot coastal deserts are the most extensive in the world. Their approximate length is 16,000 kilometers (10,000 miles). The warm coastal deserts are 7,700 kilometers (4,800 miles) in length; the cool coastal deserts 7,000 kilometers (4,400 miles); the cold coastal deserts 1,300 kilometers (800 miles).

The hot deserts are distinguished by extremely hot summers. The average temperature of the hottest month is above 30°C (86°F), which means that summer afternoons are usually above 32°C (90°F) and often above 38°C (100°F). Owing to the coastal location and tendency for breezes to blow from the sea during the daytime, high humidity accompanies the high temperature, resulting in uncomfortable and sometimes almost unbearable physiologic heat stress. The southern coasts of the Red Sea and of the Persian Gulf are notorious for their oppressive humid heat. The equatorward portion, which I have called the tropical subtype, is warm even in winter, with winter temperatures getting above 25°C (80°F) in the heat of the day and dropping only to 21°C (70°F) in the coolest part of the night. This hot tropical climate prevails only along the southern half of the Red Sea, and the adjacent coasts of Arabia and the Somaliland horn of Africa (fig. 1-1). The poleward or subtropical subtype has mean winter temperatures between about 10°C and 22°C (50° and 72°F), with daily minimums getting down even lower. This subtype prevails along the Persian Gulf and the Indian Ocean margins of Iran, Pakistan, and northwestern India; the northern Red Sea; the Gulf of California; and northwestern Australia.

Most of the hot coastal deserts have relatively little cloud or fog and, hence, are especially suitable for desalination processes employing solar radiation. An example of such utilization of a hot coastal desert is the sophisticated experimental multiple-effect pilot solar desalination plant of the University of Arizona–University of Sonora, tried for a period at Puerto Peñasco on the Gulf of California (Hodges, Groh, and Thompson, 1967). Hodges (1969) has written about a new and exciting experiment at Puerto Peñasco, which lends itself to encouraging agriculture in coastal deserts. Instead of depending on solar radiation for power to distill seawater, use is made of the waste heat given off by a diesel-electric plant. The crops are grown in inflated plastic greenhouses instead of in the open air, and seawater is introduced into the greenhouses in impervious canals. The humidity of the air is kept high by direct evaporation from the seawater, and enough of this moisture condenses on the plastic roof to water the plants, without the use of the

TABLE 1-1

Principal Types of Coastal Deserts

Type	Mean Monthly Temperature	
	For Warmest Month	For Coldest Month
1. Hot		
Tropical	> 30°C (86°F)	> 22°C (72°F)
Subtropical	> 30°C (86°F)	10°–22°C (50°–72°F)
2. Warm	22°–30°C (72°–86°F)	10°–22°C (50°–72°F)
3. Cool	< 22°C (72°F)	10°–22°C (50°–72°F)
4. Cold	< 22°C (72°F)	< 10°C (50°F)

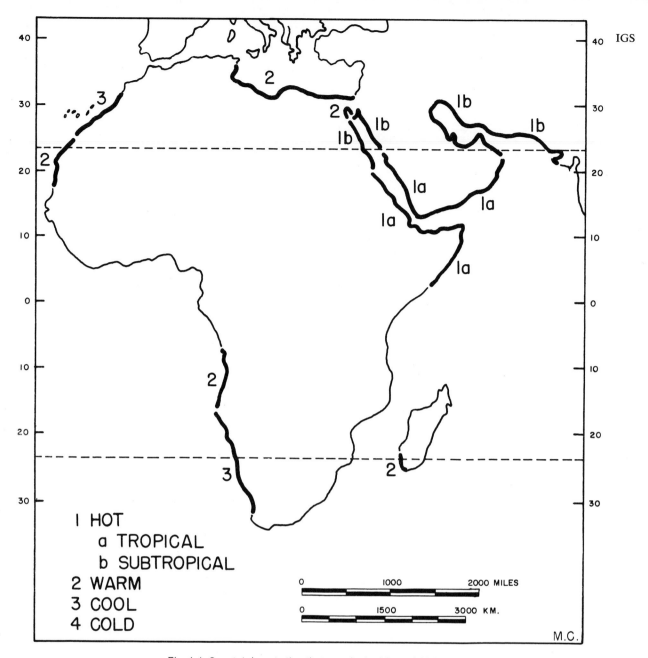

Fig. 1-1. Coastal desert climatic types in Arabia and Africa.

expensive distilled water from the power plant, except in the heat of summer, when insufficient water condenses in the greenhouses.

The hot coastal deserts have an advantage for steam, particularly along the Red Sea and Persian Gulf. Owing to restrictions at the mouths of these bodies of water and, in the case of the Red Sea, a shallow sill at the mouth, the water does not mix freely with the colder water of the open ocean, and as a result the water temperature is higher by several degrees. In summer the surface temperature of the water in the southern part of the Red Sea is above 30°C (86°F); in winter, over 25°C (77°F). The Persian Gulf is hotter than the Red Sea in summer, but much cooler than the Red Sea in winter.

The second type of coastal desert, I have called warm rather than hot because summers are less hot, with mean temperatures even of the warmest month being less than 30°C (86°F). The longest single stretch of this climate, about 2,600 kilometers (1,600 miles), is along the Mediterranean Sea, where the Sahara reaches the coast. Though occasional short rainstorms occur in winter, most of the days are bright, sunny, and mild at that season. The climate, plus the nearness to densely populated Europe, makes this one of the potentially important winter seaside resort areas of the coastal deserts.

Warm coastal desert climate exists also along the west coast and south end of Baja California, Mexico, with winter rainfall in the north, summer rainfall in the south, and nonseasonal very slight rainfall in the center. All the other west-coast deserts of the world, too, have warm climates in their equatorward portions, in Peru, Angola, Mauritania, and Australia. Like the southern end of Baja California, these areas all have their rainy season, such as it is, in summer, in harmony with their marginal tropical locations. A small but distinct area of this type of warm desert occupies the southwestern coast of Madagascar. Tourism is developing in some of these areas, notably the southern end of Baja California, where visitors from the United States and mainland Mexico come by air or sea. Near their equatorial ends, these west-coast warm deserts are relatively free from fog or cloud. Toward their polar ends, especially in Baja California, Peru, and southwestern Africa, there is frequent fog or low cloud in summer, and the warm deserts merge into the cool deserts.

A particular biologic and tourist resource of this part of Baja California is the gray whale, which comes thousands of miles from the North Pacific to breed in Scammon's Lagoon, in 27°45′ north latitude (Gardner, 1960; Scammon, 1970). It is an illustration of a natural resource that must be constantly protected against indiscriminate hunting. By 1861 the whalers practically had exterminated the whales of Scammon's Lagoon. Now that the whales have again increased, there has been some talk of exploiting them again, but so far the Mexican government has prevented it.

Type 3, the cool deserts, form the poleward portion of all five west-coast deserts. They occur nowhere else, if we consider the shores of the Great Bight of Australia as an extension of the west-coast desert. The longest stretches are along the Peru-Chile coast 3,400 kilometers (2,100 miles) and the west coast of South West Africa

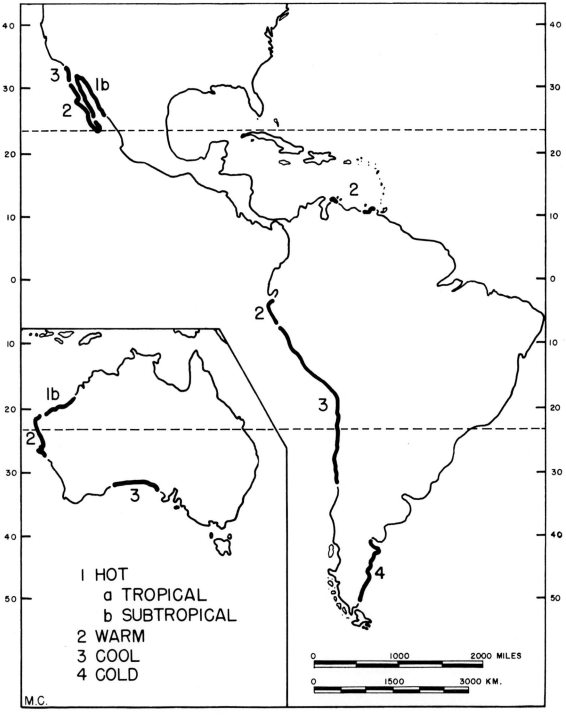

Fig. 1-2. Coastal desert climatic types in Baja California, Mexico, South Africa, and Australia.

and South Africa 2,000 kilometers (1,200 miles). Northern Spanish Sahara and southern Morocco have a smaller segment, and Baja California has a short section, just south of the California border (figs. 1-1 & 1-2). These are fog deserts, where fog forms over the cool water offshore and drifts onto the adjacent land either directly along the ground or as "high fog" or low cloud. To the cooling effect of the cold water offshore is added the reduction of solar radiation by the fog blanket. There are regional differences in the season of maximum fog. Along the Peru-Chile coast the season of greatest fog abundance is winter; along the Baja California and northwest African coasts there is a summer maximum of fog. The receipt of solar radiation also is affected by high clouds associated with the seasonal rainfall regimes of the neighboring humid areas: winter in most of the cool coastal deserts, but summer in the main part of the Peruvian coast. In any event, the cool deserts of the coast provide attractive relief for people from the hot interior deserts or valleys. The special circumstances of cool ocean water offshore has resulted in great fishing resources.

The coldest coastal desert of all, type 4 of the classification, occurs only along the southeast coast of Argentine - Patagonia (fig. 1-2). Here a cool ocean current reinforces the dry rain-shadow effect of the Andes on the westerly storms that normally prevail in these latitudes. This desert extends beyond 50° south latitude: farther poleward than any other coastal desert in the world. The mean temperature of the coldest month is below 10°C (50°F), and of the warmest month between 10° and 22°C (50° and 72°F). With such limited heat, only the hardiest crops can be raised in this region, and other resources must provide the principal economic basis for settlement.

We are presently considering only the connecting oceans and seas. The inland lakes and seas are not considered, not even the Caspian Sea. The Caspian Sea has a climate quite different from those of other coastal deserts. Most of its climate (except for the southwest borders) is continental: colder in winter than the coldest of the coastal deserts (type 4), averaging below freezing; and ranging with the warm deserts (type 2) in summer. Thus field crops could be raised in summer, but no frost-susceptible trees or shrubs. The sea affects the temperature of the land immediately adjacent to it, as for other coastal deserts.

Other forms of climatic classification might be drawn on a more sophisticated basis. For example, the climate suited to date-raising might be delimited on the basis of the temperatures of the warmest and coldest months. The trouble is that each crop would justify its own division. An ideal system would make allowance for the rapidly changing temperatures and other climatic factors within a few miles of the coast. The mean daily maximum temperature of summer is often 10° to 20°F (6° to 11°C) more at twenty-five miles in from the coast than at the coastline; and the daily minimum during winter is lower back from the coast. Thus there are parts of the western Sahara where dates cannot be raised near the coast. The trouble is that it is hard to find how far an existing weather station is from the coast, and at only a few places do coastal deserts have enough weather stations to make a cross-section of temperature (Logan, 1960; Meigs, 1935). Hopefully, other more definitive systems will be drawn up. In the meantime, the presently outlined simple system will give the main climatic characteristics that distinguish the coastal deserts from one another. A series of maps showing these coastal desert areas is included here.

Bibliographic References

GARDNER, E. S.
   1960   Hunting the desert whale; personal adventures in Baja California. Morrow, New York; McLeod, Toronto. 208 p.

HODGES, C. N.
   1969   A desert seacoast project and its future. *In* W. G. McGinnies and B. J. Goldman (eds), Arid Lands in Perspective, p. 119–126. American Association for the Advancement of Science, Washington, D.C.; University of Arizona Press, Tucson. 421 p.

HODGES, C. N., J. E. GROH, AND T. L. THOMPSON
   1967   Solar powered humidification cycle desalination; a report on the Puerto Peñasco pilot desalting plant. International Symposium on Water Desalination, 1st, Washington, D.C., 1965, Proceedings 2:429–459.

LOGAN, R.F.
   1960   The Central Namib Desert, South West Africa. National Academy of Sciences/National Research Council, Washington, D.C., Publication 758. 162 p. (ONR Foreign Field Research Program, Report 9)

MEIGS, P.
   1935   The Dominican frontier of Lower California. University of California, Publications in Geography 7. 192 p., esp. p. 77.
   1953   World distribution of arid and semi-arid homoclimates. *In* Reviews of Research on Arid Zone Hydrology. Unesco, Paris. Arid Zone Programme 1:202–210. [Maps dated 1952 were revised in 1960]
   1966   Geography of coastal deserts. Unesco, Paris. Arid Zone Research 28. 140 p.
   1967   Coastal deserts: Prime customers of desalination. International Symposium on Water Desalination, 1st, Washington, D.C., 1965, Proceedings 3:721–736.

SCAMMON, C. M.
   1970   Journal aboard the bark *Ocean Bird* on a whaling voyage to Scammon's Lagoon, winter of 1858–1859. Edited and annotated by David A. Henderson. Dawson's Book Shop, Los Angeles. 78 p. (Baja California Travels Series, 21)

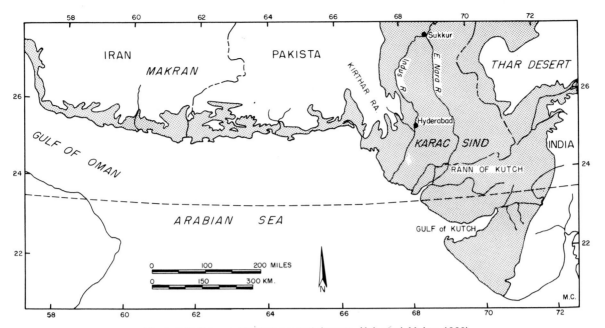

Fig. 1-3. Kathiawar-Makran coastal deserts. (Adapted, Meigs, 1966)

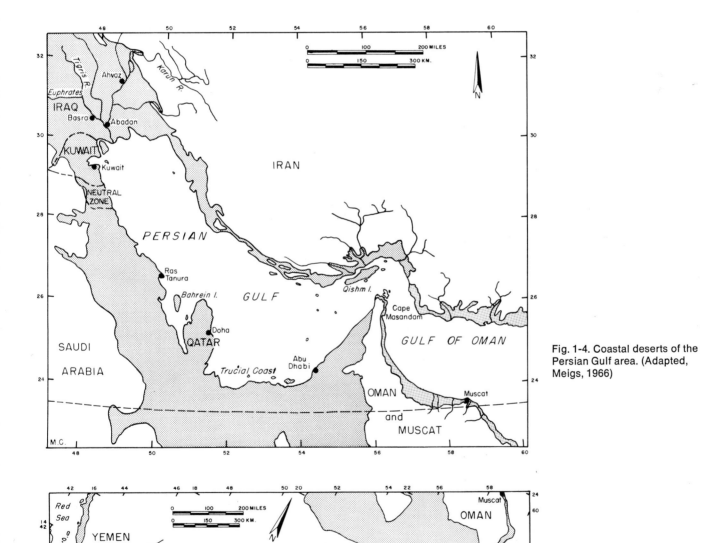

Fig. 1-4. Coastal deserts of the Persian Gulf area. (Adapted, Meigs, 1966)

Fig. 1-5. Coastal deserts of southeastern Arabia. (Adapted, Meigs, 1966)

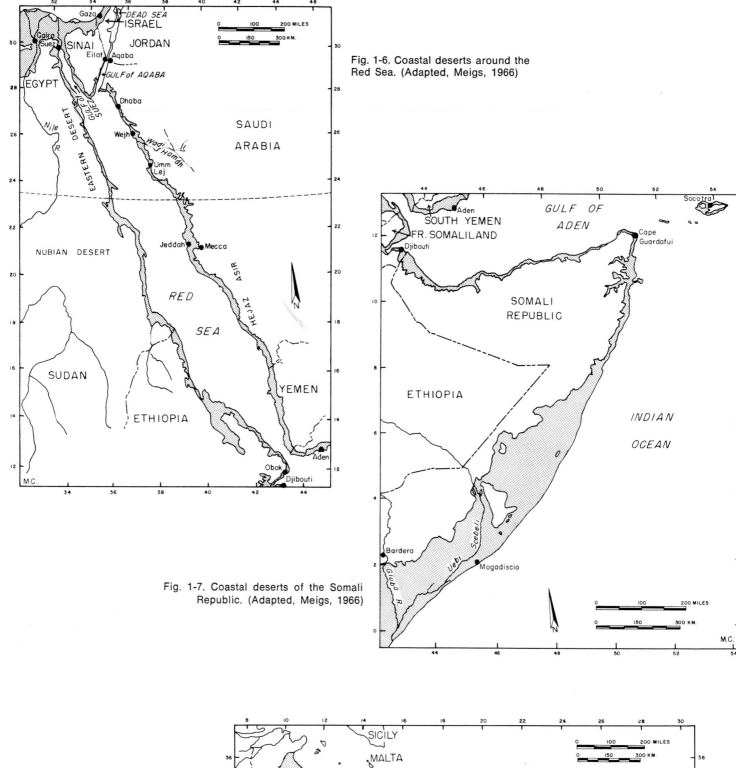

Fig. 1-6. Coastal deserts around the Red Sea. (Adapted, Meigs, 1966)

Fig. 1-7. Coastal deserts of the Somali Republic. (Adapted, Meigs, 1966)

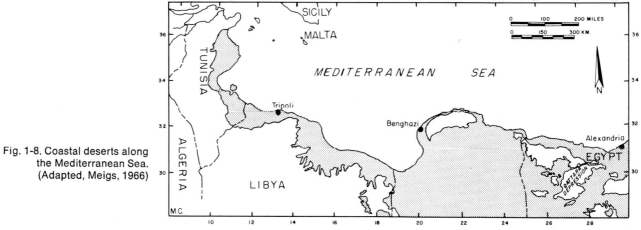

Fig. 1-8. Coastal deserts along the Mediterranean Sea. (Adapted, Meigs, 1966)

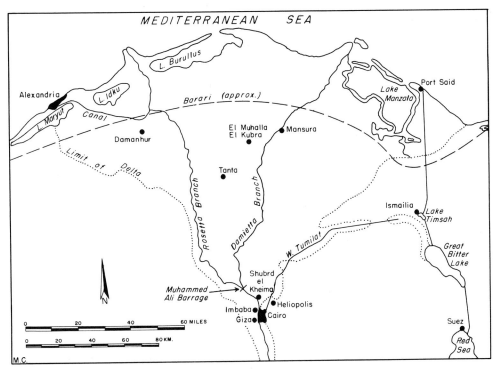

Fig. 1-9. Coastal deserts of the Nile Delta. (Adapted, Meigs, 1966)

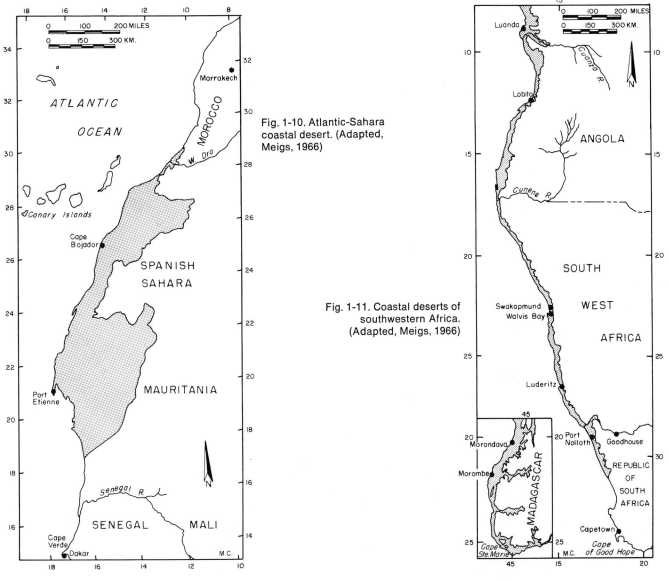

Fig. 1-10. Atlantic-Sahara coastal desert. (Adapted, Meigs, 1966)

Fig. 1-11. Coastal deserts of southwestern Africa. (Adapted, Meigs, 1966)

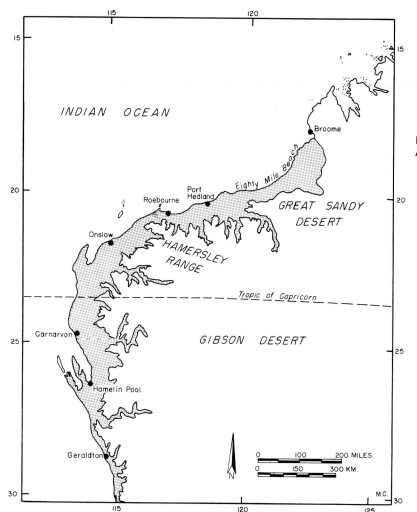

Fig. 1-12. Coastal deserts of western Australia. (Adapted, Meigs, 1966)

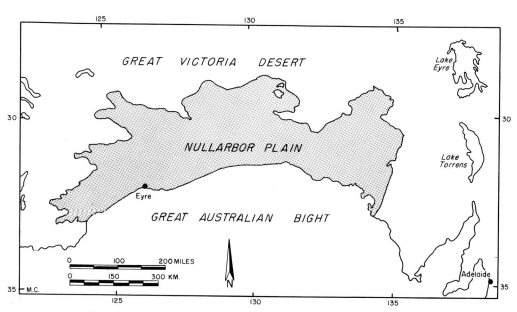

Fig. 1-13. Coastal deserts of southern Australia. (Adapted, Meigs, 1966)

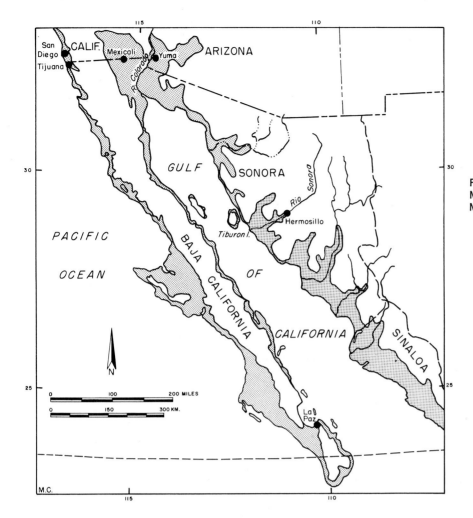

Fig. 1-14. Coastal deserts of western Mexico and Baja California. (Adapted, Meigs, 1966)

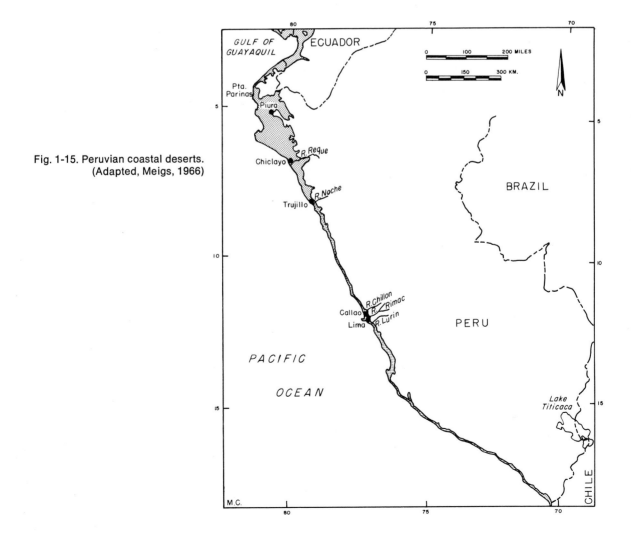

Fig. 1-15. Peruvian coastal deserts. (Adapted, Meigs, 1966)

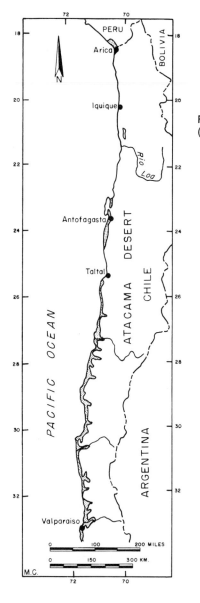

Fig. 1–16. The Atacama desert. (Adapted, Meigs, 1966)

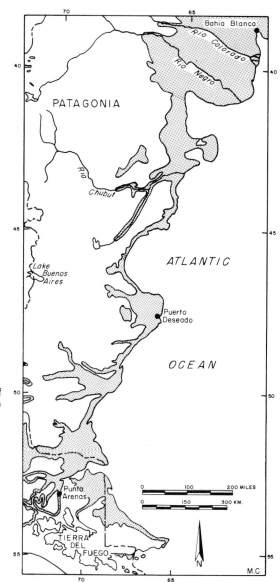

Fig. 1-17. Coastal deserts of Patagonia. (Adapted, Meigs, 1966)

CHAPTER 2

REMOTE SENSING OF THE WORLD'S ARID LANDS

Leonard W. Bowden

Department of Geography, University of California, Riverside, U.S.A.

Man is born with sensors that allow him to perceive the environment around him. His eyes respond to the spectrum between 300 and 700 nanometers, his ears detect sound between about 16 to 20,000 cycles per second, and his skin senses mild temperature changes or textures which are touched. Other than man's limited range in taste and smell, most of what goes on in the world is unheeded or beyond evolutionary man's land-based sensing capability (Parker and Wolff, 1965).

In an attempt to better understand the constant turmoil of the environment, man has developed a large group of sensing aids that help him perceive his surroundings and detect previously unnoticed activity. Cameras, radiometers, seismographs, radio receivers, geiger counters, magnometers, and similar instruments are used to penetrate the environment that normal sensing does not allow man to grasp. By remotely sensing our surroundings, we can acquire information about phenomena not in intimate contact with or in the observable range of man's normal sensors (Parker and Wolff, 1965).

One driving force toward developing environmental remote-sensing equipment has been the military in search of intelligence information. Secondly, there has long been an awareness among scientists that remote sensors could greatly increase our ability to view the earth and to examine the earth's surface in the confines of an office or lab. In the same way that a surgeon studies an x-ray photograph before diagnosing or operating, the Earth scientist can extract data at a much greater speed than if the landscape were actually traveled. Aerial photographs, taken in the panchromatic range of the spectrum, have become an indispensable and commonplace research tool for numerous disciplines.

In the 1950s, a new realm opened up for the Earth sciences and related disciplines when two things happened. First, various military organizations released to the scientific community access to new and refined remote-sensing gear; and second, spacecraft or satellites began to actually orbit the Earth.

Spacecraft offer new sensor platforms for looking at the Earth and have the advantage of being above the major portion of the atmosphere as well as passing over a large amount of the Earth's surface in a short time. Never before have scientists had such a platform for viewing and studying the Earth at a macro-scale or micro-scale with seemingly unlimited options in between. Nor has there been such a challenge to the ingenuity, ability, and capacity of the Earth scientist to absorb data on a global scale and convert it to usable, scientifically proper applications.

The applications are apparent. Demands for water, food, minerals, and energy are outrunning our "know-how" for meeting them by traditional methods. Clamor for more and better working conditions, living and recreational space, and an affluent life are being stifled by overcrowding, pollution, erosion, mismanagement, starvation, disease, and political unrest. It is necessary to look at any and all new techniques that will help the Earth scientist help his fellow man. Remote sensing of the environment is such a technique.

Remote sensing, itself, is not new. It simply means being able to measure or sense the properties of something without being in actual contact with it. A remote sensor may range from a compass to a camera to a geiger counter and beyond. The sensor may be land based, water based, air based, or space based, and may cost a pittance or multimillions of dollars. Some sensors function only in specific locations and only in narrow limits of the electromagnetic spectrum; other sensors have broad spectral bands or universal application or both (fig. 2-1). Take, for example, the camera and its recording device — film — which has for years been one of the world's most important sensors, yet operates only in that narrow visible light or near visible light range. The camera's outstanding contribution is that man's perception of his environment is also in that range, and most data need somehow be converted to man's perceptual range. (Do not misinterpret the above to mean the camera is limited only to visible application. It is invaluable in recording or storing other spectral data on film — for example, x-ray photography or laser holograms.)

Nevertheless, an infinite amount of activity is going on in the world that is out of man's visual, audio, or physical sensitivity. The activity ranges from less than 0.03 nanometers to more than 300 meters, and probably far beyond. There are objects that absorb, transmit, or reflect, or any possible combination, so that the infinite properties possible to sense are almost beyond comprehension.

One can assume, with reasonable certainty, that everything in nature has its own unique distribution of reflected, emitted, and absorbed radiation. By actually measuring a sample or by simulating the characteristics, one looks for the windows in the spectrum that are the

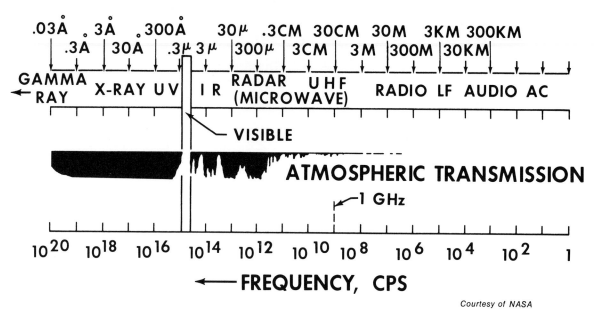

Fig. 2-1. Characteristics of the electromagnetic spectrum significant to remote sensing.

best wavelengths to sense the physical and chemical properties of certain objects and to sense or identify the properties remotely, be it from a rooftop, a balloon, or an orbiting spacecraft. The problem is, however, that few natural or man-made materials are perfect radiators, or even near perfect, and the shape of the curve of wavelength distribution depends not only on temperature but also on emissivity — and there are significant gaps in our knowledge of how to sense radiation and emissivity (Colwell, 1963).

Therein lies the challenge of mating practical, applicable, useful sensors in any or all parts of the spectrum to devices that will provide local, regional, or global coverage on a continuous basis. What is potentially possible at present, and how the sensors are or will be applied, is of prime concern.

The single greatest interference to remote sensing the Earth's surface is the atmosphere and the atmospheric path through which the sensed images or data must pass. Second to the atmosphere, of course, is the day-night contrast in which the "windows" available change dramatically in a 24-hour period. Another prime consideration is sun angle and latitude, which affect the "windows" in similar ways. One finds, for example, that sensors working in the radar range give day-night, all-weather capability, but unfortunately radar information is sometimes not as useful for detailed study as the infrared or panchromatic ranges, because of gross pattern, mechanical limitations, or high operating expense. The decision about which instruments to choose when there is a size and space limit (and there always is one in an aircraft or spacecraft) often means picking one or two and being satisfied with less than the ideal.

The case for remote sensing the Earth from space is easily defensible, both academically and economically. All Earth scientists are much aware of the limitations in understanding and properly analyzing the physical and cultural environment of the Earth. Already we are finding ourselves in the position of knowing more about areas on the surface of the moon than we know about half of the surface of the Earth. Our maps are inadequate; resources are uninventoried; populations and land uses of the world are only roughly estimated; air and water pollution and misuse are undetected; and doubtless many mineral resources are undiscovered or inaccurately measured. Our knowledge of location, temperature, precipitation regimes, vegetation types and zones, heat and water balance, ocean currents and salinity, forest and grass fires, and soil moisture and groundwater availability are but a few of the items that beg for discovery and quantification. Space data has resulted in many islands and mountain peaks being relocated cartographically, correcting earlier errors in mapping and surveying. Continuous surveillance of the Earth's surface by either manned or unmanned spacecraft offers an almost unlimited potential in data collection, information sources, and aerial overviews for the Earth scientist (Conference on the Use of Orbiting Spacecraft in Geographic Research, 1966).

One area of study that has come to the forefront in space photography is that of the world's arid lands. The reason is straightforward: most of the unclassified data released for general use has been from equatorial orbits (roughly 30° north and south latitude) which cover three-fourths of the world's arid lands. In addition, the scant cloud cover and water vapor has made arid landscapes available for photographing during both planned

Remote Sensing

and random times even when the primary mission was not to remotely sense the Earth (figs. 2-2 & 2-3). Sensing humid areas of our planet, except for weather satellites designed to examine cloud cover, still awaits more sophisticated imagery in such ranges as radar, infrared, or ultraviolet before a great deal of data can be collected from the surface. In fact, to the observer in space, most of the Earth most of the time looks like a cloud-enshrouded blue and white planet. Even when there is a lack of cloud cover, the penetration of haze and mist from evapotranspiration and other sources create a special film, filter, or recorder problem — as does the air pollution from cities. However, some of the results from the Gemini and Apollo missions were magnificent despite the limitations (figs. 2-4 & 2-5). A random but as yet unpredictable clearing may occur, such as the historical accident on Gemini IX when both the Peruvian coast and the Andes were clear at the same time and the astronauts were able to film the entire region (fig. 2-5).

Fig. 2-2. Indian subcontinent. Photographed in the visible range from Gemini XI, this image was taken from 410 miles in space. A surprisingly cloudless region surrounds the subcontinent, while the monsoonally influenced clouds swirl over the land. In the original color image, color differences from brown to green delineated humid from arid regions. Adams Bridge to Ceylon is visible both above and below the water surface.

Fig. 2-3. Cloud-enshrouded Earth. Imaged from 23,000 miles in space by the NASA Applications Technology Satellite. Note the lack of cloud cover over two of the world's coastal desert areas, that is, the southwestern portion of the North American continent (A) and the west coast of South America (B).

Photos courtesy of NASA

Photo courtesy of NASA

Fig. 2-4. The Nile Delta photographed in the visible range from Gemini V spacecraft. One original image showed more than 1,200 separate villages and settlements. The view covers nearly 500 square miles, and the dark irrigated cropland contrasts sharply with the lighter surrounding desert. In addition to Cairo (at arrow), the larger cities of Banha and Zifta-Mit Gamr lie close to the Nile in the north, with Tantá in the upper left corner. The photograph surprised urban geographers in that the hierarchical concept of Christialler's central place theory was not only applicable to a delta but was so clearly visible on a spacecraft image. Cairo is one of the few cities photographed with any clarity because many cities are blanketed by atmospheric pollution which interferes with photographic clarity in the panchromatic range.

Photo courtesy of NASA

Fig. 2-5. Gemini IX photo of the Peruvian coast, showing coastal desert area, Andes Mountains, Trujillo, and Chicalyo. A series of photos such as this one allowed the United States Geological Survey in cooperation with the Peruvian government to mosaic the entire Peruvian complex and has led to a new insight into the geologic structure, vegetation zonation, and resource wealth of the country.

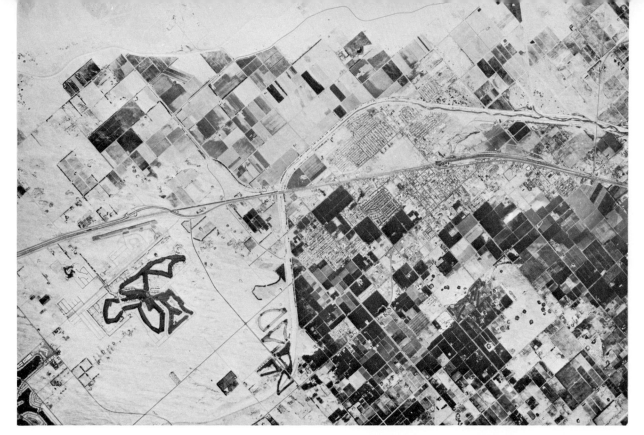

Photo courtesy of NASA and Southern California Test Site

Fig. 2-6. Black-and-white copy of a high-altitude Aero Ektachrome Color Infrared photograph of the Indio area in the Coachella Valley of the Salton Sea region, California. In the original color image, scale 1/60,000, resolutions of 10 centimeters or less were available, and a 14-category land-use map was only one of the many data extracted. The clarity, haze penetration, and high resolution of such imagery makes it invaluable as a sensor image as well as a reference from which to compare more experimental sensors that are not fully developed.

Figures 2-6 through 2-18 are examples of remote-sensing imagery that is potentially possible of any of the world's arid lands. All examples are from the Salton Sea Region of the Colorado Desert in Southern California. The Salton Sea area is part of a test site operated by the Geography Department, University of California, Riverside, for the Earth Resources Program of the National Aeronautics and Space Administration and the United States Geological Survey. Impressive as these images may seem, one must keep in mind that they are only a minuscule sample of what to expect in the future. Some are from hand-held cameras shooting through a space shield with little attention given to the proper filter or exposure and with rather ordinary film. For example, from the earlier Gemini images taken in 1965 of the Salton Sea and the Colorado Desert region of Southern California and Mexico, one can extract data about things down to 200 meters in size (fig. 2-7).

Color infrared photography taken on Apollo IX (March 1969) had a resolution of 80 meters (fig. 2-8) and could

Fig. 2-7. Gemini V imagery of Salton Sea region, California. Note extent of coverage (Mexican border to Colorado River), high resolution, land tenure, international boundary, oases, field pattern, gyrol in Salton Sea, and classic examples of desert geomorphology (for example, fans, bajades, and inselbergs). Specific locations are: (1) Mojave Desert, (2) Colorado River Valley, (3) Little San Bernardino Mountains, (4) Chocolate Mountains, (5) Coachella Valley (note San Andreas Fault which trends between 3, 5, and 4; 5 also locates Indio shown in figure 2-6), (6) San Jacinto and Santa Rosa Mountains (Palm Springs is between 3 and 6), (7) Salton Sea, (8) Peninsular Ranges, (9) Anza-Borrego Desert, (10) Imperial Valley, and (11) Colorado Desert.

NOTE: Figures 2-8 through 2-20 are from experiments conducted in the area shown in figure 2-7.

Photo courtesy of NASA and Southern California Test Site

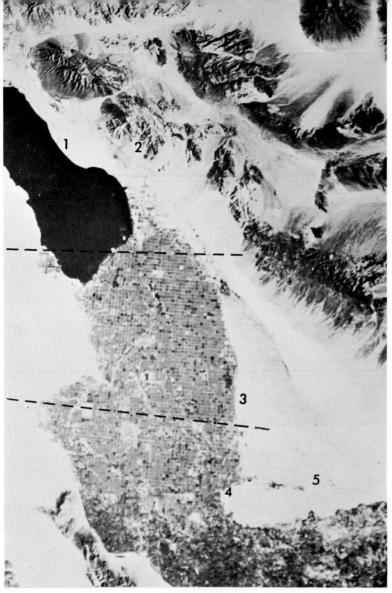

Fig. 2-8. Black-and-white print of the Salton Sea, Imperial Valley region, taken from the Aero Ektachrome Infrared image acquired by Apollo 9, March 1969. Numbers refer to (1) Salton Sea (2) Chocolate Mountains, (3) Imperial Valley, (4) California-Mexico border, and (5) All American Canal. The original 70-millimeter image covered approximately 100 miles on the side and had a resolution of 80 meters. Shades of gray in the Imperial Valley are reproductions of red tones which recorded the near infrared reflectance. Vacant fields appear as a light color, while dark patches indicate irrigation of bare fields prior to seeding. The Imperial Valley contained mostly alfalfa, cereals, and sugar beets at the time of overflight, while the irregular-shaped fields of Mexicali Valley in Baja California, were either wheat or fallow awaiting cotton seeding. Dashed lines set off the area appearing in figure 2-9.

Fig. 2-9. A detailed, five-level, digital land-use map (digits one and two are not shown because they are the same for the entire valley, that is, 8 [Resource Production], 1 [Agriculture]) was prepared as a model of semiautomatic land-use mapping from spacecraft imagery. The map (of the area between the dashed lines in figure 2-8) is a combination of pre-flight and post-flight sampling and laboratory photo interpretation. A digital computerized code is used as the first step in planned future machine-readable systems. The data management and information system problem is well exemplified in this map. Pertinent questions might be: (1) Was the quality control on the film such that similar results will be obtained on future missions? (2) Can these data be telemetered to avoid film recovery problems with as good or better results? (3) Was the diurnal and seasonal timing proper, and does continued imagery require sequential data, and, if so, under what control? (4) Is the map the best method to portray the data, or should such items as magnetic tape be introduced? (5) What additional information other than land use can be extracted? (6) Are the method, model, and classification applicable locally, nationally, or world wide? (7) Who are the users, and what are the benefits and costs?

Photo courtesy of NASA and Southern California Test Site

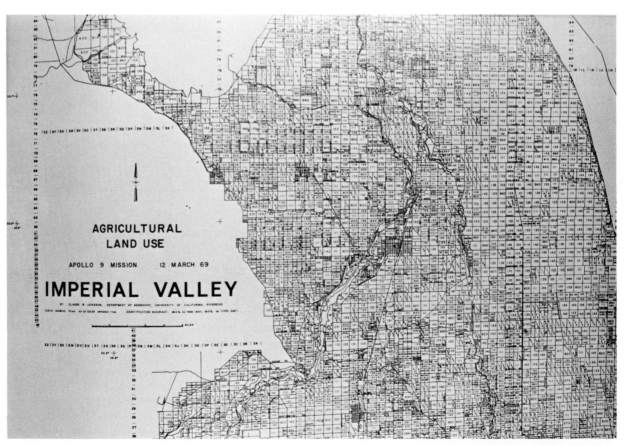

Courtesy of Southern California Test Site

be analyzed to identify fifty-four categories of agricultural land use (fig. 2-9). There are also examples of experimental pictures taken of the Salton Sea area from 13 kilometers altitude with an effective focal length of 5.0 meters (200 inches) that have a resolution of 0.2 meter. Evidence exists that one can expect a photographic coverage range of 8 square kilometers with less than 100 meters resolution with sensors presently available or planned for the forthcoming earth resources satellites. Combining space acquired information with digital telemetry, such as is available from operational weather and communication satellites, suggests that sophisticated data about the world's arid lands and their occupants will be acquired in great quantity over the next few years. This is true especially if one adds the more exotic thermal infrared, radar, and passive microwave sensors that are becoming available.

In addition, the lid has hardly been opened on ultraviolet, laser, gamma, and audio sensors. All these offer the potential of differentiation of properties (both physical and man-made) of the atmosphere, land surfaces, or water bodies.

Figures 2-10 through 2-20 illustrate a variety of experimental sensors that have potential (mostly unproved at present) for adding considerable data for solving arid-lands problems. Figures 2-10, 2-11, and 2-12 demonstrate extremes in photographic sensing in the panchromatic and near infrared range. As illustrated in figures 2-10 through 2-20, the sensing systems have a great potential, accompanied by many drawbacks.

Unfortunately remote sensing of the environment by spacecraft is not without problems and conflicts. Despite the multidisciplinary nature of the technique, there is often disagreement on the most desirable sensor when it is designed to serve such diverse fields as oceanography and forestry. The data the geologist wants are not always the same as those which the urban geographer or the meteorologist desire. Critical decisions have to be made as to whether it is best to provide the ultimate for one or two disciplines or a little bit for everybody. There is always a weight limitation and a benefit-cost factor to consider. In addition, there is the question of whether overflying another nation is spying or infringing upon that nation's sovereign rights.

Maybe most important of all is the time-lag between acquisition of data and its availability to the user. The data management problem is unbelievably complex and will only get more complicated. The burden lies directly on the Earth scientist to develop a system to manipulate and analyze the large amount of data that spacecraft can furnish. We are in the position of being overwhelmed by data we have desired for years because we were unprepared to handle the tremendous volume available. It would be tragic, however, to let the opportunity pass and the data go undeciphered and unused.

NOTE: Imagery in this chapter is shown through the cooperation of the National Aeronautics and Space Administration, United States Geological Survey, Office of Naval Research, and Southern California Remote Sensing Test Site.

Bibliography and References

ALEXANDER, R. H.
    1964   Geographic data from space. Professional Geographer 16:1–5.

BOMBERGER, E. H., AND H. W. DILL
    1960   Photo interpretation in agriculture. *In* Manual of Photographic Interpretation, p. 561–666. American Society of Photogrammetry. 868 p.

BOWDEN, L. W. (ED)
    1967   Studies in remote sensing of southern California and related environments. Status Report 1, Technical Report 1, USDI Contract No. 14-08-0001-10674, Departments of Geography, University of California, Los Angeles and Riverside. 105 p.

COLWELL, R. N. (CHAIRMAN)
    1963   Basic matter and energy relationships involved in remote reconnaissance: Report of Subcommittee I, Photo Interpretation Committee. Photogrammetric Engineering 29(5):761–799.

COLWELL, R. N., AND D. L. OLSON
    1965   Thermal infrared imagery and its use in vegetation analysis by remote aerial reconnaissance. Symposium on Remote Sensing of the Environment, 3rd, 1964, Proceedings, p. 607–620.

CONFERENCE ON THE USE OF ORBITING SPACECRAFT IN GEOGRAPHIC RESEARCH, HOUSTON, TEXAS, 1965
    1966   Spacecraft in geographic research, report of a conference. National Academy of Sciences/National Research Council, Washington, D.C., Publication 1353. 107 p.

JOHNSON, C. W., L. W. BOWDEN, AND R. W. PEASE
    1969   A system of regional agricultural land use mapping tested against small scale Apollo 9 color infrared photography of the Imperial Valley (California). Status Report III, Technical Report V, USDI Contract No. 14-08-0001-10674, Department of Geography, University of California, Riverside, 76 p.

PARKER, D. C., AND M. F. WOLFF
    1965   Remote sensing. International Science and Technology 43:20–31.

PASCUCCI, R. F., AND G. W. NORTH
    1968   Mission 73: Summary and data catalog. NASA, Manned Spacecraft Center, Houston, Texas, NASA Technical Letter 132. 296 p.

PEASE, R. W., AND L. W. BOWDEN
    1969   Making color infrared a more effective high altitude sensor. Remote Sensing of the Environment 1(1):23–30.

SIMPSON, R. B.
    1966   Radar: Geographic tool. Association of American Geographers, Annals 56(1):80–96.

Fig. 2-10. Black-and-white reproduction of Aero Ektachrome color transparency of the Alamo Delta, Salton Sea (location is directly above number 10 on figure 2-7). The image was photographed with an RC-8 camera with a 6-inch lens from 5,000 feet (1.3 km).

Photo courtesy Phil Slater, University of Arizona and Southern California Test Site

Fig. 2-11. The quarry in the boxed area in figure 2-10 becomes apparent in figure 2-11. Individual trucks can be distinguished despite the fact the photo was taken from 40,000 feet, or eight times as high as for figure 2-10. Systems such as used here, the Automatic Lock-on Tracking System (ALOTS), demonstrate that small-area, high-resolution photography is obtainable from high-altitude aircraft and spacecraft. Such systems are expensive and often heavy, which means the decision to use them must be scrutinized carefully. A system like ALOTS in space could easily produce imagery of large-scale, small-area character comparable to conventional aerial photography.

Fig. 2-12. Three multicamera images of narrow bands in the visible and near infrared spectrum show small-scale, large-area photography as the other extreme from figure 2-11. Taken from Apollo 9, of the Imperial Valley, these are images of the green (left), red (center), and near infrared (right), reflectance. Multiband photography from space offers many advantages in that black and white spectral images can be transmitted electronically and managed digitally or reconstructed as images in panchromatic true or false color. Machines can reconstitute and display data from such images at the whim of the user.

On the other hand, similar data can be obtained by color separation, densitometric tracing, and other processing of multi-layer film images such as the color infrared photo in figure 2-8. One of the greatest problems in the photographic process, despite the overall utility and the advanced techniques, is obtaining some idea of user needs before the sensing occurs. Often the hardware and the platform exist, but the uses — whether applied or theoretical — have not been defined adequately enough to assure that the right data are gathered.

Photo courtesy of NASA and Southern California Test Site

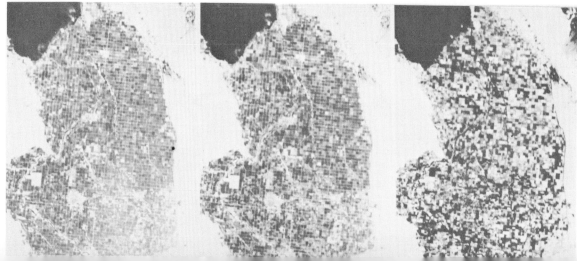

Photo courtesy Westinghouse, NASA, and Southern California Test Site

Fig. 2-13. This reading by Westinghouse's Side-looking Airborne Radar (SLAR) was gathered at night over the Imperial Valley (note Alamo Delta in right margin). Polarized and cross-polarized images are scanned and photographed simultaneously. SLAR has been used successfully for topographic mapping in cloud-covered Panama and offers great potential as a sensor in all-weather day-or-night circumstances. Generally not as useful as high-quality photography, SLAR does add a new dimension because it operates on a controlled signal transmission, reflection, and reception principle. Unlike many other sensors that must measure terrestrial reflected or emitted energy, radar measures the strength of the return of its own generated signal. Radar (and SLAR) is well known for use in navigational aids and polar research applications. Studies indicate its high potential for urban area analysis. One suspects that radar's potential for arid-land research will be subdued for years to come because of its greater utility in humid and polar regions. Some radar systems (see fig. 2-18) have some limited potential for soil-moisture detection. Many radars penetrate vegetation cover of certain plants (the bright reflectance in this photo is from sugar beets, while cotton and cereal grain have lesser returns). To date most attempts to use radar in arid-lands research have been to study geologic formations, lineaments, faulting, and vulcanism.

Fig. 2-14. Thermal infrared image of the Alamo Delta, Imperial Valley. Beyond about 1,200 nanometers (1.2μ) infrared reflectance can no longer be recorded on conventional films. In order to record infrared reflectance and emission, specially cooled or filtered devices are necessary. These sensors are designed to record in a variety of ways, but two types useful for remote sensing of the landscape are those that record profiles and those that scan and produce a photographable image. The image shown here is a thermal infrared scan in the 8 to 14μ range taken at night over the Alamo Delta. The shade or tone of the image indicates the relative degree of warmth or coolness of the target area. For example, note that the water is light or warm, while many of the fields (bare soil) are cool. The same area imaged 12 hours later would produce a reversal with cool water and warm soil (see fig. 2-16). Scanners and profilers can be designed for narrow, wide, or multiband detection; they generally have an automatic gain control working on both sides of an ambient temperature and giving interesting relative information but little quantitative temperature data. Figures 2-15 through 2-17 show additional applications.

Photo courtesy of NASA and Southern California Test Site

Fig. 2-15. Multi-sensor remote sensing. Often there is an advantage to simultaneously sensing a target in two separate parts of the spectrum. The advantage comes only when specific needs are present, although the unexpected appears surprisingly often. This thermal infrared scan (8-14μ) was taken at the same time (11:05 a.m. PDT) as the photo in figure 2-16. The image, while highly distorted and static scarred, shows clearly the extent of flood irrigation taking place.

Photo courtesy of NASA and Southern California Test Site

Fig. 2-16. This black-and-white visible-range photo shows the same field patterns as figure 2-15. Note the uniformity in tone and texture of the dark fields on the right side, with only the slightest clue to water activity.

Photo courtesy of NASA and Southern California Test Site

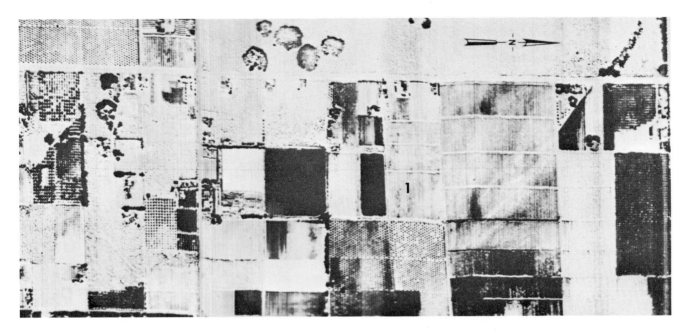

Fig. 2-17. *Above:* Daytime thermal infrared scan along Jackson Street in the Coachella Valley north of the Salton Sea. Dot-patterned fields are citrus groves, while fields with linear texture are vineyards. Solid dark fields are alfalfa, and light fields are bare or fallow. Surface moisture from recent irrigation shows plainly in many of the open fields.

Below: Ultraviolet scan taken simultaneously with above photo. Note the correlation of high thermal emission and high ultraviolet reflectance in a field by field comparison. However, not every field correlates, as can be seen at number 1. The solution to the contrast is probable difference in crop age, leaf cover, and irrigation practice.

Photos courtesy of NASA and Southern California Test Site

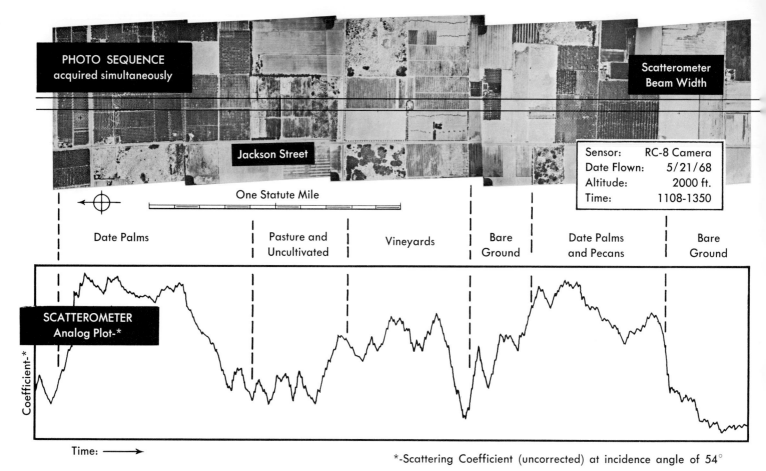

*-Scattering Coefficient (uncorrected) at incidence angle of 54°

Photo courtesy of NASA and Southern California Test Site

Fig. 2-18. This composite shows a graphic correlation of radar scatterometer return with land use and vegetation along Jackson Street in the Coachella Valley. The radar scatterometer produces a profile instead of a scan image as with SLAR in figure 2-13. The scatterometer is still very experimental and requires considerable training to operate, computer time to correct scattering coefficient, and qualified technicians to interpret. On the other hand, the scatterometer has proved successful as a day-night, all-weather, digital sensor and is most often used in conjunction with more conventional sensors.

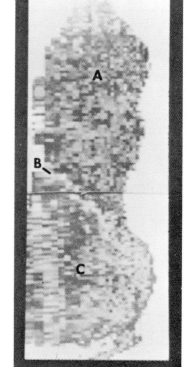

Photo courtesy of NASA and Aerojet General, Inc.

Fig. 2-19. Microwave image (19.35 gh) of Salton Sea acquired at 37,000 feet. The portion around A was a smooth sea and had an average microwave brightness temperature of 158° K, while the lower portion (C) averages 174° K. B points out a very sharp east-west wind-shear line dividing A and C. Passive microwave, unlike radar, does not measure the strength of its reflected wave but measures the Earth's own emission in the microwave range. In the example shown a television image has been produced from scanned digital data. The data can be recorded on film or machine analyzed or both. Passive microwave was first found to be an effective atmospheric sensor, but more recent experiments indicate a variety of uses in measurement of sea state (or roughness), urban energy balance, snow depth, and surface moisture. The original images, reproduced here in black and white, used color differences to indicate emission differences. Each cell, while distorted, is approximately 1,700 feet on a side.

Fig. 2-20. Microwave imagery along the border between Imperial and Mexicali Valleys. Difference in land tenure plus the distinctive canals produce a diverse pattern. Both the cities of Calexico and Mexicali show low brightness temperatures in contrast to surrounding irrigated fields.

Photo courtesy of NASA and Aerojet General, Inc.

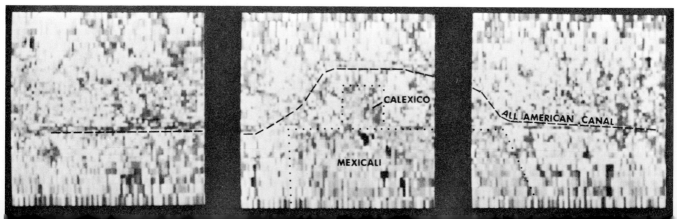

CHAPTER 3

PROBLEMS AND IMPLICATIONS IN THE DEVELOPMENT OF ARID LANDS
David H. K. Amiran
Department of Geography, The Hebrew University, Jerusalem, Israel

The Frame: General and Regional

Arid lands hold special significance in the geographic fabric of the world. They occupy a large sector of the globe's land surface, amounting to 36 percent of it. Of this area over 20 percent is fully arid and 16 percent semiarid. Due to the global pattern of climate, much of this desert area is interposed between the more habitable and thus more populated areas of the world's ecumene. They, therefore, form barriers to communication and limit the extent to the spread of normally populated areas. Deserts, however, are not all alike. There is a significant difference between the deserts of the eastern and western hemispheres. Due to the basic orographic pattern of Asia and Africa, the deserts there are truly interposed between the humid areas to the north and south of them. They form areas of separation or disruption often no less effective than the seas.

By contrast, the Cordillera aligned along the Pacific Coast of both the Americas generates a very different pattern. Here, alpine topography modifies the distribution of arid areas. In the special case of South America it places one of the world's most extreme deserts right along the coast. This creates special problems for the countries from Ecuador throughout the whole length of Peru and into Chile, but it also gives them singular advantages only feebly mirrored in Africa and west-coast deserts elsewhere. It is the same cool current which stabilizes the air and prevents rainfall that nourishes the legendary wealth of marine life in coastal waters sustaining one of the world's largest fisheries. Here then is a rare opportunity for positive use of a desert environment.

In general, arid lands pose difficult problems to man. Their extent is large, their nature difficult, their opportunities limited or insignificant, considering the effort and investment needed for their development. They normally have but a scant population, and, more often than not, for reasons rooted in their history, this population does not command the advanced technological and managerial competence required for the *mise en valeur* and the utilization of so difficult an environment. Under the stringent limitations of present-day economic standards, the result is often rather the planned nondevelopment of arid areas and a gradual voluntary thinning-out of their population. Economics impose a common limitation on arid areas. Water, the main problem-child of arid lands, is generally deficient and often nonavailable. But sometimes tremendous amounts of flood-water deluge these lands; rains in the wake of *el niño* are a case in point. However, the force and volume of these desert floods is generally such that it is not only technically difficult to harness them even in part, but considering the rareness of their occurrence it is often economically not feasible to construct the extensive works needed to turn this tremendous but short-lived resource to beneficial use.

Good resource bases available in fully arid areas, such as mineral or fishery resources, are strictly localized. Man has no choice open to him as to the range and location of their development. He can utilize the particular type of resource at the particular place of its occurrence, or he can defer its use, or he can abstain from using it. Possibilities range much wider and encompass a much more normal line of development in semiarid areas. As a matter of fact, these areas have considerable potential value for specialized and specially timed agricultural production, recreation, and tourism. Here, a reduced humidity and an already considerable aridity combine to a clear and usable climatic advantage. Here, too, populations often are rather dense, and, as they live in a difficult environment, have acquired considerably advanced skills. The *karezkan* of Afghanistan or the *moghanis* of Iran, masters in constructing and servicing kanats, the well-known subterranean water conduits, are a case in point. Another example is that of the ancient Nazceñas, constructors of the puquios of the Peruvian South.

Whereas stability is a keynote of an arid environment, and man has but little power to unbalance such stability, nature's instability introduces an element of risk into efforts to develop semiarid areas. In too many instances a developmental project has been started with the best of intentions but without the necessary knowledge of the intricate ramifications and interrelations of nature's mechanisms, so delicately balanced in semiarid areas. This has resulted in damage to vegetation and wild life, in soil erosion, air pollution, and water depletion. Here, competent knowledge of the ways in which geographic forces operate in nature is essential even more than elsewhere.

A particular combination of potential uses obtains in coastal deserts. These deserts are accessible, and most of them are endowed with resources of one type or another. The special case of the coastal deserts of South America has the added advantage that rich resources of good-quality water supplied by the Andes are available in certain valleys, providing the base for intensive agricultural development.

Incentives Toward Arid-Zone Development

The accelerating increase in world population is bringing about a distinct consciousness of man's increasing need for additional land. In the mind of most people, these additions to the ecumene, the land area actually used by man, are to satisfy two demands. The first is the requirements for land to physically accommodate the increasing population, the cities in which they are to live, the industrial plants to produce for them, and the open space to satisfy their recreational needs. The second, and this would require considerably more land, is the additional agricultural land to grow the food needed by this increasing population.

It is for this specific use that the eyes of many are directed to the world's arid lands, which include 36 percent of the terrestrial area of the globe. This area is but sparsely populated and used, large tracts of it containing much of the world's uninhabited lands.

A common line of thought is, therefore, that by making water available it will be possible to irrigate substantial parts of the arid lands and turn them into major food-producing regions. Persons entertaining such hopes tend to assume that present technological advances should make the desalination of seawater possible at acceptable prices within the next decades. Furthermore, the low overall population numbers in arid lands do make these one of the primary target areas for accommodating our expected population increase — in the view of these people.

Population and Agricultural Land. There cannot be the slightest doubt about the acutely serious nature of this problem. It should, therefore, be appropriate to critically examine the geographic premises and the available evidence in order to evaluate the potential usability of the world's arid lands in trying to solve, or solve in part, the problems created by the steep increase in the world's population.

The need to accommodate and supply more men in this world is by no means new. Accepted demographic analysis provides the figures in table 3-1. This shows that the world population increase has been steeply accelerating since about 1850. However, contrary to prior expectation, this increase has not been accompanied by a corresponding areal expansion of land occupied. Neither has the number of settlements increased significantly, nor is the total area onto which human habitation extends significantly larger than it was two generations ago. Nor, indeed, has there been a significant increase in the area of agricultural land. What did happen was a tremendous shift of population from rural to urban places of settlement and, as a no less distinct secondary feature, a shift from the smaller towns to the larger ones. As a complementary feature there is today, in the early 1970s, a significant number of settlements so sparsely populated as to be partly or even totally defunct.

With all the inadequacies of the world's food supply, substantial increases in food production have taken place since 1850. However, essentially, these have not been achieved by a proportionate expansion of the area cultivated but rather by a technological and genetic revolution in agriculture that made it possible to obtain from the same fields much higher returns. In actual fact, in a number of instances where increases in agricultural production have been most significant, these have been achieved while the area cultivated actually *decreased*. The introduction of more efficient agricultural techniques made it not only advisable but mandatory to withdraw marginal land from cultivation (cf. *Atlas of Israel,* 1959, sheet XII/2, maps B and C, and table A). Comparative statistics relating to use of irrigation are given in table 3-2.

Taking due recognition of the disparity between the considerable increase in the world's population and the world's food supply, and between the rather insignificant increase in area occupied by man and cultivated by him in bringing about this increase, there appears to be no historic evidence to support the theory that the further increase of the world's population forecasted for the next half century can be accommodated only by a roughly commensurate, or even very substantial, increase in area utilized.

Demineralization of Seawater: When? The contribution of the arid zone to the future land requirements of man may essentially serve two purposes: agricultural uses and nonagricultural ones.

TABLE 3-1

Increasing World Population

Year	Estimated Population (in millions)	Average Annual Increase %
1650	470	—
1750	694	0.5
1850	1091	0.6
1900	1571	0.9
1950	2517	1.2
1967	3420	2.1

Sources: W. F. Willcox: **Studies in American Demography.**
United Nations: **Demographic Yearbook 1966.**

TABLE 3-2

Comparative Irrigation Statistics Relating to Cultivated Lands in Israel

	1944–45*	1965–66
Total cultivated area (1,000 hectares)	571	407
Unirrigated area	531	249
% of total cultivated area	93.0	61.2
% of unirrigated area, 1944–45	100	46.9
Irrigated area	40	158
% of total cultivated area	7.0	38.8
% of irrigated area, 1944–45	100	395

* Refers to area within 1965–66 borders of Israel only.
Sources: **Atlas of Israel**, sheet XII/2, Table A, 1959.
Central Bureau of Statistics: **Statistical Abstract of Israel**, 18, 1967, p. 319, table L/2.

The prospect of cultivating the large areas of uncultivated and idle land in the arid region by applying irrigation to them holds great attraction to many a casual observer. Linking this to the expectation of achieving the desalination of seawater at cost levels acceptable for agricultural purposes seems, therefore, to open up revolutionary vistas.

At the present time it would appear wise to delay further discussion of such eventualities and not to include them in present planning, for the demineralization of water from the sea can be achieved only at prices that preclude use in agricultural production, except in special local circumstances. However, once technology provides demineralized seawater at cost levels suitable for agriculture, our whole spatial pattern of agriculture will have to be re-planned. Contrary to the present situation, where farmers obtain their water by gravity flow, areas close to sea level will be then at a premium, and any higher elevation will have to bear cost penalties for lifting the water. Once such a development enters the stage of reality, coastal deserts might become preferred areas.

The Arid Environment: Strictures and Limitations

Climate, no doubt, is the basic factor determining the natural environment in arid lands. Aridity of climate brings about a number of additional factors inherent in desert nature, generally the more pronounced the more arid the climate. It may be useful to list these factors once more to emphasize the strictures they impose on man's activity in arid lands.

First and most obvious, arid lands are poor in water. At most places no water is normally available. Subsurface water is available for development in certain areas, especially in semiarid ones; but it is often limited in quantity if used within the range of natural recharge rates as it should be, often expensive to supply, and sometimes poor in quality.

Second, the very low amount of soil moisture and consequently of soil fauna and microfauna resulting from low humidity and high evaporation brings about a considerable scarcity of developed soils, and especially of soils usable for agriculture in arid lands, especially in the more arid areas. Much of this area has no *soil* in the common sense of the term. Furthermore, the same aridity factors bring about the widespread development of surface crusts in arid areas.

Third, those arid lands that have the advantage of rich mineral resources have to balance this against the disadvantage of a considerable and often injurious mineralization of the soil. This creates severe problems of alkalinity, or worse, even of excessive concentrations of boron, if irrigation is introduced. There is ample evidence of such problems in South America. Generally, therefore, arid areas suffer from one of two natural disadvantages, mutually exclusive: either problems of soil mineralization when irrigating, or a lack of mineral resources.

Fourth, heavy rains are rare occurrences at any specific place. They may occur but once in many years. If they do occur, they may be of great intensity. But wide distribution of crusts sealing the surface prevents the absorption of rainfall into the soil and this leads to maximum runoff, resulting in flash floods which often cause heavy damage and even loss of life.

Fifth, as a result of these conditions of the natural environment, arid areas generally are but sparsely populated. In striking contrast to settlement in humid lands settlement in arid lands is discontinuous. The availability of water only at a restricted number of places, and the accompanying restricted availability of other environmental conditions enabling settlement, are the basic reasons for this. The more arid the climate of an area the larger is in general the extent of nonsettled land between points or nuclei of settlement. As a result it has often been necessary to bring people from outside the arid area to staff development projects that demanded a considerable amount of manpower.

Sixth, as a further result of the poor environment, arid-zone populations rarely command modern technological and managerial skills. This, therefore, is an additional reason necessitating the staffing of the higher ranks for advanced modern projects with personnel from outside the arid area. Not only does this add additional expense to the project, but by necessity it involves delays and a certain amount of trial and error as the staff concerned learns to adapt its skills to the conditions of an arid environment. Furthermore, inadequate appreciation of the possibilities and limitations of the arid environment, and of its nature and economics, leads sometimes to an exaggerated investment, either not fitted to the environment of the region or considerably beyond its economic capacity.

Seventh, in contrast to possibilities in humid lands, the range of environmental possibilities in arid lands is limited — that is, there is a narrow range of choices for development. In general, the more arid the climate the more limited the environmental possibilities. As a result, semiarid regions are at a considerable advantage compared with the arid ones. Whereas arid areas are in a stable environmental balance, nature in semiarid areas is in a rather delicate state of balance, for reasons which have been stated elsewhere (Amiran, 1965). Most damage done by man to arid areas has affected their semiarid parts.

Some Conclusions Concerning Arid-Zone Development

A proper appreciation of the various factors mentioned above as inherent in the nature of arid lands forms the background to certain conclusions which should be carefully considered in planning development in arid areas.

The Arid-Zone Inferiority Complex. Many people and even populations are not properly aware, or even are unaware, of the basic difference in environmental conditions between arid and humid lands. Their scale of values is modeled on those of humid areas which they consider as *normal*. They, therefore, consider as normal and as desirable an environment closely resembling the imaginary normal environment of humid lands. Arid-zone people will sometimes go to great efforts to engage in agriculture, to supply locally their own food, and to create greenery in their settlements. This is the result of what might be called the "arid-zone inferiority complex," a widespread and serious disease of arid-zone people. By trying to live the life of the humid zone, they attempt to wrest from an unfavorable environment results it is not fitted to produce. At the best, such results are achieved at high cost, because one has to work against the environment instead of with it, and the products are obtained at much higher cost than under humid conditions.

A similar manifestation of exaggerated attention to matters concerning arid lands, out of proportion to their real importance, distinguishes many scientific discussions (cf. Ward, 1968, p. 124).

Expensive Infrastructure. The infrastructure required for the provision of regional services at levels considered adequate by the population concerned depends on two basic factors: the quality of services required and the extent of area for which these have to be provided. At any given standard of quality, the cost per unit of consumer served (number of residents, households, or settlements) generally will be lower the greater the population density of the area as a whole.

As arid areas are characterized by low overall population densities — with infrequent exceptions, such as river oases — and especially by wide spacing of occupance with long empty distances between discontinuous areas, or even places of settlement, the regional infrastructure will have to pay for the installation and maintenance of these non-revenue-generating empty areas and thus be unduly expensive, or if this is economically impossible it will have to compromise at a lower level of quality. The result is either poor services or services rather expensive per unit of consumer use as compared with areas of a *normal* settlement pattern as characteristic for humid lands. Here is an important factor of regional economic inequalities where arid lands are at a relative disadvantage.

The Danger of Regression. By its very nature regional development introduces new elements and factors into the landscape pattern and, therefore, modifies the balance of the environment. The new balance which will be created will operate satisfactorily only as long as it is actively maintained. Furthermore, as the modification concerns the environment, to succeed in restoring a viable balance it has to affect all factors shaping the particular environment, and not just one of them. Failure in either or both of these two procedures will result in a negative change in environmental development, thus leading to regression. Examples unfortunately are by no means lacking. Irrigation without simultaneous introduction of adequate drainage has led to alkalinization in many an arid-zone irrigation project; or sizable irrigation projects, which after some time are no longer properly maintained, deteriorate and in extreme cases become inoperative. Before the inauguration of such projects the area concerned was probably but sparsely inhabited. The execution of the irrigation project brought about its settlement. Its deterioration leaves the population involved with a severely depleted environment to which they have to respond either by emigrating from the area or by continuing to stay there under conditions of regression of various degrees of severity.

The same sequence of events might affect development other than irrigation projects or agriculture. Mining towns are a frequent case in point. The exhaustion of the mineral resource, a consumer shift to a different commodity (as, for example, the substitution of other types of fertilizer for nitrate, which affected the economy of the Atacama Desert of Chile), or critical shifts in pattern and cost of transportation can remove their *raison d'être*. In view of the limited range of alternate possibilities available in the arid environment, the town may undergo rapid depopulation and turn into a "ghost town," the few inhabitants of which represent a negative selection of those least adaptable to change. It is by no means accidental that ghost towns are a more prominent feature of arid areas than of nonarid ones.

It appears that the limited range of economic development choices available, which makes many arid-zone towns single-purpose towns from a functional point of view, is the basic reason for the relative frequency of ghost towns in arid lands. A subsidiary factor is the element of size. Not only is there an upper limit of town size beyond which special services and organizational measures have to be applied to make the large modern metropolitan city livable, there obviously is, as well, a lower limit below which a settlement established as town cannot continue to function as such. This is especially true if the remaining population of the town in question must adapt to a considerably smaller number of inhabitants: regression will become critical once population decreases below a minimum number and, therefore, capacity to maintain essential services above a minimum standard is lost. As previously stated, the population fabric of arid areas is generally discontinuous and brings about relatively strong concentrations of population at a few centers or in a few areas. The result is that both at the traditional and at the modern technological level, urban settlements in arid areas contain a large proportion of the total population.

Mutually Exclusive Possibilities for Development. Every project in regional development changes the operation of the formative factors in the natural environment of the region. Humid regions have a wide range of active factors which can to a degree correct one another. In arid regions this range of factors is much narrower, and so their corrective ability is much less.

It might occur elsewhere that various projects of development are mutually exclusive. However, the limited activity of corrective factors in an arid environment, and particularly the absence or dearth of rain and river flow which could remove injurious residue, make this a particularly acute problem in arid lands. As an example, one might mention the discussion of introducing a petrochemical industry versus the interests of the resort industry at the desert port of Eilat, discussed elsewhere herein. And there is the case of the resort hotel at Callao, Peru, which was put out of operation by the stench of a fish-meal plant. Finally, the blueprints for the seaside development of "Puerto Grau" near the southern end of the coast of Peru proposed an industrial port, a refinery, a fishery port, and a seaside resort in this order from south to north, the direction of flow of the Peru current. Since this obviously involves a number of mutually exclusive elements, it is to be hoped that it will not be implemented in this form.

Guidelines to Arid-Zone Development

Principles for Arid-Zone Planning. The basic principle in planning for arid-zone development, as in planning elsewhere, should be a realistic appraisal of the area's possibilities and its potential for development. The primary concern must be a realistic appreciation of the weaknesses resulting from the limited availability of water, soil, and often of manpower and capital resources as well. In short, one should treat deserts as deserts and not try to apply nonarid standards to their development except under very specific circumstances. An arid environment imposes definite physical limitations and therefore a strictly limited range of choices. In arid areas, therefore, even more than elsewhere, congruous use must be made of the resources available, and efforts at incongruous development must be avoided. *Incongruous development* might be defined as development bringing about a critical depletion of the area's resources, thereby diminishing its potential for future inhabitants, or as a development that will not create economically and socially successful communities. Development must be designed in a way that will not preclude the future use of the area by populations of more advanced technological standards. This principle holds true equally for arid as for nonarid areas (Higbee, 1952).

Considering the partial deficiency in environmental resources and conditions of any arid environment, there is nothing easier than wasting great amounts of money in arid-zone development without positive results. Due to the dearth of capital usual in these regions, such funds will generally be available but once.

Great responsibility devolves on planners of arid-zone development projects: it might take from 5 to 15 years to find out that a project area will not provide its population with a base for living. With the dearth of choices available in the arid environment, most of the people involved will move elsewhere if they can afford to do so.

Arid lands are an outstanding example of the situation where one should avoid creating or conserving what Watson (1967) called conditions of *uneconomic geography*.

Environmental Planning. Since arid-zone development is development under difficult environmental conditions, no project should be initiated without the most careful and comprehensive planning. Planning teams should include geographers trained to study natural environments and the changes which will occur in them as a byproduct of the development planned. Their task will be to assess the extent to which a given project will fit both the environment and the natural resources that can be made available. Economists will evaluate the economic feasibility of projects, cost of the necessary infrastructure, and the economic potential and feasibility of marketing. Sociologists will evaluate the human aspects of the populations involved, their social approach, habits, and adaptability, both of populations already in the area and of those who would join them from elsewhere. Finally, and definitely last in order, engineers will plan and implement technical installations within the framework of the project. Implementation of projects should be made under constant review and re-evaluation by the planning team. In view of the practice in many a developing arid country (including some in South America) it should be stated explicitly that one cannot commit a greater error than trying to *engineer* arid-zone development projects. Jackson (1962) when dealing with the dry areas of the USSR stated this in unequivocal terms in saying that arid climate "grants little margin for error, delay, or inefficiency." Arid-zone projects do not admit of quick technological fixes.

Diversity and Priorities. The restricted choices for development of arid areas should not lead to a unidirectional approach in planning; in particular one should avoid putting all one's eggs into the basket of arid-zone agriculture. To the contrary, development should aim at diversification. The planner should be open minded and try to make the best use of the positive assets of the development area. In very general terms these would include the following, and possibly in the following order of priority.

1. Mineral resources: Oil, iron, and copper ore and a considerable variety of other resources are found in arid lands.

2. Fishery: The arid west coasts of the Americas and of Africa contain some of the world's richest sea fisheries.

3. Urban and industrial development: The warm and dry climate, which has the additional advantage of necessitating less expenditure in plant construction, and the ample amount of unoccupied land available, form powerful incentives for the growth of arid-zone cities. Urban and industrial development in arid areas require particularly careful planning and control to avoid despoliation of the natural environment, including the avoidance of air pollution, a particular danger in arid areas with frequent inversions.

4. Resort industry and tourism: The present socioeconomic structure permits an ever-growing number of people to enjoy extended vacations at a considerable distance from home. The almost certain dry and sunny

weather of arid areas, the scenic beauty of many parts of them, as well as the historic significance of certain areas, bring about a steady increase in their attractiveness to tourists. The time element of some such tourism is rather interesting, as it permits people to get away from home for their vacation at the climatically most unpleasant period of the year. The considerable volume of winter-tourism in the United States and that from Northern Europe, especially Scandinavia, to the southern Mediterranean lands clearly shows such a trend.

5. Agriculture: In an overall view, agriculture here has been ranked last, on purpose. The possibilities for development of arid-zone agriculture are very uneven. Small-scale development is possible at many a place, but it will be of rather local significance. In rather limited and sometimes exceptional cases, there is, however, the possibility of large-scale highly-intensive agricultural development. These require careful planning for optimal use of the resources involved.

Climate Farming: Some Considerations of Advanced Uses and the Optimal Use of Groundwater. Arid areas (at least their warm parts) have two great advantages from the point of view of the farmer. First, a high level of temperature exists during most of the year, at least for a considerable number of hours around noon. Second, the usually clear skies and the low level of humidity make pest control rather easy. As against these advantages there are very definite disadvantages: dearth of suitable soil, lack of water or of water of acceptable quality, high rates of water consumption due to high rates of evaporation and transpiration, long distances to major markets and ensuing high cost of marketing, and sometimes restricted availability of labor.

Of all the items mentioned, normally water is the scarcest commodity. It should be utilized, therefore, so as to give the optimum return. Due to high rates of evaporation loss, the economic return from water used in arid-zone agriculture is comparatively low. Careful consideration, therefore, should be given to deciding whether to allot the available water to industry or resorts or to agriculture. As compared with humid areas, the distribution network or pumping required for irrigation agriculture is an additional item raising the cost of agricultural production.

For these and other considerations, agriculture in arid lands has but a slim chance of success in growing the average range of products that command normal prices. Its real chance is in concentrating on products that find no competitor on the market from other areas, at least at the time the arid-zone farmer can market his products. Here the year-around high temperatures can assist the subtropical or tropical arid-zone farmer who is willing to concentrate on winter vegetables, fruit, and flowers. These command prices high enough in metropolitan and other markets that they render a fair profit, notwithstanding their high production and transportation cost. This specialized type of agriculture will, to a considerable extent, concentrate on winter crops; therefore it requires careful and expert operation to avoid frost damage. In many cases winter vegetables, fruit, and flowers are grown today under a plastic cover, both to diminish frost damage and to decrease evaporation; therefore their production requires a considerable investment of capital. Many such winter products are flown to European markets from North African oases and from the arid borderlands beyond the Mediterranean.

Geographical Implications of Seawater Demineralization. Although modern technology makes it entirely feasible to demineralize ocean water, the water so treated is still too expensive to be used for irrigation. But if costs can be brought down sufficiently to make crops grown with demineralized seawater competitive, the cultivation pattern, and by implication the settlement pattern, in arid areas will require an entirely new assessment (*Focus*, 1968; Meigs, 1966).

Contrary to the present, *natural* patterns, the most bountiful source of water available, and the water of best quality, will be at sea level instead of at higher elevations. Consequently water will no longer be delivered to the area of consumption by gravity flow but by lifting it to the respective elevation. This will place a premium on the low elevations of coastal deserts and make water and commodities produced by it more expensive the higher the elevation at which it is used.

It appears, therefore, likely that for a considerable time to come demineralized seawater will be used preeminently at the lowest elevations only. To many a coastal desert this will mean not only an increase in quantity available but the substitution of a good-quality water for a mineralized one. Since such a pattern of water availability is distinctly different from the natural one, its utilization for agriculture will depend on the availability of suitable soils in areas that are very arid by nature. Moreover, demineralized water will not be usable economically unless the product or service obtained commands a sufficiently high price to pay for lifting of water and for the installation of a delivery network. In summary, Meigs' (1966, p. 17) assessment appears still valid that the use of demineralized water for domestic and industrial consumption appears close at hand, whereas agricultural use is still a rather distant and none too certain possibility.

Development of Semiarid Areas and Their Problems. For obvious reasons semiarid areas are more favorable for development than fully arid ones. They approach humid conditions to a certain degree in level of moisture, availability of productive soils, relatively suitable numbers of population (and sometimes of a fair level of regional competence), continuity of agricultural settlement, and therefore a less expensive regional infrastructure. Their drawbacks are the marginality of these various elements, the more so the closer they are to the borders of the fully arid areas, particularly in relation to their climatic marginality. Here, interannual changes in rainfall may seriously affect a generally low-rainfall area that can hardly bear a series of drought years.

To maintain populations at satisfactory economic levels in semiarid lands requires their integrated, advanced, and regionally complete development. It does not permit any temporary lessening of regional management, such as the neglect of an irrigation system. The semiarid environment is particularly delicately balanced by nature, much more so than the relatively stable arid one (Amiran, 1965, p. 208). The extensive damage sustained by parts of many a semiarid environment by such factors as removal of natural vegetation, destructive grazing, soil erosion and partial removal of soil, and deterioration of water resources, is amply demonstrated unfortunately in the Mediterranean lands, California, the Norte Chico of Chile, and elsewhere. The semiarid area, therefore, makes special demands for competent planning and implementation, and it will flourish only if it is operated by a population of adequate educational, technical, and managerial competence.

Urban Development and Conservation of the Quality of the Environment. One may assume as a definite forecast that urbanization of arid countries will increase in volume and speed, and that, in common with other areas, the large town will be a prominent feature of such development.

The nature of the arid environment provides certain favorable elements for this, especially relatively ample space and dry climate with often clear skies. But at the same time it lacks certain features and corrective processes that regulate the proper integration of the city in its environment. The wide distribution of atmospheric inversions makes large cities in arid areas particularly prone to air pollution, from Los Angeles to Santiago de Chile and elsewhere. The low level of general biologic and microbic activity aggravates problems of the disposal of sewage and refuse. Unfortunately the ecological problems and the hazards to health created in this field are clearly evident around many a large city in arid lands, and the surrounding countryside is disfigured with discarded trash. As urbanization grows these problems will get worse. Satisfactory solutions are likely to be expensive, but the expense must be met.

For reasons of principle and as an important aid for the development of tourism — the increasing importance of which for arid-zone economies has been noted above — special attention should be given to the preservation of the natural scenic beauty of arid regions. The earth's dry lands contain some of its most interesting and spectacular landscapes, which are at once of great scientific interest and have great potential value for attracting tourism. In the United States, for example, Grand Canyon, the Petrified Forest, White Sands, and many other spectacular features have been admirably preserved as National Parks and National Monuments, and they attract large numbers of tourists. Similar developments in other arid areas should be encouraged at every opportunity, both to preserve scientifically important features and to assist more people to enjoy the wonders of nature.

Multiple Development. An attempt has been made here to discuss the limitations imposed by the arid environment as well as some of the more important prospects for the development of arid lands. Technological advances, as well as a widespread rise in standards of living, permit an ever-increasing number of people to avail themselves of the products and the facilities of the world's arid lands.

For a situation like this, which in part becomes technically or economically possible only by modern technological advance, and in part caters to affluence, and is subject therefore to changes in market conditions and social fashion, flexibility and multiple development are of utmost importance. Any resource base on which a given area can create a most-favored position of supplier, or even a monopoly, should be developed to the full. Mineral resources, offshore fishery, tourism, and out-of-season agriculture are all cases in point.

However, as all such commodities and services are subject to changes in demand and/or supply, any arid-zone development that aims at permanency should be diversified and not unidirectional. Unless this can be achieved, arid areas will remain backward, and to an increasing degree. Furthermore, arid environments also must be managed with a high degree of flexibility, switching major attention from one branch of activity to another as development and general trends demand. In addition, one should not commit the error of overestimating the limitations of an arid environment and develop just the most favorable kind of activity in any one area. The disastrous results of such a unidirectional approach were clearly demonstrated following the collapse of the nitrate economy in the Atacama Desert of Chile.

Because of these environmental limitations and the need for intelligent and quick use of the various alternatives available to arid-zone man, the best results will be achieved by the population which applies the greatest amount of creativity and agility to its affairs. It appears, therefore, that education holds the key to arid-zone development as it does elsewhere.

The Importance of Education in Arid-Zone Development. The arid zones contain some of the world's poorest environments. Yet, interspersed are small areas endowed with particularly rich environmental resources, as for instance an oasis or, on a different level, an oil field. By and large, however, arid lands are inhabited by people distinguished by a rather low level in education, at least in the conventional meaning of the term. These people are but poorly fitted to meet the difficult and complex problems involved in the improvement of an arid area to a state where it can compete successfully with other areas in attaining a standard of living acceptable in the late twentieth century. Unless such a standard can be met, one of the following two developments (or both) will occur. Either, the indigenous population will drift into economically more advanced areas to seek employment and decent wages, or the economic management of the area concerned will be taken over by outside interests and the indigenous population will be absorbed in the process.

This is the typical case with oil or mineral developments. The first of these alternatives is particularly serious, as, for obvious reasons, it will drain the arid area concerned of its most enterprising and valuable manpower, the loss of which it cannot afford if progress is to be made. Arid-zone life today, more than ever before, needs people with a particular competence to solve its problems. Therefore, it can least afford to lose its best-fitted men and women.

Clearly it is those arid areas in which irrigation, sometimes highly advanced, has been developed which are endowed with managers and technicians who are particularly competent. These people have to their credit some of man's most outstanding achievements.

It is worth noting that irrigation development has not always been restricted to river oases and to projects deriving their water from groundwater which could be made available either in natural oases, by using artesian aquifers, or by conducting water in *kanats* (known as *puquios* or *socavones* in South America). In various areas of the Near East, including some of its most arid parts, irrigation cultures existed in the past based entirely on utilization of the erratic winter stormwater runoff. This example of adaptation to an extremely limited environment, and its highly successful utilization under completely balanced adjustment to the mechanism of nature, must be rated among man's outstanding intellectual achievements in environmental management.

Contradictory trends are discernible in the arid zones today, parallel to trends in development in general. On the one hand many areas are losing out in the battle for the retention of their manpower, and at an accelerating pace as far as the younger generations are concerned. By contrast, certain areas turn to highly advanced, specialized, and sophisticated types of development, such as exploitation of mineral resources, the resort industry, and out-of-season specialty agricultural products for out-of-region marketing. There is no doubt that the first type is moribund and that it is the second which holds the positive prospects for the future of arid lands For such a development a proper educational standard is a prerequisite, a field in which, unfortunately, many arid-zone populations are deficient today. Very limited advantage is derived from technical assistance rendered to arid or other areas as long as the local population cannot provide counterpart manpower for the operation of such projects and eventually their management. As elsewhere, education holds the key to arid-zone development (Amiran, 1965, pp. 208–10). As elsewhere, too, the need must be the achieving of adequate levels of education as soon as possible in order to counteract the negative effects of the first alternative in arid-zone development — gradual depopulation. But one cannot "buy time" in education; the achievement of adequate educational standards takes time, time which many arid-zone people can but ill afford today.

It is with this in mind that the symposium at Lima in 1967 formulated its fifth resolution: "That investments now in education may in the long run be more productive in raising living standards than investments in irrigation," or, one may add, investments in any other technical or economic field.

Bibliographic References

AMIRAN, D. H. K.
    1965   Arid zone development: A reappraisal under modern technological conditions. Economic Geography 41:189–210.

FOCUS
    1968   Ceylon. Trends and techniques: Nuclear power to desalt water. Focus 18(9):12.

HIGBEE, E. C.
    1952   The three earths of New England. Geographical Review 42:425–438.

ISRAEL, CENTRAL BUREAU OF STATISTICS
    1967   Statistical abstract of Israel, no. 18.

ISRAEL, MINISTRY OF LABOR, SURVEY OF ISRAEL, JERUSALEM
    1970   Atlas of Israel; cartography, physical geography, human and economic geography, history. Sheet XII/2, Agriculture: Irrigation, maps B (Main pipelines, 1965), C (Soil and water balance), and Table 1 (The water supply, 1948 1965). 2d ed. Elsevier Publishing Co., Amsterdam.

JACKSON, W. A. D.
    1962   The Virgin and Idle Lands Program reappraised. Association of American Geographers, Annals 52:69–79.

MEIGS, P.
    1966   Geography of coastal deserts. Unesco, Paris. Arid Zone Research 28. 140 p.

UNITED NATIONS, STATISTICAL OFFICE
    1967   Demographic yearbook, 1966. United Nations, New York. 716 p.

WARD, R. C.
    1968   Review of *The Problems of Water*, by Raymond Furon. Geographical Journal 134:123–124.

WATSON, J. W.
    1967   Uneconomic geography. [Guest Editorial] Economic Geography 43(2): facing p. 95.

WILLCOX, W. F.
    1940   Studies in American demography. Cornell University Press, Ithaca, N.Y. 556 p.

CHAPTER 4

THE LARGER URBAN CENTERS OF THE COASTAL DESERTS

Andrew W. Wilson
Department of Geography, Area Development, and Urban Planning,
University of Arizona, Tucson, U.S.A.

Much of the discussion of the human use of desert lands in the past has been centered on agriculture and other types of primary economic activities. Irrigated agriculture in particular requires the use of large amounts of water, a scarce resource, the shortage of which provides the reason the areas are called deserts in the first place. On the other hand, desert areas are often pleasant places in which to live or visit if one is not desperately dependent on a limited water supply. Water often goes much farther in supporting humans when they are not engaged in irrigated agriculture, thus more can be paid for it. These nonagricultural activities generally center on urban living. This chapter is a look at the large-scale urban developments of the coastal deserts.

For identification of the areas of coastal desert, reliance is placed on Meigs' discussion herein. To limit the study to the most important centers, only cities or metropolitan areas of more than 100,000 population are included. Population data are from the *United Nations Demographic Yearbooks* (United Nations, Statistical Office, 1968 and earlier). Furthermore, not only is the latest available population figure included, but where possible an average rate of population growth has been determined. Most of the listed cities appear on the maps of coastal deserts at the end of Chapter 1.

The largest city of the world's coastal deserts is Cairo, in the Nile River delta of the United Arab Republic (table 4-1). A dozen other cities in the United Arab Republic also are large enough to qualify; one, Alexandria, is also a millionaire city. In fact, the United Arab Republic, when compared to any other nation, has not only the largest city but also the largest number of cities in the coastal desert realms.

The second largest desert city is Karachi, in Pakistan. It has only about two-thirds the population of Cairo, even though in Karachi's case the figure represents metropolitan rather than city population. Like Cairo, it is listed among the fifty largest urban areas in the world (World Almanac, 1968). Two more Pakistani cities are included in table 4-1.

Not far out of the "fifty largest" category is Lima, Peru, which ranks third in population among the coastal desert cities. It is somewhat ahead of Alexandria, United Arab Republic, which is fourth. Four other Peruvian cities are over 100,000 population, although Callao may be included in the figure for Lima's metropolitan area.

The fifth and smallest of the millionaire coastal desert cities is San Diego, in the southwest corner of the contiguous United States, and the only representative from that country, since the other large desert cities of the United States are situated more than 1,000 feet above

TABLE 4-1

Coastal Desert Cities of Population Over 100,000

Rank	City, Country	Type of Area*	Population and Year†
1	Cairo, U.A.R.	CC	4,219,853 (C–1966)
2	Karachi, Pakistan	MA	2,886,000 (E–1968)
3	Lima, Peru	MA	2,072,824 (E–1968)
4	Alexandria, U.A.R.	CC	1,801,056 (C–1966)
5	San Diego, U.S.A.	CC	573,224 (E–1967)
		MA	1,198,100 (E–1967)
6	Hyderabad, Pakistan	MA	658,000 (E–1968)
7	Giza, U.A.R.	CC	571,249 (C–1966)
8	Mexicali, Mexico	CC	386,760 (E–1968)
9	Tijuana, Mexico	CC	323,251 (E–1968)
10	Basra, Iraq	CC	313,327 (C–1965)
11	Callao, Peru	CC	306,976 (E–1968)
12	Kuwait City, Kuwait	CC	99,609 (C–1965)
		MA	295,273 (C–1965)
13	Port Said, U.A.R.	CC	282,977 (C–1966)
14	Abadan, Iran	CC	272,962 (C–1966)
15	Suez, U.A.R.	CC	264,098 (C–1966)
16	Tripoli, Libya	CC	247,365 (E–1968)
17	Tanta, U.A.R.	CC	229,978 (C–1966)
18	El Mahalla el Kubra, U.A.R.	CC	225,323 (C–1966)
19	Aden, South Yemen	CC	150,000 (E–1964)
		MA	225,000 (E–1964)
20	Luanda, Angola	MA	224,540 (C–1960)
21	Ahvaz, Iran	CC	206,375 (C–1966)
22	Jeddah, Saudi Arabia	CC	194,000 (E–1965)
23	Mansura, U.A.R.	CC	191,459 (C–1966)
24	Hermosillo, Mexico	CC	181,532 (E–1968)
25	Shubra el Kheima, U.A.R.	CC	172,902 (C–1966)
26	Imbaba, U.A.R.	CC	182,000 (E–1965)
27	Mogadiscio, Somalia	CC	172,677 (E–1967)
28	Zagazig, U.A.R.	CC	151,186 (C–1966)
29	Damanhur, U.A.R.	CC	146,079 (C–1966)
30	Ismailia, U.A.R.	CC	144,163 (C–1966)
31	Trujillo, Peru	CC	142,197 (E–1968)
32	Chiclayo, Peru	CC	141,131 (E–1968)
33	Benghazi, Libya	CC	137,295 (C–1964)
34	Antofagasta, Chile	CC	127,234 (E–1968)
35	Sukkor, Pakistan	CC	123,100 (E–1967)
36	Hilla, Iraq	CC	84,717 (C–1965)
		MA	110,795 (C–1965)
37	Piura, Peru	CC	101,585 (E–1968)

*CC = central city; MA = metropolitan area.
†C = census; E = estimate.
Source: **United Nations Demographic Yearbook, 1968.**

sea level. Near San Diego is Tijuana, Mexico, one of three cities from that country on the list.

Most of the remaining cities of the thirty-seven listed are in Southwest Asia, either near the Red Sea or the Arabian (Persian) Gulf. Exceptions are Tripoli and Benghazi in Libya, Mogadiscio in Somalia, Luanda in Angola, and Antofagasta in Chile.

Perhaps of even more interest than recent size is an examination of growth rates, where these can be calculated. However, there are difficulties involved. In some cases the number of years spanned is small, and thus the rate for those years may not be representative. This is especially true if one or both of the population figures is an estimate rather than the result of a census. In spite of these caveats, the growth rates are exhibited in table 4-2.

The fastest growth rate is that for Giza, United Arab Republic, with an average yearly increase of almost 19 percent. This is based, however, on a span of only four years, and with a beginning figure which is an estimate.

Second fastest growth is for another middle-sized city, Callao, Peru, which again is for a short span (five years), and with the recent figure an estimate. Callao's rate is about twelve and one-half percent.

Just over 12 percent is Mogadiscio, Somalia, but here the span is only three years, and both population figures are estimates. In the 11 percent bracket are Aden, in South Yemen, and Kuwait City, in Kuwait. Aden has grown substantially since a census in 1955, according to a seemingly rough estimate (150,000) made in 1964. Kuwait City's growth rate, however, is based on two censuses eight years apart, and probably represents the most reliable percentage among the faster-growing cities. Tijuana and Mexicali, the two cities of Baja California, Mexico, show almost equal growth rates of 10.7 and 10.5, respectively. Considerably behind is Basra, Iraq, at 8.4 percent, and then follow Cairo, United Arab Republic, at 7.4, and Jeddah, Saudi Arabia, at 7.1 percent.

San Diego, in the United States, and the third Mexican city, Hermosillo, are in the 6 percent bracket, while Ahwaz, Iran, along with Trujillo and Lima, Peru, is in the 5 percent group. Suez, in the United Arab Republic, is close behind at 4.9 percent.

Other cities have growth rates under 4 percent and thus are growing, in many cases, little faster than their national population growth rate. This includes many of the Nile delta cities in the United Arab Republic. Presumably, these are, for the most part, agricultural trade centers. One of these cities has actually lost population. Ismailia shows a 2.0 percent loss for the years 1962 to 1966. Since 1967 there has been an as yet unrecorded

TABLE 4-2

Comparison of Population Growth Rates for Selected Coastal Desert Cities

Rank	City, Country	Type of Area*	Population and Year†	Later Population and Year†	Percent Increase Per Year
1	Giza, U.A.R.	CC	287,200 (E–1962)	571,249 (C–1966)	18.8
2	Callao, Peru	CC	155,953 (C–1961)	279,500 (E–1966)	12.4
3	Mogadiscio, Somalia	CC	120,649 (E–1963)	170,000 (E–1966)	12.1
4	Aden, South Yemen	CC	54,995 (C–1955)	150,000 (E–1964)	11.8
5	Kuwait City, Kuwait	MA	125,929 (C–1957)	295,273 (C–1965)	11.2
6	Tijuana, Mexico	CC	59,952 (C–1950)	339,737 (E–1967)	10.7
7	Mexicali, Mexico	CC	64,609 (C–1950)	350,170 (E–1967)	10.5
8	Basra, Iraq	CC	164,623 (C–1957)	313,327 (C–1965)	8.4
9	Cairo, U.A.R.	CC	1,090,654 (C–1947)	4,219,853 (C–1966)	7.4
10	Jeddah, Saudi Arabia	CC	80,000 (E–1952)	194,000 (E–1965)	7.1
11	San Diego, U.S.A.	MA	836,175 (C–1960)	1,168,000 (E–1965)	6.9
12	Hermosillo, Mexico	CC	132,324 (E–1963)	167,735 (E–1967)	6.1
13	Ahvaz, Iran	CC	119,828 (C–1956)	207,011 (C–1966)	5.6
14	Trujillo, Peru	CC	100,130 (C–1961)	129,470 (C–1966)	5.3
15	Lima, Peru	MA	1,436,231 (C–1961)	1,833,700 (E–1966)	5.0
16	Suez, U.A.R.	CC	107,244 (C–1947)	264,098 (C–1966)	4.9
17	Antofagasta, Chile	CC	104,559 (E–1963)	117,187 (E–1966)	3.9
18	Zagazig, U.A.R.	CC	131,200 (E–1962)	151,186 (C–1966)	3.6
19	Alexandria, U.A.R.	CC	919,024 (C–1947)	1,801,056 (C–1966)	3.6
20	Mansura, U.A.R.	CC	101,965 (C–1947)	191,459 (C–1966)	3.4
21	El Mahalla el Kubra, U.A.R.	CC	198,900 (E–1962)	225,323 (C–1966)	3.2
22	Tanta, U.A.R.	CC	139,926 (C–1947)	229,978 (C–1966)	2.7
23	Luanda, Angola	MA	164,340 (C–1948)	224,540 (C–1960)	2.6
24	Port Said, U.A.R.	CC	177,703 (E–1947)	282,977 (C–1966)	2.5
25	Damanhur, U.A.R.	CC	133,200 (E–1962)	146,079 (C–1966)	2.3
26	Abadan, Iran	CC	226,103 (C–1956)	270,726 (C–1966)	1.8
27	Imbaba, U.A.R.	CC	164,500 (E–1962)	170,300 (C–1966)	.9
28	Ismailia, U.A.R.	CC	156,300 (E–1962)	144,163 (C–1966)	—2.0

*CC = central city; MA = metropolitan area.
†C = census; E = estimate.
Source: **United Nations Demographic Yearbook, 1967.**

(by the United Nations) loss of population on a grand scale by Suez and Port Said, as a result of the occupation of the opposite bank of the Suez Canal by Israel.

Some reasons for urban growth in the coastal deserts are more or less clear. The cities of the United Arab Republic are all in the desert, since the entire country is desert. It is impossible, therefore, to have the largest cities in humid areas, as in the United States of America. However, most of these cities are associated with the valley of an exogenous river, in contrast to several large arid lands cities in North America, coastal or otherwise. In fact, the two largest cities in the United Arab Republic are associated with the fertile Nile delta. Cairo, of course, is the national capital and therefore an administrative center for government as well as much of the business world. Like all giant cities, it is also a manufacturing center because it is a large market in and of itself. Some of its manufactures are sold in the rest of the country, too, or even exported. In all, 42 percent of Egypt's industrial establishments in 1952, and 27 percent of workers in manufacturing, were located in Cairo (Berger, 1963). Cairo also is an educational and information center and is in fact multifunctional, as is typical of a great metropolis. Giza, the fastest-growing of the cities listed in table 4-2, is a suburb of Cairo, and its rapid growth is a product of propinquity to the growing giant. Most likely its faster rate of increase is a result of its smaller base figure.

In a similar way Callao's growth is tied to that of Lima, the capital of Peru, for which it is port and suburb. The two cities almost have grown together in recent years. Lima and other Peruvian cities are the magnets for migration of surplus population from the rural areas in the same way that Cairo is in Egypt. Again, the principal cities of Peru are all in the desert areas, since the humid East is even less developed and, in fact, less attractive. The final target of migrants looking for better economic opportunity is Lima and its environs. Here is the manufacturing, distribution, and administrative center of the country.

Karachi, in Pakistan, is the dominant port of that country, and a major air terminal as well. Because it is an important port, much manufacturing of imported raw and semifinished materials is carried on there. As the former national capital and a regional center, it is of importance in administrative functions.

San Diego, in the southwestern corner of the continental United States, is a multifunctional city with a relatively high standard of living. In 1968, out of a total work force of 412,200 in the Standard Metropolitan Statistical Area, the employees on nonagricultural payrolls (which also excludes the self-employed) were 340,900 (U.S. Bureau of the Census, 1969). The latter were distributed, compared with overall national data, as shown in table 4-3.

San Diego is typical of arid-land cities in the United States in its relatively low percentage of employment in manufacturing and its relatively high percentages in services and government. Much of the manufacturing it does have is aerospace and electronic in nature, some of it what has been called "footloose" because transportation costs are an unimportant portion of total costs of production and marketing. This is a type of industry commonly found in most of the other United States arid-land cities. In contrast, however, per capita income in San Diego is above that for most other arid-land urban centers and for the country as a whole.

Across the border from San Diego, the cities of Mexicali and Tijuana, Mexico, are becoming manufacturing centers under the Mexican Border Development program, which provides a free zone into which United-States-made components may be imported for assembly, with the finished product then exported back into the United States. This may be done without Mexican duty and with United States duty only on the value added

TABLE 4-3

Non-agricultural Employment in San Diego and in the United States in 1968

Employment Categories	San Diego*	United States
Manufacturing and Mining	19.0%	30.0%
Wholesale and retail trade	21.6	20.8
Services	18.2	15.6
Transportation and public utilities	5.4	6.4
Contract construction	5.0	4.8
Finance, insurance, and real estate	4.5	5.0
Government	26.3	17.4
Totals	100.0%	100.0%
Percent of total work force unemployed, 1968	3.9	3.6

*Standard Metropolitan Statistical Area.
Source: Adapted from **U.S. Bureau of the Census, 1969.**

TABLE 4-4

Population, Employment, and Income Data for Arid-Land Cities of the United States, 1970*

City, State	Population in 1000s†	% Empl. in Mfg.†	Per Capita Income‡ Amount	Rank§
United States overall	207,976	26.3	$3,920	–
San Diego, Ca.	1,358	15.6	4,003	77
Albuquerque, N.M.	316	8.3	3,532	174
Bakersfield, Ca.	329	8.6	3,600	162
Colorado Springs, Co.	236	–	3,578	168
Denver, Co.	1,228	17.1	4,255	45
El Paso, Tx.	359	23.6	2,982	239
Fresno, Ca.	413	14.9	3,931	94
Las Vegas, Nv.	273	–	4,411	27
Phoenix, Az.	968	19.9	3,800	121
Ogden, Ut.	126	–	3,361	203
Provo, Ut.	138	–	2,487	250
Pueblo, Co.	118	–	3,371	202
Salt Lake City, Ut.	558	–	3,425	195
Tucson, Az.	352	7.8	3,586	166

*All data for appropriate Standard Metropolitan Statistical Area.
†**U.S. Bureau of the Census, 1972.**
‡**U.S. Bureau of Economic Analysis, 1972.**
§Rank in Standard Metropolitan Statistical Areas.

during assembly. As of 1967 Mexicali had 25 such free-zone industrial plants, and Tijuana had 13. One-third of the wage income in Mexicali and almost one-third in Tijuana came from these factories. The other major sources of wage income in Mexicali were 22 percent from agriculture and 21 percent from work in the United States across the border, again mostly agricultural. In Tijuana, 33 percent of wage income was from work north of the border (in the San Diego area) and 27 percent was from recreational (tourist) services (Dillman, 1970).

In Kuwait the major source of wealth, of course, is the fabulous oil reserves of this small country. All else is minor. No refining is done in Kuwait City. In Abadan, Iran, on the other hand, one of the world's largest oil refineries is responsible for the major economic stimulus.

As the technological revolution continues to play an always enlarging role in the coastal deserts of the world, the movement of surplus agricultural workers from country to city certainly will continue. This means that urban activities will play an increasing role in these countries (Wilson, 1960). Other chapters in this book illustrate this general statement. The problems of urbanization need to be analyzed continually as solutions to the disturbing realities of urban growth are sought.

Bibliographic References

BERGER, M. (ED)
    1963   The New Metropolis in the Arab World. Papers prepared for an international seminar on city planning and urban social problems, sponsored by the Egyptian Society of Engineers, and Congress for Cultural Freedom, Cairo, Dec. 17 22, 1960. Allied Publishers, New York. 254 p.

DILLMAN, C. D.
    1970   Recent developments in Mexico's National Border Program. Professional Geographer 22(5):243–247.

UNITED NATIONS, STATISTICAL OFFICE
    1969   Demographic Yearbook, 1968 and earlier. New York: United Nations.

U.S. BUREAU OF THE CENSUS
    1972a   Statistical abstract of the United States. Government Printing Office, Washington, D.C. 1032 p.
    1972b   General population characteristics, U.S. summary. Government Printing Office, Washington, D.C.

U.S. BUREAU OF ECONOMIC ANALYSIS
    1972   Survey of current business 52 (5):42.

WILSON, A. W.
    1960   Urbanization of the arid lands. Professional Geographer 12(6):1–4.

CHAPTER 5

THE CHANGING ROLE OF WATER IN ARID LANDS
Gilbert F. White
Institute of Behavioral Sciences, University of Colorado, Boulder, U.S.A.

Arid-land research is one of the deeply exciting intellectual adventures of our generation. Cutting across established scientific disciplines, this research seeks to relate the findings of pure science to technological development in its social applications: radiation physics in Tucson is seen in relation to energy collection devices in New Delhi and to the political structure and daily rhythm of a remote Mexican village. More than two dozen disciplines are involved, from pure physics and chemistry into biology and the complex realm of human ecology. The efforts at correlation span fifty nations and embrace not only halting steps to fashion research institutes in the young educational systems of Cairo and Pakistan but sophisticated ventures in transferring the established research machinery of Rothamstead or Wageningen to new situations. To try to work across these many boundaries of outlook, scientific disciplines, and nations, is at best courageous and possibly foolhardy.

Perhaps these attempts are the more challenging because they center their interest upon a meager environment: the arid lands in which the margin between human failure and success commonly is narrow and in which delicate changes in rainfall and land use can trigger profound shifts in soil, vegetation, sediment movement, and water flow. These are risky lands. Much of the table is bare, but where chances are taken the stakes are high.

In any arid landscape with its thin soils, sparse vegetation, and exposed erosion forms, the green-fringed evidence of water long has been taken to mark the land's most valued asset. Lack of water defines the arid lands. The presence of water in oasis well or ephemeral stream bed or massive exogenous river has been regarded as a stamp of resource wealth, an essential key to economic growth. It has been commonplace to say with Jean Brunhes, "Water is pre-eminently the economic wealth; it is, for men, more truly wealth than either coal or gold." And a great state can wage a bitter battle at the polling booth on the simple argument, "Water for people, for progress, for prosperity."

We now are moving into a time when the more traditional formulations of the role of water are challenged and altered. New pressures upon the arid lands inspire this challenge and set the scene for social change on an unprecedented scale. In the western fringe of the Australian Desert, the dry wastes of Turkestan, the sweeping plains of the north-central Sahara, and the southwestern United States, both society and landscape are undergoing drastic transformation. Against this shifting backdrop of denuded hills and sprawling cities, water continues to play an important part, but one that is beginning to differ in radical degree from its part in the past.

Unique Aspects of Arid Lands

A geographer must at once register the reservation that no two points of an arid zone are alike in all respects: each has its unique combination of terrain, soil, vegetation, and moisture. For this reason, generalities about arid lands are to be approached with the caution of a hydrologist assessing a drainage basin, knowing that one steep sector may be perennially dry while a nearby valley fill is usually saturated. Potential water supply in a stretch of the Egyptian desert where fresh water accumulates chiefly as a thin layer on a brackish underground supply is not to be regarded as the same as in an area of the central Sahara crossed by seasonal streams rushing from a rocky massif to dissipate themselves in salty playas. A plan or policy for resource development that fails to take explicit account of this tremendous diversity in local environment is likely to go astray.

Nevertheless, from the spotty experience of the arid one-third of the earth's land surface, a few trends may be noted. To describe those trends is not to venture a prediction. The current vogue of projecting our resource future by extrapolating past trends in the use and supply of natural wealth has heuristic value in trying to pierce misty uncertainties of the future, but it carries the danger of these trends being accepted as possible guides for that growth. Sometimes such trends obscure the realization that a present view of resources in itself shapes future use of those resources.

This is my excuse for reviewing familiar experience with water management. Today's vision of the role of water in the Santa Cruz Valley, in the Colorado Basin of Arizona, and in the arid zone as a whole, seems likely to affect the complex of individual decisions determining how remaining waters will be allocated or reallocated for human use in the decades ahead. According to the framework of public attitudes in which both familiar and unfamiliar hydrologic facts are arranged, water may be

---

NOTE: This chapter has been adapted from the Riecker Lecture, which was reprinted in the *Arizona Review,* March 1967.

cast in roles that will either inhibit or advance social aims. These public attitudes may be expected to reflect in some measure the sweeping changes in progress, and the forces shaping them.

Social Changes in Arid Lands

Arid lands now are places of remarkably rapid social change. In the great reach of dry mountains, plateaus, and plains stretching from Central Asia and Rajasthan to the Atlantic coast of Morocco, the cultures of centuries of nomadic life are being battered and warped.

The first and most noticeable social change pertains to evolution of nomadic life into a sedentary life. Sedentarization has been a watchword of recent social policy in many sectors. On a massive scale the nomads of the Soviet Union have been attached to permanent bases and prescribed ranges, their flocks limited, their institutions altered to meet the needs of sedentary agriculture. A population of something more than ten million people, divided among at least one hundred ethnic and language groups, has been taken from its tents and seasonal migrations and settled on farming plots by government plan and force. Iran and Iraq carry out calculated programs to impede the nomad, to settle him on cultivated lands, or to run him into the labor market of the city: the wandering Kurd becomes the taxi driver of Baghdad. Along its Mediterranean coast, Egypt has sought to make stable homes for the Bedouins who had roamed the Western Desert.

Even where government policy has not sought to eliminate nomadism, the turn of political and economic events combines to do so. While agricultural agents in disputed Saharan territory work to improve the flocks, water points, and pastures, other forces render nomadic life less tenable or desirable. Air and truck transport eliminates caravan trade and cripples the caravanseries. Law enforcement and administrative regulations curb the movements of pastoral peoples. Markets for animals become less encouraging. Jobs in petroleum and other mineral exploitation and in growing urban centers attract the nomadic men, offering them income and amenities and a different kind of insecurity. Throughout the Old World deserts nomadism is dead or ill. Possibly the only growing edge of nomadism in a fresh context is the restless invasion of North American deserts by trailers and pickup trucks.

In the dry lands of the New World as well as in nomadic territory, other aspects of social change have special significance for water management. Perhaps as evident in the Santa Cruz Valley as in any arid area, these aspects need only to be enumerated to suggest the scope of their impact.

One of these additional aspects of change is the rapid advance of technology for finding, lifting, storing, transporting, and treating water. Experts have pointed out the implications of new methods for locating and pumping underground water, storing and moving water on a large scale, improving water-application techniques and dryland plants, treating raw water and waste water, transmitting electrical energy, and scheduling water use. Within six or seven decades the concept of where and how water could be stored, moved, and treated has changed dramatically. The technical means are at hand to go deep and far for water at low cost, to increase the returns from the volume withdrawn, and to change its quality.

An important feature of the technology of water management is its rate of advance. Opportunities to regulate and treat water are unfolding much more speedily than they are being used. While there are conspicuous examples of applications of the new knowledge — nets of tube wells where gas engines are novelties, massive concrete arches where baskets are the common means of moving earth, and intricate hydrologic models for educational use where most of the citizens are illiterate — much water use lags far behind. Irrigators in a Peruvian valley are less open to innovation than the young engineer who designs a new dam. Fifty years ago a farmer or manufacturer could be less profligate with water in field or plant than today because he could not then have been aware of opportunities to line ditches or use brackish waters or recirculate industrial coolants. Tragically, a number of water developments in Southwest Asia are running on to rocks of inept management as soon as they are launched; the first division of a project begins going out of cultivation before the others are complete. Thus, technological advance enlarges the possibilities of wise water management at the same time as it widens the margin between knowledge and performance.

Another significant aspect of social change is the increased demand upon arid lands for recreational and industrial use. Throughout the arid zone, cities are growing more rapidly than total population. In Egypt where an increase of 1.5 percent per year took place in total population during 1927–47, urban population increased at 3.0 percent per year.

Urbanization and the Changing Land

Cities are becoming the dominant feature of all lands. In countries of Southwest Asia a principal element in the strategy of economic development is industrialization; manufacturing is seen as the great expanding sector of the economy, but an essential aid to such expansion is strengthening agricultural production, and so irrigation development is a step toward intense urbanization. Rather than representing competition between city and country for water, as in the southwestern United States, irrigation expansion is seen as an aid to ultimately preponderant manufacturing activity.

With urbanization, water use changes in several ways. Industrial uses assert themselves and claim larger volumes of water than does the entire urban population. In industrial centers of the United States the manufacturing intake may amount to one-third to fifty times the municipal intake. Total water needs therefore may increase more rapidly than total population. Manufacturing commonly leads to a rise in per capita income, which in turn is reflected in increased per capita water use for household purposes. Industry is more demanding upon water for waste disposal than for direct use: dilution requirements mount tremendously with manufacturing aggregations and may quickly attain troublesome proportions where streams are thin or ephemeral. Growing urban populations reach out for recreation amenities, either extending living quarters to the dry lands or going to them for holiday use. This in turn raises the standards of water quality for the few available stream courses. Workers create more rigid demands for clean water in picnic areas, residential areas, and fishing grounds at the very time their plants are requiring enlarged volumes of water to dilute industrial wastes.

These phenomena of urban growth are familiar signs in the southwestern United States, but they are proliferating in other sectors of the arid zone, and they give every indication of becoming more intense in the decades ahead.

Hand in hand with social transformations have gone changes in the land itself. The gashes of new highways have exposed fresh terrain to grazing and recreational use. Increased population has intensified the destruction of natural vegetation with concomitant alterations in fauna and in soil. Just as agricultural stages of occupance of the older arid lands had cleaned out fuel-wood supplies in brush and gallery woodland for miles around, the new cities are invading scenic landscapes, appropriating them for building and recreational sites, and stretching along improved roads beyond former limits of exploitation. The net effects of these physical changes are difficult to assess. If the ideal drainage area for catching water is a paved surface, some of the growing edges of the arid zone are approaching the ideal. Broader influences upon infiltration rates, sedimentation rates, and evapotranspiration are less certain. Much of the arid zone — and particularly its semiarid fringes — is in a delicate balance that, once triggered by modified vegetation or drainage, may be severely dislocated. Knowledge of fundamental processes in the hydrologic cycle is growing but still does not permit refined estimates of effects of changed land use and technology upon extreme flows, water quality, and total water output.

Water Use Versus Supply

In this setting of social flux and physical development and degradation, a dominant fact is that for the first time in many areas the volume of consumptive water use has approached the potential supply. Because it has been common to describe arid lands as places that are deficient in precipitation, there is a tendency to think of them as being short of water. If potential water supply is taken as the volume of water available to an area from surface or underground sources, regardless of the amount effectively available through engineering works, then there are only a few drainage areas in the arid lands where use yet approaches potential supply. Shortage stems from political ineptitude in developing supplies, or from the heavy cost of doing so, rather than from limits of potential supply. Paradoxical as it may seem, in the arid zone large volumes of water annually evaporate from playas or flow into the sea without having served human needs.

To be sure, basins like the Ili and the upper Salt already are used virtually to full capacity: no flood spates go uncontrolled, no water stands until evaporated purely by chance. On a larger scale, the entire southwestern United States is approaching that condition. Consumptive use has not come close to potential supply, but withdrawals are nearing the mean runoff and, with re-use of irrigation waters, are exceeding it in some areas. On the other hand, great basins such as the Nile and the Tigris-Euphrates still discharge 40 percent or more of their mean flows without economic return. Indeed, the prevailing condition in much of the arid zone is one of large-scale losses by evaporation and transpiration of water that has served no human use. This margin generally is dwindling. Under the impact of new technology and new uses, the physical limits are in sight.

Always before in arid areas the threat of water shortage has been accompanied by knowledge of unused potential supplies awaiting exploitation. Now, as limits loom ahead, only the most heroic of schemes for long-distance transport or for new technology can stand in the way of compulsory choices among alternative uses for a given supply. Engineers and ranchers and civic leaders have known that physical supply has clear limits, yet much of their action, as in both Arizona and California, presupposes that somehow, somewhere, the supply can be augmented.

A sobering aspect of any attempt to assess arid-zone resources and potential is our state of knowledge as to basic natural processes. At many critical points it leaves us in doubt as to the grounds upon which new technologies can be developed. Ignorance of some of the vital mechanisms still is great. The mechanism by which rain droplets form in a cloud, the precise fashion in which evaporation takes place from land surfaces, and the means by which mineral nutrients in the soil become available to plants are among the baffling unknowns. Descriptions of hydrologic relations are largely empirical. To the extent that scientific understanding of such processes can be deepened, the way may be paved for more efficient water use plans, and perhaps, for drastic innovations in such directions as weather modification and irrigation agriculture.

Revision of Public Attitudes

Important as it is to fill the gaps in scientific knowledge, it seems likely that even more urgent is a drastic reorientation of public attitudes toward water. Unless water can be seen in a different role, much of the new research and investment may yield meager and frustrating returns. The prevailing view of water in the social and economic growth of arid lands faces revision in at least three regards.

Needed: Water Budgeting

The most obviously needed revision of attitude, and probably the most difficult to achieve, is the recognition that using more water may not be the panacea for water ills. The typical solution for water problems in the past has been to go deeper or farther for water or to irrigate more land or to build a bigger dam. Many an area has been like the proverbial drunkard who thinks the best corrective for excess use is another little drink, or a large one early the next morning.

One of the more startling dilemmas in the arid lands is the rate at which old irrigated lands are lost while new ones are wrested from the sagebrush and mesquite. Salting and waterlogging are the two most active thieves in the ancient lands of southwestern Asia and northern Africa.

While ambitious programs for harnessing the Indus, the Tigris-Euphrates, and the Nile are commanding both public attention and public funds for the enlargement of irrigated acreage under new storage dams and canal systems, equivalent acreage is quietly going out of cultivation. Without fanfare and with only limited police measures to stop the robbery, salt accumulation is taking as much as 40,000 acres a year in Pakistan. More than 60 percent of the irrigated lands of Iraq are seriously affected by salting, the product of over-application of water and unsuitable soils. In the fruitful delta of the Nile, as much as one-quarter of the land is threatened by a high water table resulting from inadequate farm drainage. Declining yields of cotton testify to the curtailed productivity. There now is doubt that the investment in massive new works is any more than offsetting the inconspicuous dissipation of old investment in farms that are abandoned to salt-crusted fields and soggy soils.

Cities are also beginning to steal irrigated land, as in Southern California where the demands of urbanized areas are accompanied by a retreat of irrigation and an intensification of agriculture on the urban fringe.

In the western United States, according to a recent report by the Agricultural Research Service, it would be practicable to enlarge the irrigated acreage from 29,600,000 to 55,500,000 acres by increasing the total intake of water by 307 million acre-feet if no improvement in efficiency were to be made, and by less than 200 million acre-feet if there were acceptance of known and tested methods of conserving water as it flows in streams, canals, and distribution ditches so as to cut the losses from seepage, phreatophytes, and excessive irrigation applications. Irrigated land could be increased greatly at the cost of making better use of the water already in effective supply. Intelligent use of water budgeting, for example, might achieve tremendous savings in water without altering the usual methods. It would be sanguine, however, to think this would come soon. Much easier in a political sense than altering the irrigation practices and canal system in the Gila or the Sind is constructing a new project to apply more water to another area by the same inefficient methods.

Much the same situation applies to industrial use of water in arid lands. Although there are successful instances of low water intake per manufacturing product, as in the case of steel production in the Great Basin or petroleum refining along the Texas coast where intake per unit of product may be as little as one-thirteenth of intake in some other plants, the prevailing tendency in national as well as international projections of water use is to assume that new industry will continue its present habits of lavish water use. Projections of water need, assuming currently wasteful practices will persist, may become self-fulfilling prophesies. They predict shortage and encourage the water user to think that his expanded activities inevitably lead to shortage. The common solution where potential supplies seem limited is to seek new supplies rather than actively to encourage widespread economies in water use. Moreover, much water law permits the careless user to remain entrenched in his indolence.

In disposing of waste a similar attitude prevails. During recent decades relatively little attention has been given to improved methods of treating waste; much more concern is expressed for regulating streams so as to provide adequate dilution for the growing volume of untreated or partially treated waste.

Several conditions help explain this widening gap between knowledge and application in water use, and the tendency to rely upon more water rather than better management. Illiteracy, poor adult education facilities, water law and administration, pricing systems, inflexible social organization, lack of capital, marketing and production methods, and the high risks of transition are among the reasons.

To put the argument more sharply, the cure for threatening water shortage is not necessarily more water. Indeed, some predictions of shortage may hasten or assure the shortage by discouraging remedial action which may be taken with present supplies. The farmer who is running short may find the needed supplies in seepage from canals, in excessive applications, in evaporation from reservoirs. A manufacturer may decide to move into a water-short area, knowing that by waste reduction, recycling and reuse, he can operate as effectively and even as profitably as at a site where water is abundant. As technology for water use advances, and as the possible gap between knowledge and application enlarges, the opportunities to benefit from water-saving measures will increase. The cure for water shortage may be the use of the available knowledge of how to deal with water.

Needed: A Fair Price on Water

A second public attitude facing revision has to do with value of water. It is shifting, but so slowly as to promise little effect before the pattern of water development in many basins has been frozen into dams, canals, and court orders that sturdily defy revision.

Like air, water is accepted as a God-given commodity outside the ordinary pricing system. Throughout large sectors of the arid zone initial water rights are acquired at nominal charge, and no tax is levied on water holdings or water use. There is a widespread assumption that water will be provided by public agencies at cost. Market mechanisms rarely are used to determine a fair price for water, and when a substitute for cost is sought it generally is in the direction of estimates of what the user can bear. Implicit in much of the public policy toward irrigation development in arid lands is the belief that water regulation and the cultivation of new lands are inherently good, and that water, while an item of inestimable value for national growth, should be treated as an item of lowest possible value for computing costs and returns.

This is illustrated by the rate schedules for municipal water supplies. In the days when the germ theory of disease was still not widely accepted, and the back-yard well and the outdoor privy seemed good enough, public and private water companies sought to promote new connections and the installation of water-using equipment by rates favoring large use at low cost. The more a customer used the less the unit cost. These promotional rates still prevail, even though leaders in the water industry decry the under-valuation of their commodity with its resultant restrictions upon service and plant expansion. A city such as Denver, while compelled to ration water for lawn sprinkling during dry periods, encourages maximum intake among its customers, reaps the financial benefits of enlarged revenues, and earnestly seeks additional supplies. Tel Aviv and Karachi, rapidly expanding cities in arid situations, until recently based their water charges upon the assessed valuation of property. Under this scheme the owner of a large establishment is impelled to use very large quantities of water since he is paying for it anyway. In Chicago, with annual rainfall of thirty-seven inches and a tremendous source within a mile of the shore line, a middle-class family pays as much for water as does a similar family in Boulder, Colorado, with annual rainfall of seventeen inches and a glacier source twenty miles away.

Recent pioneering studies sponsored by Resources for the Future give fresh perspective on the economic productivity of water. The San Juan investigation suggests that in terms of returns in employment and volume of production, an acre-foot of water used in industry may be a hundred times as valuable as if used in agriculture. Recreational use of water flowing in a fishing stream may yield returns ten times larger than those from the same volume of water in an irrigated farm downstream.

Obviously, these values presume a demand for the industrial and recreational use, although it seems ridiculous to envisage the deserts blanketed with factories, homes, and playgrounds. To the extent, however, that national economies may favor such development, the systems of planning and regulating new water development may require heavy revision, for irrigation will have to give way to other claimants, and pricing systems will have to take account of other demands if wise allocations are made. It does not seem too much to expect rate schedules that charge least for the first gallons used and increase rapidly as usage exceeds a reasonably efficient level for the particular type of use. Nor is it unlikely that means will be found to expedite the transfer of water rights for new uses with adequate compensation for the present owners. The full social consequences of this type of transfer are not yet apparent.

The people of areas such as the Santa Cruz Basin or the lower Indus must recognize sooner or later that water may be more useful in the service of industry or recreation than of agriculture. Their cities may grow literally at the expense of fields, and farming operations of crops which can be grown elsewhere at similar levels of cost may be transferred out of water-short areas, letting the transport of the product substitute for transport of water into the area.

Current pricing systems, water rights laws, and public attitudes are largely opposed to readjustments in use priorities. They can be expected to change slowly at best, and then only in response to severe cases of misallocation or to persistent public education that paves the ground for revisions in public attitudes and institutions.

Needed: A New Look at Water Programs

A third shift is needed in attitudes toward investment in water facilities. They have come to enjoy a kind of sanctified status in arid lands that sometimes belies the facts of their effects. From the Helmand to the Tigris-Euphrates and the Zambezi there exists a sort of magic about investment in water storage and transportation, a spell of solid confidence that to be sure does not necessarily fall over the international banker or the taxpayer of more humid climes, but that inspires enthusiastic public action within the arid lands.

A simple example is in public investment in flood protection, a program claiming heavy federal outlays in cities of the arid United States. Since the national flood-control policy was initiated in 1936, more than seven billion dollars have been expended on reservoirs, levees, and channel improvements to curb the annual flood losses. Twenty-four years later the toll of flood losses remained high. The explanation lies in part in methods of counting losses and in the incidence of floods, but in large measure it rests in continuing encroachment upon natural flood plains. Even in basins with elaborate reservoir systems, new building in hazard zones has tended to offset greater protection so that while the Corps of Engineers valiantly beats back the enemy on one front their position is threatened from the rear.

This experience has lessons for other water-regulation programs. By assuming that only engineering works were needed to curb the cost of unruly streams, other possibly effective means were neglected. Little or no attention was paid to such alternatives as land-use regulation or flood-proofing of buildings. By assuming that the engineering works would do what the benefit-cost calculations had solemnly estimated they would do, without attempting to verify the practical results in land use, the public reaped quite different effects. A single-purpose channel improvement may invite further encroachment upon natural flow ways; a single-purpose levee may set a confident scene for later catastrophe; a single-purpose reservoir may appropriate a unique dam site without assuring complete reduction of flood losses.

Another case is public investment in irrigation as a means of stabilizing grazing. For a long time new federal reclamation schemes in the western United States have been justified in part as measures to protect the economics of nearby grazing areas from fluctuations in forage supplies. When one of the older projects in this category, the Uncompahgre in western Colorado, is examined, it is found that the grazing industry in fact was dislocated by the investment: the combination of reclamation and forest land policies attending the project made the ranching economy less stable than before and probably supported further deterioration of the range.

A prevailing attitude toward water in arid lands is that public investment, usually framed to stabilize the economy and promote growth, will yield the expected returns. Water, having been vital to past development, is taken as the easy key to the next stage. But water, although necessary, may not be sufficient, and the other requirements of sufficiency may be hard to define. It is much easier to get one hundred thousand dollars to investigate a new project than to get one thousand dollars to find out what happened after an earlier project was completed. Methods of assessing the effects of investment in water now are being refined, and while much remains to be learned as to the redistribution of income gains and losses, rough estimates are practicable.

When questions of this sort are raised about the impact of water-development projects, the traditionally complacent view of "the program" for a given basin is shaken. To re-establish confidence that public aims are in fact being met requires more searching studies and a wider range of choice than ordinarily is available. In practice, new water projects are put forward as the best judgment of an engineering staff which has investigated hydrology, geology of dam sites, soil conditions, structural requirements, and expected demand for water. A single plan judged most feasible is recommended for legislative or administrative approval. The choice is between "yes," "no," or "yes" with minor modifications. Only passing reference is made to the hundreds of other water-control plans that might have been designed as alternatives, or to the possible strategies of investment, promotion, or regulation in other sectors of the economy which might achieve similar results.

If the public welfare is taken as the broad, albeit vague, aim for an arid basin, and if industrial and recreational production are seen as principal features in sound economic growth, it may well be that water development may be a necessary though secondary aspect of investment promoting cheap transport of raw materials, efficient marketing facilities, and amenities conducive to a stable labor force. We can only speculate at this stage as to what might be the effects of investing in such facilities in lieu of certain of the great dams under construction in the upper Colorado Basin. It must be speculation because there have not yet been full plans which provide the executive and legislative agencies with a range of choice. The nearest approach has been in some of the International Bank studies of investment needs in under-developed countries.

Analysis of the New Technology

At this point my argument may be challenged on the ground that having begun with the tremendous repercussions of new technology in water development I do not properly credit the possibilities of future advances in that direction. It may be asserted, for example, that saltwater conversion or evaporation repression or weather modification may so enlarge the potential or effective supplies of water as to render invalid much of what has been said about shortage of supplies. There is some support for this, but it is shaky.

Take the case of salt-water conversion. A few years of international experimentation has spawned what is almost a modern mythology about brackish water. Important improvements have been made in methods of removing minerals, and experimentation with both pilot and laboratory plants is proceeding on a wide front with fruitful collaboration among the technologists of a score of nations. The costs thus far do not promise salt water as a source for more than a few situations where urban needs can support costs of conversion that exceed the total cost of delivering fresh water supplies. They are still far above common irrigation water costs. Demineralization for agricultural use seems more likely to develop where a slight reduction would bring brackish water below toxic limits for irrigation or stock use. Undoubtedly brackish water will be cheaper to convert and more widely used in the future, but evidence does not yet warrant expectations that its availability will work miracles in the use of most arid lands. Possibly more promising would be a scientific development that would permit a reduction in transpiration, a process that accounts for more than two-thirds of water use in some areas.

Even if the more optimistic assessment of new means of removing salt or increasing rainfall were to be proven sound, their proper application would be hampered by the very conditions that hamper the use of technologies already in hand. Until they have been developed so as to be practicable alternatives to present water sources, the public emphasis upon them diverts attention from basic scientific research and from applications of other water knowledge. The prospect of going to salt water for additional supplies induces complacency toward current shortages and toward the exploration of other solutions.

We must find ways to prepare the patient and the dosage so that the promise of new technology becomes a stimulus to intelligent water use, rather than a sedative.

If the three previously discussed modifications of view are to be achieved, there will be need for a much deeper understanding of social processes in the arid zone than we enjoy now. It is not enough to know the available techniques or the combinations of them which would optimize economic returns. Only as there is understanding of the elements that enter into decisions governing water use by farmers or manufacturers or public servants can there be adequate programs to cultivate new patterns of use. We must find out, for example, how the resource managers view their land and water, what uses they think they can make of it, how they estimate the demand for and return from their products. In this fashion we may move toward recognition of conditions in which the farmer will stop over-irrigating his fields, the manufacturer will install recycling devices, and the state legislator will ask more about a proposed dam than how much it will cost and who will get the water from it.

By way of analysis, let us examine two cases of water use on quite different scales.

First, consider the lawn of one house in one oasis town of the western United States. Generally, the owner applies much more water than a water-budget analysis would show it needs. When in doubt he applies a little more. This is because the owner either doesn't know about or doesn't care about water-budgeting. He tends to use as much water as he is entitled to, and, indeed, the local water department hopes he will, so long as he doesn't exceed restrictions at times of peak demand. He resents any increase in rates as an infringement upon his inherited rights, and he thinks the more he uses the less he should pay per gallon.

An alternative would be to have no lawn at all and to cultivate a patio and border vegetation that gave him cooling relief with less water. He is slow to consider this or any other ways of dealing with his arid climate. The house is unsuited to such a change because it was designed as though for a Connecticut seacoast village. His midwestern tradition tells him a neat lawn is a mark of respectable status. Moreover, watering, in either the homely tug-of-war methods of yesteryear or the missile-control-board method of tomorrow, has therapeutic value after a trying day. Here is the water user who thinks more water is the sure solution for a dry environment, consistently undervalues his commodity, and is reluctant to consider any alternatives to his heavy usage.

Next, consider the state water plan for California. The plan essentially provides for moving water from surplus areas to deficit areas by a system of reservoirs and aqueducts. What I find lacking is a careful appraisal of how much water might be gained for the state by applying other means of using or re-using water, or what effect differences in the price of water could have upon its use, and of alternative measures which might be taken to guide the redistribution of income, land use and water in the state. Perhaps such appraisal would reveal the approved plan to be the best. Whether or not it did, a state facing gigantic expansion in its use of water deserves to know the best available answers to these questions before making its decision.

Because the arid lands are currently marked by widespread social change, it is not unreasonable to expect that views toward water may be revised drastically. To deal intelligently with conditions created by unfolding technology, new uses and dwindling margins of supply, will require new orientation as to the role of water where the potential supply is small relative to natural demands. The gap between technology and application must be narrowed, the standards of value reappraised and the prospective impacts re-examined. This would add to the already long bill of needs for physical and biological research, another set for social science research. On the scale of a single lawn or an entire state, these new views of the role of water constitute a basic challenge to both research and public education. We cannot expect sudden revolution in attitudes toward water, but we can expect that as they are revised, many promising opportunities for making effective use of the limited potential supplies will be realized.

CHAPTER 6

DESALTED SEAWATER FOR AGRICULTURE: IS IT ECONOMICAL?

Marion Clawson, Hans H. Landsberg, Lyle T. Alexander
Resources for the Future, Washington, D.C., U.S.A.

During the 1960s and beyond, there has been mounting advocacy of desalting seawater for use in commercial agriculture in various locations of the world, especially in the Middle East. The process, it is contended in the early 1970s, is both technically and economically feasible, or soon will be, and its application on a large scale can produce additional volumes of food at competitive prices — the desert will blossom like a rose — and at a profit.

Although research on desalting techniques had proceeded for many years, supported by modest funds, in the Office of Saline Water of the Department of the Interior, and the Atomic Energy Commission had been exploring the role that nuclear energy might play in desalting, the entire matter suddenly acquired international interest after the six-day Israeli-Arab war in June 1967.

Within days after the war the London *Times* published two letters recommending desalting schemes in the Middle East. A detailed letter from Edmund de Rothschild suggested three nuclear desalting installations in Israel, Jordan, and the Gaza Strip, respectively. This provoked comments and questions in the House of Commons, which generally approved the idea or at least further exploration of it. On the other side of the Atlantic the United States Senate in December 1967 passed Resolution 155 without a dissenting vote. It says in part:

Whereas the greatest bar to a long-term settlement of the differences between the Arab and Israeli people is the chronic shortage of fresh water, useful work, and an adequate food supply; and
Whereas the United States now has available the technology and the resources to alleviate these shortages and to provide a base for peaceful cooperation between the countries involved:
Now, therefore, be it
Resolved, that it is the sense of the Senate that the prompt design, construction, and operation of nuclear desalting plants will provide large quantities of fresh water to both Arab and Israeli territories and, thereby, will result in
1) new jobs for many refugees;
2) an enormous increase in the agricultural productivity of existing wastelands;

NOTE: This chapter is a slight modification of an article appearing in *Science* 164 (3884) (June 6, 1969): 1141–1148. Copyright 1969 by the American Association for the Advancement of Science.

3) a broad base for cooperation between the Israeli and Arab governments; and
4) a further demonstration of the United States efforts to find peaceful solutions to areas of conflict...

The resolution was a direct descendant of the "Strauss-Eisenhower Plan," a proposal by former AEC Chairman Lewis Strauss for which he obtained Eisenhower's backing. The proposal gained its greatest popularity through an article written by Eisenhower (1968), which, with reference to the Middle East, states the proposition in its most optimistic form:

Now it looks as if we are on the threshold of a new breakthrough — the atomic desalting of sea water in vast quantities for making the desert lands of this earth bloom for human needs.... Since we now know that the cost of desalting water drops sharply and progressively as the size of the installation increases, it is probable that sweet water produced by these huge plants [billion gallon daily capacity] would cost not more than 15 cents per 1000 gallons — and possibly considerably less.... There is every reason to suppose that it could be a successful, selfsustaining business enterprise, whose revenue would derive from the sale of its products — water and electricity — to the users.... The purpose... is... to promote peace in a deeply troubled area of the world through a new cooperative venture among nations.

These and other basic documents, including Strauss' memorandum outlining the proposal, have provided not only a flood of newspaper stories and magazine articles, but have also accelerated government-sponsored efforts. Of the engineering studies, two are directed specifically to foreign areas. Following President Johnson's meeting with Israel Prime Minister Levi Eshkol in June 1964, the Kaiser company was commissioned to make an engineering study of the feasibility of seawater desalting in Israel (Kaiser, 1966). The Oak Ridge National Laboratory has produced a report on nuclear energy centers and agro-industrial complexes to be established in various arid areas of the world (Oak Ridge, 1968). The tone of these two reports is cautiously optimistic, but a more careful review of their assumptions leads to quite opposite conclusions (Clawson, 1968). Others have written uninhibitedly — as if sweet water were already flowing into the desert at low costs (Nikitopoulos, 1968).

The Kaiser study was specifically concerned with a desalting plant in Israel; the contemplated location of the plant was on the Mediterranean seashore about

9 kilometers south of Ashdod. The power plant would be a nuclear steam-generating facility using a conventional light-water nuclear reactor with a thermal rating of 1,250 megawatts. Essentially all of the steam generated in the reactor passes through the generator without condensation; the steam exhausted from the turbine is condensed in the shell side of the brine heaters of the desalting plant. The evaporator structures consist of heat recovery stages, heat reject, and heat reject-deaerator stages. A multiple seawater intake structure would be located 450 meters offshore and 7 meters deep. An outfall facility would consist of a buried concrete box culvert with transition to an open channel beyond the desalting plant limits. The plant would have a capacity of 100 million gallons of desalted water daily; a plant operating factor of 85 percent is assumed. The generator would produce 200 megawatts salable electrical power at an estimated price of 5.3 mills per kilowatt-hour and an 85 percent power plant operating factor. The total capital costs are estimated from 187 to 210 million dollars depending on interest rate. Annual operating costs vary from 16.8 to 28.7 million dollars, for interest rates of 1.9 and 8.0 percent respectively. Crediting power sales against total costs, water costs per 1,000 gallons — conceived as the residual costs — range from 28.6 cents if interest is calculated at 1.9 percent to 67.0 cents if the interest is 8.0 percent. Several variations in structure and methods of operation are possible without major effect upon water cost.

The Oak Ridge proposal is for a major nuclear energy center, with industrial and agro-industrial complexes, as well as desalting works. It was conceived to be broadly suitable for several locations in the world, subject to specific site planning and adaptation. The proposal is based upon technologies expected to be developed over the next decade or two, not upon technologies tested and in application today — and that reason makes it less suitable for rigorous review since at *some* time in the future costs of both power and water production will presumably be lower than they are now. "Near-term" light-water reactors and "far-term" advanced breeder reactors were considered, as well as near-term and far-term desalting equipment; a number of alternative layouts were included, with industrial electrical power ranging mostly from 1,585 to 2,070 megawatts. Most of the designs would produce one billion gallons of desalted water daily. Investment costs in the nuclear plant and desalting works would range from 1.5 to 2.0 billion dollars for most designs. Various industrial processes, producing metals or chemicals and using large amounts of electricity, are considered. Alternative costs are estimated, with different interest rates and other cost factors. A highly advanced type of agriculture is assumed. The Oak Ridge complex would produce ten times as much water as the Kaiser proposal, at a cost at near-term technology of 17 cents per 1,000 gallons at 5 percent interest, 24 cents at 10 percent interest, and 32 cents at 15 percent interest, and about one-third lower at far-term technology. As pointed out in the report, these values are arbitrary, since the complexes are conceived as closed economies. They represent the incremental cost of adding one unit of water to an existing plant. But, in the size class here contemplated, these incremental costs will approximate the average cost sufficiently to stand as surrogates.

A second stage of research, which includes the outlook for marketing the expected increase in output, has been undertaken at Oak Ridge to adapt the general design specifically to conditions as they exist in a number of locations in the Middle East.

Our purpose in this chapter is to explore the economic feasibility of desalting seawater on a large scale for commercial agriculture in regions of extremely low rainfall. In preparation of the material on desalting costs we have made extensive use of the analysis of Paul Wolfowitz (1969) of the University of Chicago and a report by W. E. Hoehn (1967) of the RAND Corporation.

Desalting Seawater for Large-Scale Commercial Agriculture

Any program to desalt seawater for use in agriculture involves three closely interrelated components: (1) a source of energy, (2) a process for producing sweet water out of seawater, and (3) means for transporting the water, at the right time, to the place of its application for the growth of crops.

Each component is essential, and any part can set a technical or economic limit to the whole process. Much has been written about the first two components, but very little about the third.

Numerous sources of energy and methods of desalting exist, but we shall focus entirely upon nuclear power as the energy source and upon evaporation as the desalting method.

No persuasive case can be made for a preferred energy source. Indeed, one might simply stipulate a given cost of energy, from whatever source is locally most advantageous, and concentrate attention on the other two components. We are analyzing nuclear rather than fossil-fuel energy only because the proposals that have received most notice have been based on nuclear energy. First, the atom attracts both attention and funds. Second, there is a large well-funded atomic research establishment, certainly in the United States, staffed with imaginative and highly skilled thinkers who do not shrink from the novel and spectacular. Third, the larger the proposed installation, the greater the advantage of nuclear energy. And fourth, there are arid areas in the vicinity of seacoasts that are remote from other fuel sources and to which nuclear energy may offer less expensive access to the new technology of sweet water production. For these and perhaps other reasons the packages offered so far have contained nuclear energy as the energy source.

Regarding desalting, it safely may be assumed that

current technology offers no more feasible way than evaporation to obtain freshwater from the sea on a large scale. Even the so-called far-off, 20-years-in-the-future, technology used in the more favorable of the two Oak Ridge variants is based on evaporation. That a breakthrough — say a very efficient, stiff membrane — could change the picture goes without saying. But such breakthroughs are not now in view, despite much effort in that direction. Nor would they have a necessary advantageous association with power generation.

In short, we discuss the merits of the programs in the terms chosen by their proponents. Although costs would vary with the location of the plant and other environmental factors, it is possible to consider the problem in general and to reach conclusions which no specific application could significantly change. This is true even for the third component — the conveyance of the water once produced — provided that areas are eliminated which do not have suitable soils, or are too remote from the seacoast, or do not in other ways qualify.

Nuclear Energy

The reputation of nuclear power as a cheap source of energy understandably has reached the desalting field, once it was realized that the addition of power production (from any fuel) to a desalting plant represented a logical combination, wherever raising of steam was part of the desalting process. Nobody now doubts that electricity from nuclear power plants can indeed be fully competitive with that from fossil-fuel plants under certain conditions in certain areas. But it is also true that some of the enthusiasm of recent years was based upon circumstances unlikely to be repeated in this country or to be found at all in less-developed regions of the world. These include the large funds furnished by the government for research and development, and the initial input made by suppliers who quoted highly attractive prices for their generating equipment when it seemed essential to the spread of the new technology. For these and other reasons, some sober criticism has been directed at the evaluation of the outlook for nuclear energy (Hoehn, 1967; Sporn, 1968). This does not cast doubt on its basic competitive position but does question the extent of this advantage. There are four major considerations:

1. It will be a year or two before even one U.S. nuclear plant designed to be competitive with fossil-fuel plants has been in operation long enough to establish a performance record that would substantiate expectations. The large scaling-up in size of the equipment ordered for nuclear plants in 1967 and 1968 and the reliance placed in all cost estimates on minimum downtime make this especially significant. What little experience has accumulated from demonstration plants in the United States indicates that the high rates of availability, a *sine qua non* of low power cost assumptions, will not be easy to attain.

2. The cost of nuclear generating equipment for the long run is far from settled. Past reductions in equipment prices by manufacturers have turned out to be more in the nature of initial lures. Costs in 1967 and early 1968 were $30 or more per kilowatt installed above those of 1965 and 1966, allowing for differences in size (Hoehn, 1967). Nor has the trend abated (Brown and McGuire, 1967). The Atomic Industry Forum, 1968 ("The Nuclear Energy Industry — The U.S. Highlights of 1968," mimeographed) puts the case even more strongly: "The direct costs of constructing nuclear generating plants rose significantly in 1968. From a low in 1966 of about $100 they had increased some 30–40 percent in 1967, and there seemed to be a strong consensus that this year's increase was also 30–40 percent. While the costs of comparable fossil-fueled units also rose, the increase was apparently less abrupt." Costs of conventional equipment also have risen, but less steeply.

3. There is some uncertainty regarding the future costs of nuclear fuel, once the increased power generation begins to reduce the uranium supply and forces a diversion to higher cost sources of the mineral. But prices of competitive fuels cannot be assumed as constant either, so that uncertainty is the real problem.

4. Nuclear power plants, owing partly to the heavy cost of shielding and containment, require more capital than conventional plants do (Huenlich and Kruck, 1968). Interestingly, the same is true for the desalting phases. The far-term technology (combined flash-vertical-tube) requires more capital than does the near-term (multistage flash). Whenever a portion of the fuel is awaiting enrichment (or being enriched), being fabricated into fuel elements, awaiting loading, or undergoing cooling, it still represents capital investment and thus carries interest charges, no matter whether the utility or the supplier manages the fuel cycle. The steep rise in interest rates has been penalizing nuclear more than conventionally fired plants. No one knows whether, to what level and how soon, rates will begin to decline, but while rates are high they blunt the competitive edge of nuclear plants.

In addition to the factors cited which tend to make themselves felt in aggravated form in less-developed countries, some elements apply especially to those countries.

Economies of Scale

Electric power generation is a classic example of economies of scale; unit cost drops as size of the plant rises. This drop is especially marked for nuclear power plants, partly because the absolute cost of shielding and containment increases relatively little with reactor size.

Information available for 1967 shows a rise in capital costs per kilowatt capacity from about $130 for plants to 1,200 megawatts to about $180 per kilowatt for plants in the 400-megawatt range. This explains why no nuclear power plant smaller than 450 megawatts has been ordered in the United States since 1963, and why the smallest size plant which a utility can now order from a major U.S. supplier is 480 megawatts (Huenlich and Kruck, 1968).

However, these economies can be realized only when

certain other favorable factors are also present. In the less-developed countries they usually are absent. Chief among them are a large market for electricity and a well-developed power grid.

An engineering rule of thumb calls for reserve capacity equal to at least the largest single generator in the system to assure continued supply when that unit is out. In fact, to keep the system from collapsing in case the nuclear plant trips out, prudent engineers advise against installing in a small isolated system a nuclear plant that is larger than 10 percent of the peak load.

How many countries are there that fulfill the conditions which permit them to benefit from the economies offered by large nuclear reactors? To be competitive with power from conventional sources, "a reactor of 500 Mw, now about the lower limit in size, would . . . have to produce not less than 3.5 billion kwh per year. Presently, there is only a handful of countries outside of North America, Europe, and Oceania that consume that much electric energy per year *altogether*. And, of these, only Argentina, Brazil, Japan, India, North Korea, and the Republic of South Africa consume greatly more" (Landsberg, 1968). Even though power markets will, of course, be larger 10 and 20 years from now, it is precisely this circumstance, the "low-demand trap," that has led to the search for an adequate and reliable market, hence the recent work on agro-industrial complexes as built-in consumers.

Availability of Capital

Because nuclear power plants require a large capital outlay per kilowatt of installed capacity, the availability of capital is of great importance. Most of the countries which could best use additional power — and water — are seriously short of capital; alternative investment opportunities exist which can earn interest at much higher rates than are customary in the United States even now. Israel, for instance, permits a legal maximum of 11 percent, and the demand for loans is usually greater than can be supplied at this rate (*Quarterly Economic Review* [of] *Israel,* 1966). Higher rates are paid in various ways. It is certainly doubtful if any country which could use a large desalting project based on nuclear power should count on having to pay less than 10 percent interest per year. If for political or other noneconomic reasons the United States should decide to provide a plant in a country on a subsidized basis, any interest rate could be used in the calculations. Without discussing the merits of subsidization, however, current efforts to portray nuclear desalting as having come or about to come "of age" are based not on subsidized but on market conditions.

Costs of Equipment

Power plant costs would almost surely be higher outside the United States, especially in countries that have been mentioned as candidates for desalting plants (Hoehn, 1967, p. 165). The reasons include costs of shipment of equipment, lack of supporting industries and their production, shortage of national specialists and construction crews, and longer construction time. Only a small portion of these increases might be made up by procurement of some of the equipment from lower cost sources abroad.

Operating Performance

The cost of servicing a nuclear power plant is likely to be higher than in the United States, for reasons similar to those just cited. The intrusion of a highly complex technology into an environment not geared to it is bound to result in lessened effectiveness and higher cost of maintenance and operations generally. Although the record of power availability of nuclear plants is anything but good in the few plants that have so far operated in the United States, it is likely to be poorer in the less-developed areas of the world unless the plant can be run as a virtual enclave, and even then a good record is by no means certain. This is not to say that improvements would not gradually be attained, as they are bound to be attained in the United States. It is reasonable to believe that the economics of nuclear desalting, examined alone, would make the first plant ordered in any less-developed country disproportionately large, and it would be a long time before a second and third could be built. Thus the initial plant would for years bear the burden and cost of serving both as an economic input and as a training and experimental facility.

We do not wish to appear unduly pessimistic. Not all of the adverse factors need come true, but some are sure to be felt. And there is little in the picture that points to the emergence of unforeseen favorable elements, at least not without consideration of other energy sources. A good deal of what has been said above would not be true of fossil fuels, abundantly available in the Middle East, where vast amounts of natural gas are flared, and where the marginal cost of crude oil is extremely low. Moreover, the economies of scale would be less pronounced and size problems somewhat alleviated.

But for the moment no such proposals have caught the public fancy, although there is no generally valid technical, economic, or other connection between nuclear energy and desalting. Indeed, from the viewpoint of international complications, the association of desalting with nuclear energy probably represents an obstacle rather than an aid to achievement of the economic objectives in some parts of the world. A first indication of change in that direction could be the proposed bill transmitted to the Senate by the Department of the Interior on 17 January 1969, three days prior to the Administration changeover. It would authorize United States participation in a dual-purpose plant to be erected in Israel. An upper limit of 40 million dollars would be placed on any grant made to help finance the desalting techniques and necessary modifications in the power production of the plant. Financing of the balance, as well as the choice of energy source, would be determined by the Israeli government.

Efforts to overcome these difficulties have taken two forms. One has been to present an optimistic picture of the expected costs of water by making highly favorable assumptions for cost and output factors. The other has been to broaden the scope of operations beyond production of power and water and to test the feasibility of a large agro-industrial complex in which the large volume of electricity that cannot be absorbed by the ordinary demand of the country can find a ready market.

First, Kaiser used plant costs based upon a price schedule that became quickly outmoded, as we have pointed out, and has led to later revisions (Clawson, Landsberg, and Alexander, 1968). For a detailed review of the Kaiser Engineering proposal, see Hoehn (1967). We have not so far seen any similar careful review of the Oak Ridge study. The cost changes would matter less if prices of crops had undergone similar increases, but they have not. Second, interest charges, and fixed charges based upon that interest, have been unrealistically low. Even the highest variant has an interest charge of only 8 percent. The assumed downtime for the generating plant is 10 percent, certainly a highly optimistic assumption over the lifetime of the plant (Wolfowitz, 1969). And since the cost of water is arrived at by deducting from the total costs of the dual plant the income from selling the large amounts of electricity that are not consumed internally, a constant price of that electricity is assumed for the lifetime of the plant. This method discounts the possibility of slowly falling revenue stemming from a declining power-cost level in the economy as a whole. To the extent that the income from electricity sales is maximized, the "cost" of water, as the residual, is minimized.

The Oak Ridge study does make some allowance for higher costs outside of the United States, and it sets out a wide range of possible interest rates. But in terms of the Oak Ridge concept there is no need (and perhaps no basis for doing so) to determine separately the cost of either power or water, since it is the returns for the operation as a whole that measure the profitability of this closed complex.

We shall deal later with the various assumptions, but one needs mentioning here. In both studies joint production economies are reflected in the cost of water. While such a subsidy from one part of joint production to the other may be wholly desirable in a given case, it is apt to mislead those who are interested in the cost of desalted water regardless of its association with power production. In fact, the popular discussion has fastened on precisely the costs of water that have emerged from such studies without awareness that water cost is to a substantial degree a function of the price at which power can be sold. In this connection the Oak Ridge study straddles the fence. Although its basic concept of an integrated complex renders the costing of either power or water meaningless, it fails to exploit this advantage in a consistent way and presents both costs separately, albeit in a somewhat off-hand manner, and, it must be said, to the decided detriment of the entire exercise; for it leads the reader to marvel at the agricultural sector calculations being based predominantly on water at 10 cents per thousand gallons, when the rest of the study clearly spells out that no such water is in the offing, not even in the far-term model 20 years hence.

The same phenomenon has turned up in the case of the dual plant in the Los Angeles area. Before the plan was tentatively shelved in mid-1968, the estimated cost of water at the plant had risen from 22 cents to nearer 40 cents per thousand gallons. But "20 cents per thousand gallons" has left a lasting impression with well-intentioned but ill-informed writers and speakers.

Desalting Process

Much of the uncertainty to which we have drawn attention in the discussion of the energy-producing component is presented in aggravated form in the desalting component. Here too, the scale of operations proposed in each instance is greatly in excess of anything that has so far been tried, although in the Kaiser proposal the large capacity is reached by replication of small base-modules that are only five times the size of anything now in operation.

In each proposal the nuclear power plant and the distillation plant would be closely linked. Anything which led to a shutdown in one would force an early shutdown in the other, although planned maintenance in either process might be carried out during forced shutdowns of the other process. The schedule for a power and distillation plant in the Kaiser proposal calls for demanding availability of 85 percent jointly, or a downtime of 15 percent. There is little on which to base an appraisal of this assumption. But it may be noted that the Point Loma demonstration plant of the Department of the Interior, prior to its transfer to Guantánamo Bay in Cuba, had an availability of only 70 percent (Wolfowitz, 1969). Since that was an early plant, one would expect later ones to operate more continuously, if it were not for the type of difficulties that the Point Loma plant experienced. A serious one that the Kaiser plant does not seem adequately to have taken into account was the problem of drawing water out of the ocean. Many materials obstructed the intake pipe — kelp and sand and silt, fish, even large stones. Large and expensive stilling basins must be installed if such difficulties are to be avoided, and one may assume that the type of difficulty will vary from one location to another. "It is evident from our observations both at San Diego and elsewhere," an engineering evaluation states, "that the importance of trouble-free intake systems either does not get through to those responsible for the design of the system or that there exists a tendency to skimp in the design in order to reduce costs" (Foster and Herlihy, 1967).

Another unknown is the discharge of hot and bitter brines in volumes 100 times and more than that of existing plants. This could present awkward problems. At the minimum, the discharge point must be removed by considerable distance from the intake point to prevent even partial recirculation of ever-saltier water; expensive piping out to sea may be required (Powell, 1967, cited by Wolfowitz, 1969). Adverse ecological consequences of dumping these wastes are inevitable. Neither they nor the possible costs of dealing with them have received attention.

Although each report is concerned with the future, some comparison with present plants is sobering. The lowest cost plant operating today (providing water for Key West, Florida) produces water for 83 cents per 1,000 gallons, but with a subsidized interest rate loan from the federal government; without the loan the cost would have been close to one dollar per 1,000 gallons. In 1966 two private utilities in Southern California, the City of Los Angeles, and the Metropolitan Water District of Southern California, assisted by funds from the U.S. Department of the Interior and the Atomic Energy Commission, entered into an agreement to build a desalting plant on man-made Bolsa Island in Southern California to serve an urban area with a high demand for both electricity and water. The plant was to produce 150 million gallons of desalted water daily and have a generating capacity of 1.8 million kilowatts of electricity. The originally estimated cost, including water conveyance and power transmission, was 444 million dollars; by the summer of 1968 estimated costs had escalated to 765 million dollars, owing to a greatly lengthened construction period, increased equipment and interest cost, stricter design criteria, and, one of the smallest items, a 10 percent larger power output. The project is uneconomical at this price, and the proposal in its present form has been shelved.

If desalting of seawater is not economical in Southern California today — where alternative water must be brought long distances at high cost, where electricity surely has a ready market, and where much of the water would not go to agriculture — then where is large-scale desalting of seawater economical? If it is not economical at an interest rate of 3.5 percent, and at the lifetime capacity factor for both water and power of 90 percent, assumed for this venture, then what are the prospects under less generous assumptions?

These recent experiences may not apply to desalting costs in the more distant future, but they are at least sobering. This is particularly true when one attempts to make corrections both for the unrealistically low fixed charges, and especially the interest rate (the Oak Ridge proposal is the most realistic in that respect), and some allowance also for the optimism incorporated into the estimates at various stages. Wolfowitz (1969) has tried to make adjustments for the proposed Israeli plant. Using as his point of departure the lowest estimated cost of 28.6 cents, based on an interest charge of 1.9 percent, he demonstrates persuasively that the likely contingent expenses not included would bring the cost to 40 cents per 1,000 gallons at the farm. If adjustment is then made to a more realistic but still modest fixed charge such as 10 percent, the resulting cost of water at the farm would rise to somewhere between 90 cents and one dollar per 1,000 gallons.

Application of Desalted Seawater to the Land

The third, and most generally neglected, aspect of desalting seawater for use in large-scale agriculture is the conveyance of the water from where it is produced at the edge of the sea to the land, which may be some distance inland and at a much higher elevation. The desalting plants discussed above will produce water in a constant stream (except when shut down for repairs or servicing), but the farmer wants water in a different time sequence during the year. In some way, water must be stored and transported, from one time and place pattern to another, and substantial costs will be incurred in doing so. Much of the discussion of the economics of desalting seawater overlooks this point; someone will compare the costs of water at the plant (usually grossly underestimated) with the value of the water at the farm (usually grossly overestimated).

In an arid region, irrigation water is essential for successful production of most crops, but so are several other inputs. The farmer combines them all into the farm operation program for production of crops and livestock which, in view of prices, costs, and markets, seems to him most likely to produce the greatest net income. The resulting time sequence of irrigation-water use is usually highly seasonal in character, its exact pattern depending upon climatic factors as well as upon choice of crops and methods of crop production. Modifying the farming program to smooth out the seasonal demand somewhat for irrigation water is possible in some areas and under some circumstances, but this modification is very likely to reduce income, sometimes substantially, from the whole farm operation. By and large, for desert and arid areas where desalted water might be used, a markedly seasonal demand for irrigation water is certain, if the farmer is free to choose when he takes water; demand for off-season water may be low.

The problem of storing and conveying water from desalting plant to farm will vary greatly from one location to another, but some generalizations may be made. Desalted water, in excess of immediate need, might be stored in surface reservoirs or underground aquifers located en route or not too distant from the place of either production or application, or in the soil of the farm. In each case, some water — often a great deal — will be lost through evaporation or percolation or both; water stored in the soil may pick up salt — a great deal in most desert soils. Evaporation in most desert areas is high, often 10 feet or more annually from a water surface. There may be no suitable reservoir site; in any event, dams cost money to build. Soils and aquifers may have a low water-holding capacity or intake rate. Also, some means must be provided for carrying water by large conduits, pipes, or canals from the desalting plant or storage site to the border of each farm. In the United States, this has proven rather costly even when the water source was available by gravity flow. If the arable lands lie at some elevation, pumping costs will be considerable.

In the Kaiser report, the water-conveyance facilities and electrical transmission lines are not included. It is stated that they would add more than 15 percent to the investment. The water cost estimates are based upon 310 days annual operation of the desalting plant, but no provision is made for storing this water at times of slack demand and no allowance is made for pumping costs. There are few good surface reservoir sites in Israel. The same limestone formations which allow infiltration of natural precipitation that could later be salvaged as groundwater also are the cause of leaky reservoirs. The most suitable lands in Israel near the proposed desalting plant lie at an elevation of 500 feet or more; pumping costs, even with relatively cheap electricity, would be considerable. The cost of taking desalted water from the plant to the field includes (1) losses in transport, (2) pumping costs, and (3) costs of conveyance to the farm, including distribution canals or pipes. By far the greatest of these is likely to be water loss. A 10 percent loss of water would raise the cost of the remaining water by 11 percent, a 20 percent loss by 25 percent, and a 30 percent loss by 43 percent. The more costly the desalting process, the more costly the loss of water in storage or in conveyance.

Pumping costs depend primarily upon lift and distance. Even with high pump efficiency, lifting water requires somewhat more than one kilowatt-hour for each foot of lift for an acre-foot of water (enough water to cover an acre one foot deep, or 326,000 gallons). A 500-foot lift, as would be necessary at the most frequently mentioned Israeli site, would require about 640 kilowatt-hours of electricity; at 5.3 mills per kilowatt-hour, the rate at which the Kaiser report estimates electricity can be disposed of, this would still mean nearly $3.50 per acre-foot for energy; depreciation, maintenance, and interest on pumping equipment would probably add as much again. Finally, there are the costs of construction, maintenance, and operation of a canal or pipe system. The annual cost, including interest on capital, could hardly be less than $3.00 per acre-foot.

The Kaiser report, on the basis of 8 percent interest on invested capital, arrives at a cost of 67 cents per 1,000 gallons at the plant, or $218 per acre-foot. On the basis of the foregoing calculations, an overall loss of water of 10 percent (representing a much higher loss on the volumes actually stored), plus the other costs, would add about $34 per acre-foot to the cost, or 14 percent. If the overall loss were 20 percent, the lost water would add $55 per acre-foot to the cost of the delivered water; with the other costs, total costs incurred between distillation plant and field would be $65, or a 30 percent increase.

If all calculations in the Kaiser report were retained, but the interest rate raised to 12 percent, the costs of desalted water would be in excess of 75 cents per 1,000 gallons. If 20 percent were added for conveyance costs and losses, the delivered cost at the farmer's field, on the time schedule he wants the water, rises to 90 cents or more per 1,000 gallons.

If one accepts the Oak Ridge calculations but uses an interest rate of 12 percent, the cost of desalted water at the plant is 28 cents per 1,000 gallons; if 20 percent were added for conveyance costs and losses, the delivered price becomes 34 cents; and taking into account all the variables discussed, it seems realistic to count on a delivered cost of at least 40 cents per 1,000 gallons, or $130 per acre-foot. It should be noted that some of these additional costs, here incorporated in the cost of irrigation water, are allowed for in various ways in the Oak Ridge scheme under various capital charges of the farm enterprise. Thus, comparisons are difficult because the cost of the water remains unchanged from its cost at the outlet of the desalting plant. But primarily, it is larger size and assumptions of less costly future technology that explain the lower costs of the Oak Ridge study as compared to the Kaiser study.

Value of Irrigation Water

The value of water for irrigation, whatever its source, is affected by many variables — climate, soils, associated inputs such as fertilizer, markets, efficiency of farmers, competition from other producing areas, and many others. Throughout the whole world, water is rarely sold on a market, hence one must estimate "shadow prices" for the irrigation-water supply. It is extremely difficult to determine the *actual* value of irrigation water, but not difficult to say how it should *not* be determined.

First, the value of irrigation water to be developed by the two desalting projects cannot be determined on the basis of what a few farmers could pay to produce a highly specialized crop for a special market. There has been much loose talk about production of "winter vegetables," for instance; aside from the fact that this type of agriculture has never been the gold mine that some think it is, and that competition among producing areas in the future will reduce whatever large profits may have existed in the past (it is hardly legitimate to assume that the advantages of new technology will not be available to other, similarly situated areas), the scale of the Kaiser and Oak Ridge projects precludes this type of agriculture for more than a small fraction of the water to be produced. One hundred million gallons a day for 310 days in the year — 85 percent availability — in the Kaiser project, are nearly 100,000 acre-feet annually, or irrigation water for perhaps 35,000 acres of summer crops and much more of winter crops; the Oak Ridge project is ten times as large.

Even 35,000 acres is not much less than the total acreage of all vegetables grown annually in Israel, of which only a small fraction are exported. Such an acreage of winter vegetables could not be grown at any single location for the home market, and if exported would have disastrous results in terms of prices of products. True, tomatoes — greatly desired as a leading export — are grown in Egypt on some 200,000 acres, but exports in 1965 were the equivalent, at prevailing yields, of the harvest from 40 acres! Even in 1960, the best recent export year, exports came from the equivalent of 700 acres. The task of escalating from such levels to those appropriate to the magnitude of the desalting plants is truly overwhelming. Such comparisons and our ignorance concerning the characteristics of the specialty markets lead one to conclude that crop production from large-scale desalting works must be primarily staple, not specialty, crops.

Second, one cannot safely assume that all the increase in value of output resulting from irrigation will, or can be made to, accrue to the irrigation water; this is a trap into which economists around the world have fallen repeatedly. The quality of the labor and the management which will be required under the more intensive irrigation farming will demand, and can get, higher returns than the kind of labor and management which sufficed for the less intensive agriculture that the new irrigation replaced. Moreover, farmers and other landowners the world over have demanded and have secured some part of the increased product resulting from irrigation as a reward for their land. Further, to attract the capital needed for the new irrigated agriculture, adequate rewards must be in prospect, including a generous allowance for risk.

Some of the farm programs or budgets prepared for proposed new irrigation seem to show that very large sums can be paid for irrigation water. On closer examination, these have a fatal flaw; if the intended crop production is so profitable that very large sums can be paid for water, then it is profitable enough so that other extensive areas of the world, including those that need not pay high prices for water, can undertake such production — and the estimated price then quickly drops. Furthermore, the costs of other inputs rise rapidly, as the high yields conventionally assumed on irrigated acreage in these studies demand greatly increased applications of fertilizer, pesticides, and so forth, with attendant employment of sophisticated skills and machinery. In irrigated cotton-growing in California, for example, the other costs are so high that water costs typically constitute only 10 to 15 percent of total operating cost.

Third, it is easy to develop plans which embody a wholly new order of magnitude in farm efficiency — crop yields much higher than those obtained by farmers in other irrigation projects in the region, fertilizer inputs several times as great as now practiced, new crop varieties that lead to much higher yields, and many others. By comparing irrigation agriculture on this new higher plane of efficiency with nonirrigated agriculture (or even with present irrigation) on the older and lower level of efficiency, some very high values of water can be estimated. Irrigation does indeed open up new production opportunities, but realism is called for in estimating just how much advantage can and will be taken of those opportunities, and how soon. If the new system of agriculture is possible with new irrigation, why is it not feasible with old irrigation? What reason is there to expect that provision of irrigation water will immediately transform a backward, traditional agriculture into a modern or futuristically efficient one?

Fourth, the agro-industrial complex has been offered as an answer to the last question asked. But is it? Such complexes as sketched by their proponents employ currently unknown or untested methods in industry and agriculture, produce for unspecified markets, and appear to justify very high costs for irrigation water. The prime example here is the Oak Ridge project. Although comprehensive in the scope of things to be considered, it tends to assume optimistic outcomes, uses low costs, and fails to allow for unexpected difficulties and costs. Above all, it fails to supply a satisfactory answer to this question: if these great agro-industrial complexes are economically feasible with desalted water, why are they not feasible with natural flow or groundwater? There is nothing magical about desalted water; it is simply water.

The agro-industrial complexes of the Oak Ridge type have been defended on the ground that they would constitute a new order of technology and organization, freed of all the inhibitions of restrictive institutions, cultural values, modes of living, and so forth, which impede agricultural and industrial development in some countries. This is a dubious argument if applied to Australia, Israel, and possibly to Mexico and India. Moreover, this proposal is futuristic plantation philosophy. In many colonies of the world before Word War II, there were plantation economies, using outside capital, outside management, and producing for an export market; often they were highly efficient. Most are now liquidated as foreign enterprises; there is little reason to expect that the countries would welcome them back. The very isolation of the proposed agro-industrial complex from the mainstream of the country's culture is its most devastating weakness, regardless of the efficiency it might attain. The Oak Ridge study comments on this by contemplating that the food factory concept "would appear to be the reverse of agrarian reform programs in many countries. On the other hand, setting up an operation in a sparsely populated area might be effective in avoiding complications of existing social organizations and customs" (Oak Ridge, 1968, p. 27). One can only comment that it would save even more trouble if one were to select a less difficult geographic, social, and political setting and then find a way of letting the country to be aided share in the fruits of production by assigning to it the plant's net return.

Ignoring these broader-based considerations and insisting only that the large-scale desalting projects planned by Kaiser and Oak Ridge must produce predominantly staple crops, such as grains and cotton, for domestic and export markets, one can judge the economic feasibility from a number of recent American studies that provide estimates of the value of irrigation water for such crops. Since the contemplated farming ventures discussed above are based on highly advanced technologies and must to a large extent be competitive with world market prices, such studies are not as inappropriate a criterion as one might first think.

Young and Martin (1967) provide information and analyses to indicate that the value of irrigation water in central Arizona is less than $21 per acre-foot; Stults

(1966?), considering the situation in Pinal County, Arizona, makes analyses which imply that the value of the water is about $9 per acre-foot; Grubb (1966) estimated the ability to pay for irrigation water in the High Plains of Texas ranged from $27 to $36 an acre-foot, even in 1990; and Brown and McGuire (1967), found that the marginal value productivity of irrigation water in Kern County, California, was about $19 per acre-foot. These are all in fairly good farming areas, where the growing season is rather long, cropping patterns can be rather intensive, and crop yields are relatively high. In irrigated areas where farming is somewhat less intensive, due in part to differences in climate, Hartman and Anderson (1962), concluded that the value of supplementary water was from $1.50 to $3.00 per acre-foot; and Fullerton (1965), found that in a fairly active water-rental area, the price was about $8.75 per acre-foot. All of these examples involve rather high-level managerial competence (which is more easily hypothesized) unlike that found in some of the countries under study; the same is true of the availability of farm machinery, fertilizer, insecticides, and other inputs. It is important to note that they do not focus on the subsidized price of water but on what users can afford to pay. Thus they are directly relevant to the hypothetical cost of desalted (or any other) water. Moreover, they escape the frequent criticism that the cost of desalted water should not be compared with the actual price currently paid for water, or that the present price of water is an irrelevant object of comparison, since it must be judged in a multi-purpose use context.

On the basis of this range of American experience, it seems most unlikely that irrigation water delivered to the farm on the schedule the farmer wants it, for the production of staple crops, can attain a value greater than $30 per acre-foot (10 cents per 1,000 gallons), and a value of $10 per acre-foot (3 cents per 1,000 gallons) is a much more reasonable planning standard.

The conclusion is inescapable: the full and true costs of the proposed desalting projects, now and for the next 20 years, are at least one whole order of magnitude greater than the value of the water to agriculture. The specifics of both cost and value will vary, depending upon the location of the plant and the myriad of factors associated with that location, upon what desalting costs actually are in practice, upon crop possibilities (costs and markets, especially), and upon other variables. But it is impossible to bring planned costs and prospective values for agriculture together or even close.

Nothing we have said with regard to the prospects for desalting seawater should be construed as an argument against continued research, including the construction of a rather large pilot plant. The Oak Ridge study both merits and needs attentive reading and critical review. Such research must not stop at the farm gate nor bypass the broader implications of such programs with a few passing sentences. There is more involved here than either "truth in advertising," the discovery of a new input, or a new means of fighting hunger. The present mirage may indeed have an oasis within it, and we as a nation have the resources to pursue the matter much further. But let us not delude ourselves or the rest of the world that an early and practical solution is at hand.

Bibliographic References

BROWN, G. M., AND C. B. MCGUIRE
    1967 A socially optimum pricing policy for a public water agency. Water Resources Research 3(1):33.

CLAWSON, M., H. H. LANDSBERG, AND L. T. ALEXANDER
    1968 Economics and desalination [letter to the editor]. Environmental Science and Technology 2:648–649.

EISENHOWER, D. D.
    1968 A proposal for our time. Reader's Digest 92(554):75–79.

FOSTER, A. C., AND J. P. HERLIHY
    1967 Operating experience at San Diego Flash Distillation Plant. International Symposium on Water Desalination, 1st, Washington, D.C., 1965, Proceedings 3:57–86.

FULLERTON, H. H.
    1965 Transfer restrictions and misallocation of irrigation water. Utah State University, Logan (Thesis).

GRUBB, H. W.
    1966 Importance of irrigation water to the economy of the Texas High Plains. Texas Water Development Board, Austin, Report 11.

HARTMAN, L. M., AND R. L. ANDERSON
    1962 Estimating the value of irrigation water from farm sales data in northeastern Colorado. Journal of Farm Economics 44(1):207–213.

HOEHN, W. E.
    1967 The economics of nuclear reactors for power desalting. Rand Corporation, Santa Monica, California, RM-5227-PR/ISA.

HUENLICH, W. H. F., AND P. H. KRUCK
    1968 Future prospects for nuclear power stations of small and medium output (in German). Atom und Strom 14:149–153.

KAISER ENGINEERS IN ASSOCIATION WITH
CATALYTIC CONSTRUCTION COMPANY
    1966 Engineering feasibility and economic study for dual-purpose power-water desalting plant for Israel. U.S. Department of the Interior contract 14-01-0001-632. Report no. 66-1-RE, Job no. 6452. 183 p. [The study has been revised twice since and may be secured from Kaiser Engineers, Oakland, California.]

LANDSBERG, H. H.
    1968 Population growth and the potential of technology. In R. N. Farmer, J. D. Long, and G. J. Stolnitz, eds, World population — the view ahead, p. 138–167. Indiana University, Bureau of Business Research, Bloomington. 310 p. (International Development Research Center Series, 1)

NIKITOPOULOS, V.
   1968   The influence of water on population distribution. Ekistics 26(152):14–20.

OAK RIDGE NATIONAL LABORATORY, OAK RIDGE, TENNESSEE
   1968   Nuclear energy centers industrial and agro-industrial complexes. Contract W-7405-eng-26. ORNL-4290. 227 p. [NSA 23:11167]

POWELL, S. T.
   1967   Factors involved in the economic production of usable fresh water from saline sources. International Symposium on Water Desalination, 1st, Washington, D.C., 1965, Proceedings 3:429–443.

QUARTERLY ECONOMIC REVIEW [OF] ISRAEL
   1966   Annual supplement 1966. Economist Intelligence Unit, London.

SPORN, P.
   1968   [Remarks] *In* Nuclear power economics, 1962 through 1967, p. 2. Joint Committee on Atomic Energy, Report to 19th Congress, 2d session. U.S. Government Printing Office, Washington, D.C. 310 p.

STULTS, H. M.
   1966?   Predicting farmer response to a falling water table: An Arizona case study. *In* Western Agricultural Economics Research Council, Committee on the Economics of Water Resources Development, Economic Criteria, Water Transfer, Economics of Water Quality [and] Economics of Ground Water: Conference Proceedings, Las Vegas, Nevada, December 7–9, 1966, p. 127–141. 175 p. (Water Resources and Economic Development of the West, Report 15)

WOLFOWITZ, P.
   1969   Middle East nuclear desalting: Economic and political considerations. Rand Corp., Santa Monica, Calif. RM-6019-FF (unpublished)

YOUNG, R. A., AND W. E. MARTIN
   1967   The economics of Arizona's water problem. Arizona Review 16(3):9–18.

CHAPTER 7

PLASTIC OASES FOR ARID SEASHORES

Carl N. Hodges, Merle H. Jensen, and Carle O. Hodge
Environmental Research Laboratory, University of Arizona, Tucson, U.S.A.

Few places on earth are less arable than the United Arab emirates, a sparse scattering of shaikhdoms on the Arabian (Persian) Gulf, in the eastern elbow of the Arabian Peninsula. Dust storms, heat, and aridity preclude ordinary farming, and the low, sparse shrubs barely sustain the few camels that browse there. In 1970, nonetheless, the island of Sadiyat, a small swath of sand at Abu Dhabi, became a supplier of high-quality vegetables (Hodges, 1969a). This paradox resulted from an abundance of oil, a progressive ruler, and a technology developed in Mexico half a world away.

Abu Dhabi, largest of the seven Arab emirates, exported its first petroleum in 1962. Subsequent royalties have enabled its ruler, Shaikh Zayed Bin Sultan Al Nihayan, to bring twentieth century convenience and culture to what had been mainly an isolated nation of nomads. As one example, his government imported the techniques that made vegetable production possible on Sadiyat — a concept that evolved on the Gulf of California in northwestern Mexico. With this integrated power/water/food system, devised by the Environmental Research Laboratory of the University of Arizona in cooperation with Mexico's University of Sonora, waste heat from engine-driven generator sets is harnessed to desalt seawater; crops in controlled-environment greenhouses then are irrigated with the fresh water (Hodges and Kassander, 1967; Hodges, 1969b).

With the increase in world population, the potential of arid shorelines becomes of increasing importance. Almost everywhere, industry and residences are encroaching upon farmland. It is said that in California alone, by the year 2000, massive urbanization will have usurped at least half the prime agricultural areas. Space of any kind is to become scarce.

Excluding Antarctica, the earth's surface encompasses 13.6 billion hectares, and this includes tundra and snow-capped mountains, as well as the deserts. There now are, in the early 1970s, 3.7 billion human beings. If the available land were divided by the number of present inhabitants, each would have only 4 hectares. According to estimates by the United Nations, this total population is to virtually double by the year 2000; there will be more than 6.1 billion of us. This means that the ratio will have been reduced to less than 2.2 hectares per person.

Clearly, if man is not to suffer psychologically from overcrowding, he must migrate into places that now are thinly settled. In some of them, he must grow food and provide the other necessities and amenities. Because we will almost double in numbers, we must, of course, increase food production accordingly. Even with such an acceleration in productivity, we would merely maintain the present-day nutritional standards that already are inadequate in many nations.

If 40 percent of the population predicted for our planet twoscore years hence were distributed over the deserts, instead of there being only 56.5 meters between persons, the population could spread out to a uniform 136.2 meters. Here, then, is a vast land bank, but these regions offer more than simply space. They can be desirable places in which to live. The desert coasts can be particularly pleasant for habitation. There are some 32,300 kilometers of such shorelines (Meigs, 1969). Adjacent to them, and 300 meters or less above sea level (a reasonable lift to transport freshwater if it could be converted economically from the sea), are more than 680 million hectares. By and large, they are essentially uninhabited.

Partly because of these factors there have been growing efforts to desalt the oceans at a cost that would permit the deserts to be irrigated. The cliché is "to make the deserts bloom." The dream is almost as old as civilization.

Desalting remains relatively expensive, unfortunately. Pessimism toward technological progress may be unwise. Advances cannot always be perceived. Albeit, it is plain that marginal improvements of existing desalting processes will engender no miracles. Costs will remain fairly high, even with large-scale systems. These costs are prohibitive if compared with the water conventionally utilized for irrigation.

Efforts at the University of Arizona to lower the cost of desalting began in 1962 (Hodges, Groh, and Thompson, 1965). Shortly thereafter, that institution and the University of Sonora in Hermosillo, Mexico, agreed to construct a cooperative pilot facility at Puerto Peñasco, on the west coast of the state of Sonora.

In most small coastal communities, or, for that matter, in many remote villages the world over, small engine generator sets (generally diesel engines) provide power. A 90-kilowatt engine supplies all the electricity for the Sonoran experimental station.

Wherever power is generated this way, excess heat is a byproduct. A reciprocating diesel engine dispels to the atmosphere approximately two-thirds of the input fuel energy. At Puerto Peñasco, this surplus energy is captured to heat seawater in the desalting unit, a humidification-cycle system. The unit, the towers of which may be seen in figure 7-1, was designed by the Environmental Research Laboratory and funded first by the Office of Saline Water, United States Department of the Interior.

In this low-energy form of distillation, the outwardly most obvious features are two towers, a condenser, and an evaporator. Within the condenser are brass-lined finned aluminum tubes. Cool seawater spirals upward through the tubes. Hot air vapor envelopes the tubes and causes the seawater inside to be preheated, by the latent heat of condensation and by sensible heat transfer, to about 68°C.

At that point, the seawater moves through heat exchangers in the power plant, where it is heated further to 77°C. It then sprays down through the evaporator (packed with small plastic "saddles" to give a large heat-transfer area) countercurrent to a rising stream of air. This forms the hot freshwater vapor that moves through ducts over into the condenser. The vapor condenses there on the finned tubes and falls to the bottom of the tower.

Although our desalting process still may prove feasible for certain locations, the economics were not encouraging for open-field farming. Indeed, the more our staff pondered the problem, the more convinced it became that desalted seawater, however it may be wrested from the ocean, will be too costly for conventional agriculture in the foreseeable future.

One obstacle is that a crop in the desert normally demands, during the growing cycle, many times its own weight in water. This happens, in part, because so much transpiration occurs through the leaf stomates, and a great deal of this transpiration occurs as a result of a vapor gradient between the leaves and the outside air. If a plant were encapsulated in an environment with a high relative humidity, the vapor gradient may be reduced, thereby reducing transpiration. Thus, the amount of freshwater required should be reduced dramatically.

Fig. 7-1. Power, water, and food are produced in "package" approach at Puerto Peñasco. In the background are controlled-environment greenhouses and desalting-plant towers.

Fig. 7-2. Air-inflated horticultural enclosures in Mexico are fabricated of inexpensive plastic.

This theory led us, with Rockefeller Foundation sponsorship, to grow crops in the pairs of plastic, closed-environment greenhouses in figure 7-2. Very little air is exchanged between the greenhouses and the outside atmosphere. Seawater from a seawater well sprays at one end of the greenhouse down through a honeycomb-like packed column of corrugated asbestos. Air forced up through the spray of the packing continuously circulates around the greenhouses in a counterclockwise loop and through the column every two minutes. Because the packed column ceaselessly humidifies the air, the relative humidity hovers close to 100 percent. This greatly inhibits transpiration from leaf surfaces. As a result, comparatively small quantities of the desalted water are needed for irrigation (Hodges and Hodge, 1971).

In fact, during the winter on the Gulf of California, the greenhouses themselves become desalting systems. The greenhouse interiors are so much warmer than the outside air that moisture condenses inside the plastic film. The salt-free condensate exceeds the amount of water the crops need; so, the surplus can be collected and stored for use later.

Temperatures also can be manipulated. Whether the greenhouses are heated or cooled depends mainly upon the season and time of day. Most of the time, though, temperatures may be maintained near that of the 26°C water from the seawater well. This is accomplished by increasing or decreasing the flow rate through the packing.

From the outset of the experiments, various kinds of greenhouses have been considered, and many materials submitted to stress and cost analyses. The advantages of plastic soon became apparent. As we shall see, this evolution has not ended. The present Puerto Peñasco houses, however, are of air-inflated plastic. They were fabricated of a light-stabilized (to ultraviolet radiation) polyethylene, a film with a predictable life beside the Gulf of California of 12 to 14 months.

These plastic balloons can be erected easily by relatively untrained personnel. There are no supports to inhibit solar radiation. But perhaps the most important asset is that the polyethylene is inexpensive.

Each greenhouse combination consists of two 30-meter-long, 6.9-meter-wide half cylinders connected front and back by concrete tunnels; hence, there are 414 square meters for agriculture in each house. These greenhouses must be entered through air locks, which also contain the instrumentation. In the front tunnel are tanks in which nutrients are mixed with irrigation water. The plastic cover is attached to a redwood base bolted to concrete curbs. Widths of the houses are limited by that of the polyethylene available commercially, that is, the breadth in which the manufacturer makes the material. But the desired height and the strength of the plastic also bear upon this: the greater the radius of curvature and the wider the house, the more subject the structure becomes to stress.

Little pressure is required to keep the pneumatic envelopes taut. A primary 1/6 horsepower centrifugal inflation blower maintains a positive pressure generally of 0.01 to 0.02 psi (0.25 to 0.50 inch of water); albeit, this is a function of ambient wind velocity. When wind speeds reach about 30 kilometers per hour, the domes must be stiffened. Hence, a larger one-horsepower blower automatically switches on, and the pressure increases to about 0.03 psi (0.750 inch of water).

Temperatures, condensation, and other pertinent parameters have been measured in the Puerto Peñasco greenhouses, as shown in table 7-1. From these studies a computer program was developed that permits one to predict the performance of similar structures anywhere.

Predictably, at least in a climate akin to that of northwestern Mexico, the enclosures will be hot and humid. Sheltered from the harshness of the surroundings, then, a tropical rain-forest has been superimposed on the desert. What can be cultivated under these conditions? Economic considerations probably exclude, for the present, any widespread utilization of plastic covers for grain crops, although rice has been grown successfully in the greenhouses. But the most promising results, both agriculturally and economically, have come with vegetables.

During the first two years of experiments at Puerto Peñasco, 160 varieties of 19 different high-quality vegetables and small fruits were harvested from the manmade oases, including squash, peppers, okra, beans, and eggplant. Indeed, the controlled-environment produc-

TABLE 7-1

Measured Temperatures and Condensation in a Controlled-Environment Greenhouse, Extracted From Results of a 24-hour Performance During Experiments in Nov. 1969 at Puerto Peñasco

	Units	Hour					
		18	22	2	6	10	14
Solar Radiation	Langly/hour	0	0	0	0	46.7	42.9
Wind speed*	Meters/second	1.22	1.22	0.59	1.17	2.93	1.49
t ambient DB	°C	17	15	11	9	19	21
t ambient WB	°C	12	10	9	7	16	16
t in DB	°C	26	26	25	24	27	28
t in WB	°C	25	26	25	25	27	28
t out DB	°C	24	24	23	22	28	31
t out WB	°C	–	24	22	23	27	30
t plant at 15m	°C	–	19	24	22	28	31
Condensation	kg/hr	–	40	59	58.2	35.7	19.5
Total Condensation	kg/hr	–	155	355	472	814	909

NOTE: Data are abridged from Selçuk, 1970.
*Average air velocity: 0.598 meters/second.

Fig. 7-3. Eggplant (foreground) and peppers are among vegetables harvested from controlled environments.

tion of these foods already appears to be practical for many places in the world.

Figure 7-3 depicts some of the crops at Puerto Peñasco. All of them have been grown directly in the beach sand. Early in the investigations, plants were set out or seeded in separate plots of either the local sand or a mixture of peat moss and vermiculite. Because they grew equally well in either, the artificial medium was abandoned. Now, radishes, spinach, beets, onions, and carrots are seeded directly in the sand, while most other vegetables are started in a nursery and then transplanted.

Water-soluble fertilizer compounds are fed through the irrigation system. Several standard irrigation systems have been used, each designed to ration water sparingly. All of them release only enough water to moisten the sand to the depth of root penetration. As crops begin to mature, the frequency of irrigation increases, but not the amount of water applied. Plants that require pollination must be pollinated by hand at present. Work is being done, however, on new procedures, the application of certain hormones, for example, that may replace pollen transfer.

Most of these crops mature far more rapidly in the controlled environments, as is documented in table 7-2, than they would outside, a result of the heat and intensive light. Thus, more crops per year can be harvested. Table 7-3 lists some of the yields obtained.

There was some fear, before the experiments began, that the same hot dank conditions that promote this rapid maturity also might foster plant diseases. Actually, few problems with either insects or diseases have been encountered, possibly because the saltwater spray washes the incoming air. Not all varieties fare well; some of those routinely raised elsewhere tended at Puerto Peñasco to be beset by rooting on the stems and by extended flower clusters. Of the vegetables that have been tried, those commonly grown outdoors in hot, humid areas (tomatoes from Florida, for instance) responded best to the comparable atmosphere of the closed greenhouses. The search for adaptable varieties has continued, and a plant-breeding program was established to develop vegetables specifically for the program.

When the appropriate varieties are planted in the greenhouses, they grow very well. Yields have been far greater, predictably, than those from open fields or, by and large, when compared with harvests from conventional greenhouses (Jensen and Teran, 1971). Steps are being taken to increase all yields of vegetables. For instance, no space is left between the beds for implements. With radishes, as an example, only a center aisle is left unplanted. To further maximize utilization of the space, plants are trained, when possible, to overhead supports. A significant point is that these yields were obtained in a place where little else grows. In addition, there is another exceedingly promising ingredient in controlled-environment agriculture: carbon dioxide enrichment.

Plants must have carbon dioxide for photosynthesis, of course. If greenhouses are almost completely closed, carbon dioxide must somehow be injected. A simple but not necessarily the most efficient means is merely to introduce a small amount of outside air. An alternative that could be much more beneficial is to elevate beyond the ambient level the carbon dioxide in the enclosure. This can be done by tapping the exhaust of the diesel engine, separating out the other gases in a seawater scrubber, and then piping into the greenhouse the optimal amount of carbon dioxide for crop productivity.

TABLE 7-2

Growing and Harvest Periods, in Days, for Some Winter Crops Grown in Controlled Environments at Puerto Peñasco in 1969–1970, Compared With Average for Summer Field Crops in the United States

Crop	Puerto Peñasco Greenhouses		U.S. Field Crops	
	Growing Period	Harvest Period	Growing Period	Harvest Period
Cucumber (European type)	130	90	90	30
Eggplant	170	90	130	40
Lettuce (bibb and leaf types)	40	–	45	–
Okra	100	60	118	60
Pepper (bell)	146	41	155	55
Radish	30	–	30	–
Squash (zucchini)	135	90	80	30
Tomato	170	90	140	50

NOTE: Harvest period minus growing period equals number of days to first harvest. Growing periods and harvest periods depend greatly on local conditions.

TABLE 7-3

Yields, in Kilograms per Hectare, From Controlled Environments in Puerto Peñasco in 1969–1970 Compared With Average Yields From Field Crops in the United States

	Puerto Peñasco Greenhouses	U.S. Field Crops
Cucumber (European type)		
Fall crop	355,680	9,992[†]
Spring crop	392,864*	
Eggplant		
Fall crop	148,200*	16,055[†]
Spring crop	148,200*	
Okra		
Winter crop	89,611	11,201[‡]
Pepper (bell)		
Winter crop	33,681	10,441[†]
Tomato		
Fall crop	168,021*	15,269[†]
Spring crop	145,618	

*Based on a harvest period of 90 days.
[†] United States Department of Agriculture, **Agricultural Statistics 1969** — (data for fresh market).
[‡] James E. Knott, 1962, **Handbook for Vegetable Growers**, John Wiley and Sons, Inc., New York.

If light, water, nutrients, turbulence, and other factors are carefully controlled, the carbon dioxide level often becomes the limiting growth factor. Generally speaking, the atmosphere contains an average of slightly more than 300 parts per million carbon dioxide, but when this level has been increased by eight times, say, or to 2,400 parts per million, some crops in the controlled-environment greenhouses produced far more profusely (table 7-4). To increase unit productivity by this means should strengthen considerably the commercial viability of controlled-environment agriculture.

Components of the power/water/food system could be tailored to a variety of situations, but the total concept probably will be most applicable to such coastal desert settlements as Abu Dhabi. That Middle Eastern emirate of 80,000 square kilometers is even hotter and drier than northwest Mexico at about the same latitude (table 7-5). Although it grows some vegetables at inland oases, most of the high-quality produce comes by air from other countries, mainly from Iran across the Gulf, and from Lebanon. No lettuce is harvested in Abu Dhabi, nor are beets, cabbages, tomatoes, or cucumbers most of the year, even though all of these are in demand. There are months when okra and squash, either domestic or imported, are not abundant at the local market. Vegetable prices are high all year.

Early in 1969, Shaikh Zayed gave the University of Arizona a grant to install a power/water/food facility on Jazirat as Sadiyat (the Isle of Happiness), about 2.40 kilometers across the harbor from his island capital. By the end of that year, the site had been surveyed, and a road was being built.

A resident staff, led by James J. Riley, from the Laboratory moved to Abu Dhabi early in 1970 to oversee construction and to train the Abu Dhabians, who ultimately will assume full control of the complex. An air-inflated research greenhouse was erected; the first vegetables were harvested in the summer of 1970.

When the Sadiyat facility, now called the Arid Lands Research Center, began full operations in 1971, it included about two hectares of controlled-environment greenhouses, power generators with an output of 1,200 kilowatts, a 264,000-liter per day desalting unit, storage tanks, a packing plant, training center, and a horticultural laboratory.

The face of Sadiyat, mostly dunes and flats, will be changed in more than one way. To enhance the landscape, reduce pollution and further close the system, a small sewage treatment plant will transform the Center's waste into irrigation water for grass and trees at the Center.

Arrangement of the agricultural area will permit, for the first time, precise comparisons of the economics and productivity of inflated and relatively standard greenhouses. Half of those on Sadiyat will be more or less traditional in appearance — steel structured with plastic roofs and fiberglass siding. Crops that must be trained vertically and thus require supports, for example, toma-

toes and cucumbers, will be grown in them. The remaining, air-inflated chambers, where all other vegetables are to be planted, will be connected to two ground-level central passageways. The structured greenhouses will be placed on two sides of the air-supported houses to act as windbreaks.

This compact food factory will furnish about 1,000 kilograms of vegetables a day. As it expands later, production will increase proportionately. Crops are to be chosen, and planting planned, in such a way that the vegetables from Sadiyat should complement rather than compete with existing domestic farming. By selling its products at competitive prices, the facility soon should become self-supporting.

Supplying high-quality vegetables where and when they are needed, all year, is an achievement, but hopefully the impact of the Center will be far more encompassing than that. It also is a research station at which

TABLE 7-4

Total Season Yields for Greenhouse Cucumbers With and Without Carbon-Dioxide Enrichment for a Winter Crop, 1969–1970, at Puerto Peñasco, Sonora, Mexico

	Without Additional Carbon Dioxide					
Variety	No. 1		No. 2		Cull	
	No. fruits	Yield (kg/ha)	No. fruits	Yield (kg/ha)	No. fruits	Yield (kg/ha)
Cherokee 7	162,552	20,465	31,200	2,503	17,784	1,233
Bestseller	160,368	34,218	54,912	8,817	30,576	2,068
	With Carbon Dioxide Enrichment					
Variety	No. 1		No. 2		Cull	
	No. fruits	Yield (kg/ha)	No. fruits	Yield (kg/ha)	No. fruits	Yield (kg/ha)
Cherokee 7	180,024	19,776	39,000	2,975	24,960	1,487
Bestseller	221,832	46,701	70,200	12,119	26,316	3,229

NOTE: Plants were grown at a spacing of 4,000 plants per hectare. The last harvest was March 12, 1970.

TABLE 7-5

Comparative Climatic Data for Puerto Peñasco, Tucson, and Abu Dhabi

	Elev. (m)	N. Lat.	Yrs. of record	Mean rainfall (cm)		
				Summer	Winter	Annual
Puerto Peñasco, Sonora, Mex.	12	31°20′	20	53.64	49.99	103.63
Tucson, Arizona	729	32°15′	63	194.46	138.07	332.54
Abu Dhabi	1.8	24°25′	14	4.88	43.28	48.16

	Temperature (°C)				
	Jan. mean	Long-term low	July mean	Long-term max.	Ann. mean
Puerto Peñasco, Sonora, Mex.	11	−12	29	41	20
Tucson, Arizona	10	−5	30	46	19
Abu Dhabi	17	10	34	45	25

visiting scientists strive for new methods and new cultivars. Education is a primary objective. The first three Abu Dhabi trainees spent eight months in Tucson and in Mexico to learn English and the intricacies of controlled-environment horticulture. Back in their homeland, they are helping to teach continuing classes of other apprentices.

It would be premature, at this writing, to predict what influence the program might have on people not directly involved. But there has been one encouraging development. The only inhabitants of Sadiyat before the power/water/food facility arrived were about thirty families in two small settlements of palm-frond huts. These tribesmen are fishermen, as their forebears were. They are not rich.

With the encouragement of the Center and their own government, these people formed the Sadiyat Community Cooperative. This organization bought a boat, which it leases to the Center. It operates a canteen and plans a restaurant that will serve Center employees and guests. The profits are shared by the entire native population.

Bibliographic References

HODGES, C. N.
1969a Food factories in the desert: Accomplishments. Agricultural Research Institute, Annual Meeting, 18th, 1969, Proceedings, p. 113–123. National Research Council/National Academy of Sciences/National Academy of Engineering, Washington, D.C. 185 p.
1969b A desert seacoast project and its future. *In* W. G. McGinnies and B. J. Goldman (eds), Arid Lands in Perspective, p. 119–126. American Association for the Advancement of Science, Washington, D.C.; University of Arizona Press, Tucson. 421 p.

HODGES, C. N., J. E. GROH, AND T. L. THOMPSON
1967 Solar powered humidification cycle desalination; a report on the Puerto Peñasco pilot desalting plant. International Symposium on Water Desalination, 1st, Washington, D.C., 1965, Proceedings 2:429–459.

HODGES, C. N., AND C. O. HODGE
1971 An integrated system for providing power, water and food for desert coasts. HortScience 6(1): 30–33.

HODGES, C. N., AND A. R. KASSANDER, JR.
1966 Extending use of available supply — a system approach to power, water and food production. *In* R. G. Post and R. L. Seale (eds), Water Production Using Nuclear Energy, p. 227–233. University of Arizona Press, Tucson. 392 p.

JENSEN, M. H., AND M. A. TERAN R.
1971 Use of controlled environment for vegetable production in desert regions of the world. HortScience 6(1):33–36.

KNOTT, JAMES E.
1962 Handbook for vegetable growers. Wiley, New York. 245 p.

MEIGS, P.
1969 Future use of desert seacoasts. *In* W. G. McGinnies and B. J. Goldman (eds.), Arid Lands in Perspective, p. 103–118. American Association for the Advancement of Science, Washington, D.C.; University of Arizona Press, Tucson. 421 p.

SELÇUK, K.
1970 Technical and economical aspects of the controlled-environment agriculture in arid lands. Paper presented at Conference of the Canadian Society of Agricultural Engineers, July 8, 1970, Ottawa, Canada.

U.S. DEPARTMENT OF AGRICULTURE
1969 Agricultural statistics, 1969. U.S. Government Printing Office, Washington, D.C. 631 p.

CHAPTER 8

OFFSHORE DESERT ISLANDS AS CENTERS OF DEVELOPMENT

Homer Aschmann

Department of Geography, University of California, Riverside, U.S.A.

An almost universal feature of modern settlement in arid regions is its highly nucleated, almost pointwise, character. Often, of course, there is an overwhelming environmental basis for such concentrations, such as the presence of the only water source in a vast area, or of the vein that supports a mining camp. In other instances, the control is not obvious; a particular spring becomes a center of commerce and of intensive if limited agriculture, while similar water-source areas within the region are unoccupied or little developed.

The intent here is to take the feature of insularity and to compare the population density and degree of economic development of certain offshore dry islands with desert mainland regions opposite them. The island itself can be taken as a potential center for nucleated settlement. Demonstrating that neither the island nor the mainland has environmental advantages over the other would rarely, if ever, be possible. In the relatively small number of available examples, instances of distinctly greater development on either side exist, and they do not correlate clearly with physical conditions in the local environments. Through this focus on the feature of insularity it may be possible to elucidate the sorts of minor environmental features that affect cultural decisions and historic trains of events that ultimately create the intricately diverse and not readily rationalizable cultural landscape.

This is a tentative venture. It would have been desirable to know intimately through fieldwork throughout the world a large fraction, if not all, of the dry nearshore islands and their opposite coasts. I know only a few, and for the others my interpretation of their essential features comes from standard sources rather than intensive library research.* That I may have missed a critical environmental element is obvious. Baja California and its offshore islands, the Guajira Peninsula, Venezuela, and the Dutch Islands, the Eastern Canaries, though not the African mainland opposite, and the northern coast of Chile are the examples I know best.

Island and Mainland Comparisons

The principal criterion in the identification of examples is that the island and mainland share the same climate. Since the climate is, in large measure, the product of air masses, proximity is likely to provide substantial similarity except where there is great topographic contrast between island and mainland. The Canary Islands form a fine example. The eastern islands, Lanzarote and Fuerteventura, with relief of less than 800 meters and very modest local variation in precipitation, can reasonably be compared with the Southern Moroccan coast. The high islands to the west, whose slopes capture water that can sustain permanent irrigation, are not appropriate for comparison, even though the entire area is generally affected by the same air masses.

Islands More Highly Developed Than Mainland

The following examples are instances in which the offshore island or islands have a notably higher intensity of development than the nearby mainland:

1. Lanzarote, with about forty inhabitants per square kilometer, and Fuerteventura with about eight inhabitants, are fully desert places; only by extraordinary effort and ingenuity can they be made to support their inhabitants. The Moroccan coast south of Ifni is a wasteland supporting a few herdsmen and some fishing villages, less than one person per square kilometer. The main climatic distinction, higher relative humidity in summer, does favor the islands slightly over the coastal zone, but in other climatic parameters, island and mainland are similar indeed.

2. The island of Djerba off the southern Tunisian coast also is much more highly developed agriculturally and more densely settled than the adjacent mainland. Kerkennah to the north probably should not be considered,

---

*No effort is made to cite sources for the areas not observed directly. Standard gazetteers and regional geographies have been used, which, though not fully informative, are not in significant disagreement.

NOTE: Grateful acknowledgment is given to the United States Office of Naval Research for support, under contract Nonr 1842 (03), of field work in the Canary Islands, and earlier support of field work in the Guajira Peninsula, Colombia, and Curaçao and Aruba. The ideas expressed in this chapter were generated, in large measure, by field observations made in those areas.

since neither it nor the opposite Sfax area is truly desertic. Djerba's population density is about 120 persons per square kilometer, high indeed for its strictly agricultural and fishing base. The immediately opposite mainland is not undeveloped, but except for the regionally important trade and fishing port of Gabes, population density is about an order of magnitude lower.

3. From Cumaná to the Guajira Peninsula, the north coast of South America and the Caribbean sea for about one hundred miles to the north receives little rainfall. Despite the high humidity of the tradewinds, the constantly high temperatures mean that rainfall values of from 20 to 50 centimeters, which are typical of low-elevation stations, give rise to essentially desert conditions. Elevations above 800 meters, however, get considerable orographic precipitation. Four large and many small islands lie off the coast and share the drought. Two of the three Dutch Islands, Aruba and Curaçao, are extremely densely populated, with more than 600 inhabitants per square kilometer, though these are basically industrial populations. Bonaire, which has no oil refineries, has only twenty-five inhabitants per square kilometer. The larger Venezuelan island of Margarita, similarly dependent on farming and fishing, has about forty-five inhabitants per square kilometer.

Comparing population densities of island and mainland along this varied stretch of mainland coast is extremely difficult. There are densely settled districts on the mainland such as La Guaira and Maiquetía, the port and airport, respectively, for Caracas, and where mountains near the shore catch more abundant rainfall. The Guajira and Paraguaná peninsulas, however, may afford more accurate comparisons, climatically and because they lack exotic rivers from major mountain ranges. These areas have about five inhabitants per square kilometer.

Islands Not More Highly Developed Than Mainland

Nevertheless, by no means all offshore desert islands enjoy the distinctly higher level of development of those mentioned above. A few examples from a considerably longer list of possibilities can be mentioned.

1. The myriad tiny islands off the Peruvian coast, though of great economic worth for their guano deposits, are essentially uninhabited. The many islets off the north coast of Chile are similarly uninhabited, and they have not developed guano deposits thick enough to invite significant exploitation.

2. Two islands off the desert west coast of Baja California have somewhat contrasting development patterns. Cedros Island, close to the peninsula, has a combination fishing and fish processing center with 2,700 residents, though the rest of the island is nearly unoccupied. A similar small center, Bahía Tortuga, exists on the mainland, and other transiently occupied fishing camps are located there. Very little farming and grazing activity occurs on the mainland, and essentially none on the island. The levels of development may be regarded as essentially equivalent. Further offshore, Guadalupe Island is only slightly smaller than Cedros, and with comparably rugged topography gets at least as much moisture as Cedros Island and probably considerably more than the mainland. (While climatic records in the whole area are nearly nonexistent, these conclusions can be drawn with some confidence from the character of the natural vegetation.) Yet Guadalupe Island has only a lighthousekeeper and a tiny Mexican garrison. The myriad wild goats, introduced by mariners in the last century, are destroying the vegetation and then themselves dying and rotting.

The considerable number of islands in the Gulf of California have never experienced even the modest level of development that the opposite Baja California and Sonora coasts have. Angel de la Guarda, with an area of over 850 square kilometers, was uninhabited at contact in mid-eighteenth century and has had no permanent residents since. Tiburon, with over 1,000 square kilometers, is still visited seasonally by a few Seri Indians, but even in earlier times, when they were more numerous, the bulk of the Seri lived most of their lives on the mainland. A few of the islands in the southern part of the Gulf had aboriginal populations, but these were removed in the early eighteenth century as part of the missionization process and not replaced.

3. Socotra Island, off the East Africa Horn, shows a population density of less than four per square kilometer. Again, this is roughly the same density as in the nearest parts of Somali Republic and probably a bit higher than that in the South Arabian sultanates with which Socotra is more closely affiliated.

4. The coast of Western Australia between latitudes 20° and 27° S gets about 25 centimeters of rainfall per year, a bit more in the north, and is essentially desertic. A considerable number of islands lie off the shore, with Dirk Hartog, the largest, involving more than 600 square kilometers. A few permanently occupied lighthouses and some transiently visited sheep runs are the only developments on the islands. It must be noted, however, that recent mineral development has intensified human activity on the mainland (cf. Gentilli, Chapt. 24 herein).

Causal Developmental Relationships

From the preceding numerically small sample I shall try to identify a set of physical and cultural features that may be causally related to the islands' greater or lesser development:

Size: It is almost accurate to say that the desert islands of less than 200 square kilometers are not developed. One reason is likely to be a physical one, for such small islands are unlikely to have enough groundwater accumulation to sustain springs. The islands off the coasts of Peru and Chile fall into this category. On the cultural side such a small base cannot sustain a large enough community to provide its own social satisfactions and economic support. It would just be too lonely or socially isolated for one to want to live there.

Population size itself shows a parallel influential relationship, as does island size. The Canary Islands afford a nice example. Fuerteventura and Hierro, the inhabited islands having the smallest populations, though Fuerteventura is territorially large, show essentially stable or declining populations though they both have high natural rates of increase. Distress at the social isolation is reported by the migrants as at least as much a reason for emigration as are the economic limitations. The three still smaller islands—Alegranza, Graciosa, and Lobos—are uninhabited except for lighthouse-keepers.

Distance From the Mainland: I believe that I sense a negative impact on development resulting from increased distance from the mainland. A fairly clear example would be the islands off the west coast of Baja California. Cedros Island, which is close to the mainland, has moderate economic activity and once had a substantial Indian population. Guadalupe, farther out, had no aborigines and still is essentially unoccupied. The Galápagos Islands are really too far from the South American coast to be regarded as offshore islands. Distance from mainland Ecuador, however, is perhaps the best explanation for their lack of development. Distance from the mainland is a particularly difficult variable to measure in this context, because as it increases one tends to get into climatic environments progressively more different. I am inclined to believe, but cannot document, that Socotra would have played a far more important role in the history of the Red Sea if it were closer either to the African Horn or to the South Arabian Coast.

Political Control: More than half of the islands identified above are presently under the same political system as their opposite mainlands, but this situation has not prevailed at these places through all of recorded history. The Canary Islands had an aboriginal population that for millennia was culturally completely separate from that of the opposite mainland. Djerba has at several times been politically separated from the South Tunisian region. In each individual case, it is fairly easy to show how the vicissitudes of political history have contributed to the development or lack thereof. Generalizing from these examples, or imagining what might have been if history had taken another turn, must be highly subjective. A few examples still may be worth considering.

The Dutch Islands off the Venezuelan coast had, during Spanish colonial rule, and to some degree still have, a considerable function as bases for smuggling, although a plantation economy on Curaçao and Bonaire and a more peasant-like one on Aruba were maintained under difficult conditions. Political distinction as a requirement for smuggling is obvious. The development that made Aruba and Curaçao urban industrial centers, however, came after the development of the petroleum fields of the Maracaibo Basin. In this case it is clear that the oil companies preferred to place their great investments in refineries under responsible Dutch rule rather than risk destruction or confiscation in revolution-prone Venezuela. Bonaire, which never got a refinery, has experienced steady emigration to the more prosperous islands under the same flag. Margarita, under Venezuelan rule, has quite high development as well, notably more than Bonaire. To a considerable degree it has been enhanced by the fact that, while the overall government was no better or worse than that on the mainland, Margarita never suffered from the extremely bitter civil combat that afflicted Venezuela during the Wars of Independence and later.

It is tempting to speculate whether there would be significant development there if one or more of the islands off Western Australia were under Dutch or Portuguese rule. We can be more certain that Cedros Island, if it were part of the United States, would be a major center for both commercial and sport fishing, but that is because it affords closer access to an especially rich marine resource. In British hands, for example, it might be much less developed than now.

Social Community: My conclusion is that the critical attribute of offshore islands which have experienced markedly more development than adjacent mainland districts is that they have managed in some way to create and maintain a distinctive community of islanders who are numerous enough to supply their own social structure and attractions. The island had to be large enough and with resources enough to sustain such a community, even though with difficulty. Further, at some point in its history, the community had to feel itself a special entity with need for cooperative arrangements, especially in the area of water control and utilization and capital investment therein, in order to survive in an alien if not hostile world. Political separateness clearly fits this pattern, though it need not have been complete or maintained for very long, as Margarita and Djerba attest. Once the community was functioning it could sustain itself and create its own cultural distinctions. Even though these factors were minor, they would make the island an identifiable home.

A native, therefore, would be willing to work harder and to accept smaller material rewards than those he reasonably could hope for in a metropolitan center. Or, if forced by the limited economy to emigrate, he would plan, often with ultimate success, to return to his home island with capital accumulated elsewhere. Also, the proper mode of land development and water management was recognized by the entire populace. Typically, a reasonable level of agricultural productivity depended on heavy investment of labor and time. The returning emigré or the thrifty husbandman could make an investment in developed or in developing land that the whole community would recognize and respect as enriching itself, even though it was a low-yielding investment indeed.

Thus, intricately defined, effective social community

was or is present, to the best of my knowledge, in each of the well-developed and peopled offshore desert islands. This community may be thought of as a cause rather than as a result of development. On Lanzarote, landowners are quite individualistic, but all agree that the laborious spreading or pitting and repeatedly cleaning the lapilli (volcanic cinders) cover, which permits crops to be produced with less than 15 centimeters of rainfall, enriches the economy. A man's investment in land modification or in vines or figs is respected and socially protected. As a symptom of its decadence, there is less community agreement on proper patterns of land and water use on Fuerteventura, and some use patterns are injurious to others. On Fuerteventura, there is notably less individual capital or labor investment that might bring about a permanent, if small, enhancement of agricultural wealth.

Djerba, with its Muslim sectarianism, conservative retention of Berber speech, and heavy investment in olive and date groves, fits the pattern of a community desirous of sustaining itself. Returned emigrés bring a continuing trickle of new capital. The Dutch Islands off Venezuela now are unusual in having industrial economies and attracting immigrants; earlier, however, in addition to the distinctiveness of Dutch political institutions, they had evolved their own language, Papiamento. Margarita, although part of Venezuela, is a separate state. It receives essentially no immigrants other than national officials, and its emigrants, who will always regard themselves as Margariteños, seek to return after they have earned and saved a modest competence on the mainland.

With the exception of the Dutch Islands, it is probably accurate to state that all the intensively utilized offshore desert islands are places of poverty and difficulty. Hard and persistent effort is needed to gain a livelihood, and such effort would be likely to yield greater returns if invested elsewhere. It is, however, a dignified rather than a degraded poverty. I cannot but feel that the world is richer for containing these communities, which are cultural as well as physical islands, and that it will be poorer if they decay as younger generations are attracted to the several national metropolitan centers.

CHAPTER 9

ON THE CAUSES OF ARIDITY ALONG A SELECTED GROUP OF COASTS

Paul E. Lydolph

Department of Geography, University of Wisconsin, Milwaukee, U.S.A.

Limits of the Investigation

Dry and barren landscapes washed by the waves of the sea are not an unusual phenomenon on the surface of the Earth. Such areas extend for hundreds of miles along the subtropical west coasts of North and South America, Africa, and Australia, as well as along the north and east margins of the Sahara, much of the periphery of Arabia, and the coasts of Southwest Asia. The Cape São Roque region in Brazil and the somewhat dry promontories of the northern coast of Venezuela are also notable examples. Numerous other occurrences include such widely scattered places as spots along the dry Canterbury Plain in New Zealand, the dry northern end of the Kona coast of the big island of Hawaii, the southern coast of Cuba east of Guantánamo Bay, and so forth. A notable group of desert islands lies within a dry tongue in the eastern equatorial Pacific Ocean. This appears to be an extension of the desert coast of South America, and thus falls logically within some sort of pattern. Other small areas occur irregularly in conjunction with shielding effects of terrain features or predominant influences of other local factors.

Obviously all the dry coastal areas of the world cannot be dealt with in one short chapter. Even if this were physically possible, the treatment would have to be very uneven, since some areas have been subjected to repeated investigations while other areas have not been studied at all. This chapter does not describe any one area in detail, but rather seeks common factors applicable to a number of desert coasts which might lead to an understanding of the causes for the lack of precipitation.

The field might be narrowed by comparing only those coasts which are really dry; say, those that receive less than 100 millimeters of precipitation per year. This would restrict consideration primarily to the driest sections of the coasts along South America and southwest and northwest Africa and some restricted spots on the east African coast and south central Baja California. However, such a definition would have to include extensive sections of the coast along the Red Sea and probably spots along the coasts of Southwest Asia eastward from the Persian Gulf, although data are lacking to substantiate the extreme dryness of this latter area. The inclusion of the coasts bordering the Red Sea and extending eastward from the Gulf of Aden would seem to complicate unduly the search for common precipitation controls among a set of coasts. Therefore, they will not be included.

The scope of the study might better be limited to only coastal deserts, not all desert coasts; that is, to consider only those desert strips which seem to owe their existence to the presence of a coast, rather than to consider all dry coastal areas, many of which might be simply fringes of large interior deserts. True, some of the so-called coastal deserts that would be considered under such a definition would be the coastal fringes of extensive interior deserts, but in these cases those coasts would appear to be significantly drier than their desert interiors. In such cases, as well as in the cases of coastal desert strips, it would seem that the presence of the coastline itself somehow is a causal factor in the aridity of the coastal strip. Such a criterion would limit consideration primarily to the three world-famous cases of extreme coastal aridity which stretch over great latitudinal extent, the west coast of South America and the northwest and southwest coasts of Africa. I should like to add the coasts of southwestern North America and Western Australia, since their latitudinal and continental positions are similar to the positions of the other three coastal deserts. Mention will be made of some other coasts as well, to illustrate certain points.

The west coast of Australia, though not nearly as dry as the other four coasts, is included as sort of a norm to which the other coasts can be compared. I would like to reiterate the essentials of the procedures and conclusions of my study of the five dry western littorals that appeared in the *Annals of the Association of American Geographers* in 1957 so that they can be combined with procedures and conclusions of three or four more recent studies which have shed additional light on the causal factors for the lack of precipitation in these areas. These studies collectively present a tentative explanation of the causes of coastal deserts in areas where coasts are drier than the interiors.

Precipitation and Precipitation Controls on Dry Western Coasts

My study on the dry western littorals was prompted by the obvious similarities in location and climate among the five west coastal areas of South America, Baja California, southwest Africa, northwest Africa, and Australia. These coasts lie at comparable latitudes on western sides of continents and are strongly influenced by subtropical high-pressure cells and, with perhaps the exception of Australia, cool ocean currents. In some respects, the Somali coast of Africa is similar; the winds flow parallel to the coast throughout much of the year, and cold in-shore currents seem to prevail a good deal of the time. Separate ship reports record more than a 15°F (8°C) drop in water temperatures as the coast is approached. The dry northern portion of the coast, however, seems to be drier inland than in the immediate vicinity of the coast, and this coastal area is not influenced by a conspicuous high-pressure cell. Since in my original study I was interested primarily in establishing locational and climatic similarities and dissimilarities among a set of coasts that might lead to some general conclusions about aridity in coastal areas, it appeared that the inclusion of the east African coast would simply complicate the picture, and I eliminated it. It now appears that the east African coast might fit into a general scheme better than I had at first anticipated, and I will illustrate that idea a little later.

Although the Australian coast is not as dry as the other four coasts, and it is not driest right on the coast, as the other four coasts seem to be, its comparable location with respect to latitude, continental position, and relation to a subtropical high led to its inclusion in my study of the western coasts. This proved to be a happy inclusion, since the final results of the study hinged strongly on the uses of the Australian coast as a standard which indicated the amount of precipitation that one could expect along a subtropical west coast that did not consistently act as the eastern terminus of a subtropical high and that was not paralleled by a persistent cold ocean current. Against this standard, the other four coasts could be compared; certain conclusions could be drawn regarding relative magnitudes of influence of various aridifying factors along all the coasts; and approximations could be made of absolute magnitudes of individual effects along individual coasts.

Figure 9-1 shows these relationships very well. It is simply the graph of annual precipitation against latitude along each of the five coasts in question. If the Australian coast represents precipitation amounts along a coast that experiences only the normal subsidence within a high-pressure cell and no perceptible influence of cold surface waters, it can be stated that the annual precipitation along such a coast at the point of maximum subsidence will total about nine inches (225 millimeters) over a latitudinal distance of one or two degrees, and the precipitation will increase at the rate of about four inches (100 millimeters) per latitudinal degree in either direction.

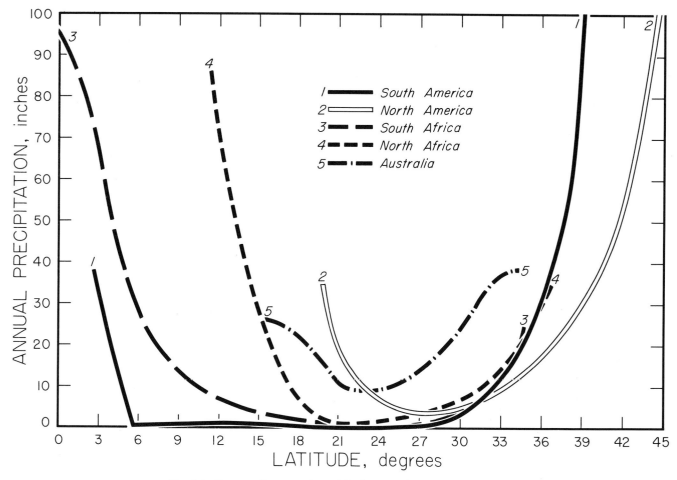

Fig. 9-1. Comparative annual precipitation values, according to latitude, along American, African, and Australian coastal regions.

At the other extreme, the plot for the South American coast indicates that the additive effects of strong surface temperature contrasts and atmospheric subsidence along the eastern periphery of a high can reduce precipitation to zero at the point of maximum subsidence and can hold it at that level against the waning effects of subsidence to low latitudes so long as active contact with the open ocean circulation is maintained. Where surface contrasts are maintained at an optimum to higher latitudes, their effects can reduce precipitation by three-quarters or more but cannot hold it at zero. Under optimum conditions, then, as illustrated by the South American coast, the absolute aridifying effects of surface contrasts are greatest at low latitudes and decrease poleward.

The northwest and southwest coasts of Africa represent cases somewhat intermediate between the two extremes represented by South America and Australia. So does the North American coast, but there the coast loses contact with the general atmospheric and oceanic circulations at such a high latitude that aridifying effects from strong surface contrasts do not coincide latitudinally with those produced by maximum subsidence, so that absolute aridity is not reached anywhere along the North American coast.

A latitudinal comparison of winter precipitation and summer precipitation separately indicates that surface effects are much more important during summer (fig. 9-2). On all of the coasts at this time, except the west coast of Australia, precipitation is nil along considerable latitudinal stretches wherever the configuration of the coast is such that direct contact with open ocean circulation is maintained. Since the configuration of all coasts is different, each coast presents a different curve at this time of year. This is not the case during the winter when, with the slight exception of Australia, all the coasts appear very similar. Surface contrasts combined with normal subsidence from weakened oceanic highs produce near absolute aridity equatorward of about 26° latitude along all the coasts during the winter, but poleward of 26° the effects of surface contrasts are nil, and precipitation increases with latitude as stormy weather prevails.

On the two American and the two African coasts it appears that the precipitation is least in the immediate vicinity of the coasts and increases along a line perpendicular to the coasts both inland and out to sea. Precipitation data are not abundant enough to substantiate this generalization in all instances, but what evidence does exist seems to point to this conclusion. And where data are lacking along transects perpendicular to the coasts, there are perceptible enough differences in precipitation amounts at coastal stations to indicate that protected embayments along these coasts consistently receive more precipitation than do exposed headlands. Thus, it appears that the exposure of a coast to the general oceanic and atmospheric circulations is a very important factor in precipitation processes. Nearly absolute aridity persists along these coasts as long as the coastline is in active contact with the paralleling flows of atmospheric and oceanic circulations, and when the coastline diverges from these circulations an abrupt increase in precipitation takes place immediately. It seems, then, that an intimate interrelation exists between the presence of the coastline itself, the eastern termination of the atmospheric high-pressure cell, the oceanic circulation along the coast, and precipitation processes in the immediate vicinity.

Moreover, there appears to be an additional amount of aridity along these coasts unaccounted for by the drought-producing factors identified above, which led me to the conclusion that there is an additional amount of subsidence immediately over such a coast which must somehow be explained by the presence of the coastline itself. Without ascribing an explanation for the cause of this apparent additional atmospheric stability, I attached to it the coined phrase "coast-wise subsidence." This has proved to be a stillborn phrase that no one has used since, to my knowledge. However, the concept itself has persisted in the minds of a number of people, and three studies in particular have carried the concept forward and attempted an explanation of this additional amount of atmospheric subsidence along certain coastlines (Lydolph, 1957).

Stress-Differential Induced Divergence

In 1958, James Lahey, then at the University of Wisconsin, completed a dissertation entitled "On the Origin of the Dry Climate in Northern South America and the Southern Caribbean." Although the Venezuelan coast is in no way as dry as the coasts previously mentioned, nor is it paralleled by conspicuous cold ocean currents, it does have aspects of climate similar to that of the drier coasts; and the raggedness of the coastline with its low mountain ranges oriented perpendicularly to the coast afforded Lahey an opportunity to observe changes in air flow across the region, leading him to the conclusion that the difference in frictional drag across the rough land as opposed to that across the smooth water set up a lateral stress in the air flow which produced a flow across the coast out to sea, which in turn induced subsidence in the vicinity of the coast. (Cf. Chapter 10 herein.)

Thus, along with certain influences of the western end of the subtropical high in the Caribbean and occasional influences of a colder sea surface, Lahey ascribed much of the aridity in this area to a stress-initiated subsidence which was due partially to differences in frictional drag over land and over water and partially to differences in turbulent motion over land and over water. Hence, he attached much significance to the reductions of wind speed over the land due to frictional drag and additional turbulent motion induced by thermal effects over the land. He substantiated his theory largely by pointing to the considerable differences in rainfall found along north-south oriented segments of the coast as opposed to rainfall along east-west oriented segments. The onshore trades along the north-south oriented segments piled up against the low mountain ranges back from those

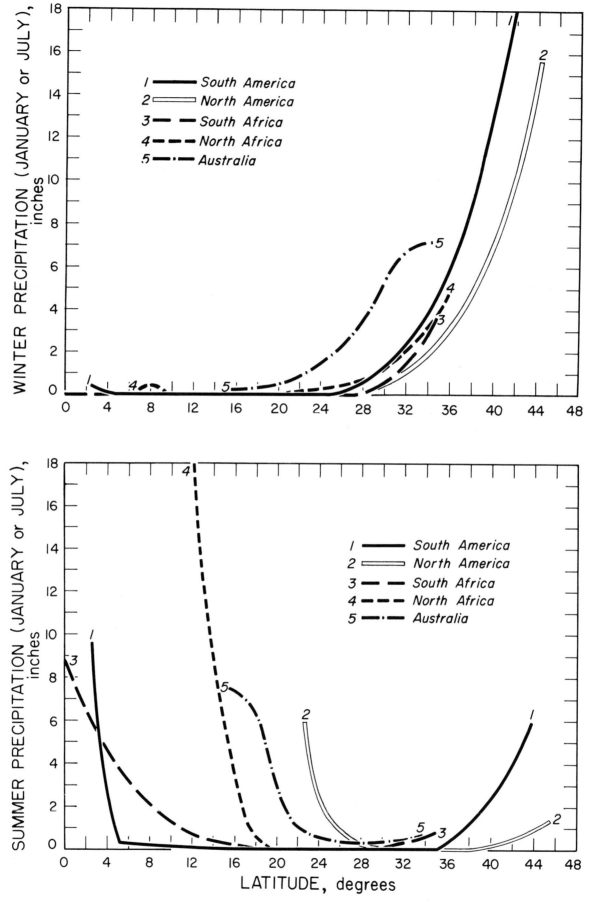

Fig. 9-2. Comparative winter and summer precipitation values, according to latitude, along American, African, and Australian coastal regions.

segments of the coast to produce considerable amounts of precipitation on the windward slopes, whereas paralleling winds along the east-west segments of the coast underwent a stress-induced divergence which suppressed precipitation on those coastal segments.

In 1961, Bryson and Kuhn, in a short article entitled "Stress-Differential Induced Divergence with Application to Littoral Precipitation," applied Lahey's concept to the various dry coasts of the world. Using an approximate equation to relate divergence with the angle at which the wind blows across the coast, they showed a good correspondence between the values of divergence thus computed and precipitation amounts along the coasts of northwest and southwest Africa, northern Honduras, and, to some extent, western Australia.

The equation that Bryson and Kuhn used to compute the divergence was $\nabla H \cdot V = (\triangle C_D / f \triangle y) [(v \sin \beta)^2 - (V \cos \beta)^2]$ where $\nabla H \cdot V$ indicates the horizontal divergence of the volume transport within the friction layer, $\triangle C_D$ is the land-sea drag coefficient difference across the coast, f is the coriolis parameter, $\triangle y$ is the distance perpendicular to the coast throughout which the divergence is being computed, v is the velocity of the wind, and β is the angle at which the wind is crossing the coast.

They also illustrated correspondences between convergence or divergence and rainfall along various segments of the coasts of New Guinea. Unfortunately, the classic aridity case, the west coast of South America, did not yield high divergence figures using this method. It is assumed that the reason for this lack of correspondence along the South American coast is that the winds used there were not representative of gradient winds in that area but were strongly influenced by the high Andes only a short distance to the east. Therefore, it appears that the ruggedness of the topography in this area negated the validity of the method.

Thermo-tidal Winds and Their Implications for Precipitation Processes

Heinz Lettau has published the conclusions of a thought-provoking piece of work which originated in connection with years of research on wind flow and weather phenomena over the Great Plains of North America and culminated in a study of wind flow and aridity along the west coast of South America. The results of his research in these two areas has led him to formulate a general equation for wind flow in the atmospheric boundary layer over meridionally-oriented extended slopes of land surfaces, which, among other things, affords a conceptional framework into which all the north-south oriented dry coasts of the world can be fitted. In this framework Patagonia and the Somali coast of Africa seem to fit into a set, along with the five west coasts previously considered as a set in my study on the dry western littorals.

Lettau's general theory is as follows: On a meridianally-oriented long-sloping terrain surface (that is, one that is sloping either to the east or to the west and the strike of which is oriented north-south) horizontal temperature gradients and, hence, pressure gradients, will develop due to differential daytime heating and nighttime cooling of the underlying surface. These thermally-induced gradients will be parallel to the slope of the land and will reverse their directions between day and night in a seesaw fashion, because at the same elevation in the air the greatest diurnal temperature variations will take place above the highest parts of the terrain slope. Thus, at one time of day the thermally-induced gradient might oppose the general pressure pattern of the area and at another time of day it might reinforce it. Moreover, this seesaw oscillation of opposition and reinforcement will induce a super-geostrophic low level jet stream to occur during some part of the day, depending upon the latitude and the east or west facing direction of the terrain slope. This low-level jet will be essentially perpendicular to the thermally-induced pressure gradient, hence, parallel to the contours of the terrain, with low pressure to the left in the northern hemisphere and to the right in the southern hemisphere. In the boundary layer there will be enough friction to cause a component of motion down the pressure gradient at all times, which for the thermally-induced gradient would mean movement down the terrain slope during the daytime and up the terrain slope at night. Hence, nighttime low-level jets appear to be associated with nocturnal thunderstorms in the Great Plains of North America, while daytime jets are associated with subsidence and aridity over the pampas of Peru and Chile.

According to Lettau, the component of horizontal air motion perpendicular to the pressure gradient can be expressed by an equation which includes the factor $f^2 \triangle V / (\omega^2 - f^2)$, where ω is the angular velocity of the earth and f is the coriolis parameter, $2\omega \sin \phi$. From this relationship it can be seen that there are three critical latitudes at which the phase of the time period of the low-level jet-stream oscillation would change sign — at the equator where the coriolis force changes sign and at 30° latitude north or south of the equator where $f = \omega$. Thus, in the northern hemisphere on an east-facing, or westward-ascending slope, low-level jets will tend to develop during the night poleward of 30° latitude and during the day equatorward of 30° latitude. On west-facing, or eastward-ascending, slopes in the northern hemisphere, low-level jets will tend to develop during the day poleward of 30° latitude and during the night equatorward of 30° latitude. In the southern hemisphere on a west-facing slope low-level jets will tend to develop during the day equatorward of 30° latitude and during the night poleward of 30° latitude, and on an east-facing slope they will tend to develop during the night equatorward of 30° latitude and during the day poleward of 30° latitude.

Lettau contends that this theory goes a long way toward explaining such phenomena as the nocturnal thunderstorms of the Great Plains of the United States, the extreme aridity of the western pampas of South America from the Gulf of Guayaquil southward to approximately

30° south latitude, the aridity of the Namib coast of South Africa from approximately 30° south to 5° south latitude, and the aridity of the Somali coast of east Africa in the northern hemisphere which terminates on the south almost exactly at the equator. Other dry coastal areas such as Baja California, northwest Africa, and Australia do not have the requisite terrain profile of a prolonged slope in one direction that would lead to thermally-induced diurnal oscillations of low-level jets in these areas. Hence, the climate of these regions is not particularly related to such phenomena, and therefore a simple worldwide pattern reflecting the effects of low-level jets, according to Lettau's theory, is not easily discernible.

New Evidence From the Peruvian Area

In 1967 Federico Prohaska spent the year at Agrarian University in La Molina, Peru, and took the opportunity to analyze in detail the day-to-day weather and radiosonde observations, which heretofore had been presented only in summarized forms in records averaged over extended periods of time and grouped about mandatory reporting levels. Thus, his work, presented separately in this volume, exposes information about the vertical structure of the atmosphere over the central Peruvian coastal area that had been previously obscured in average climatic records.

One of the most noteworthy facts that he has brought to light is the shallowness of the trade winds over Lima. The south and southeasterly winds at the surface are quickly replaced aloft by northwesterly winds generally at altitudes of no more than 1,000 to 1,500 meters. Such an air flow would seem to greatly complicate the general theories formulated earlier about aridity causes. Obviously, an understanding about the climate of the dry coasts of the world is still very incomplete. Until much more upper air data over these regions becomes available, the theories of Lettau, Lahey, and Lydolph will remain only hypotheses without the necessary corroborating evidence to establish a generally accepted axiom.

Bibliographic References

BRYSON, R. A., AND P. M. KUHN
    1961 Stress-differential induced divergence with application to littoral precipitation. Erdkunde 15(4): 287–294.

LAHEY, J. F.
    1958 On the origin of the dry climate in northern South America and the southern Caribbean. University of Wisconsin (Ph.D. Thesis). 316 p.

LETTAU, H. H.
    1964 Preliminary note on the effect of terrain slope on low-level jets and thermal winds in the planetary boundary level. *In* "Studies of the Effects of Variations in Boundary Conditions on the Atmospheric Boundary Layer," University of Wisconsin, Department of Meteorology, Annual Report, 1964, p. 99–115.

    1967 Small to large-scale features of boundary layer structure over mountain slopes. *In* Symposium on Mountain Meteorology, 1967, Ft. Collins, Colorado, Proceedings. Colorado State University, Department of Atmospheric Sciences, Atmospheric Science Paper 122. 74 p.

LYDOLPH, P. E.
    1957 A comparative analysis of the dry western littorals. Association of American Geographers, Annals 47(3):213–230.

PART TWO

LATIN AMERICAN DESERTS

CHAPTER 10

ON THE ORIGIN OF THE DRY CLIMATE IN NORTHERN SOUTH AMERICA AND THE SOUTHERN CARIBBEAN

James F. Lahey
Department of Geography, Oregon State University, Corvallis, U.S.A.

Along the northern coastal margin of South America (fig. 10-1), in Venezuela, Colombia, and the Netherlands West Indies is an area where surface trade winds move from the Caribbean Sea onto shores studded with arid and semiarid vegetation. It is an area with onshore winds located only 10° to 12° from the equator, and yet possessed of a climate with limited and erratic rainfall. Normally at such a latitude and in such a marine location, there would be abundant precipitation. The purpose of this analysis is to delineate the possible causes of the dearth of rainfall in this area.

Location and Rainfall Characteristics of the Dry Area

The dry area of northern South America (fig. 10-2) extends from the Paria Peninsula in the east to 76° W. The area has small latitudinal breadth in its eastern extremity but widens to about 4° latitude in the west. The center of the dry area is just west-northwest of the peninsulas of Paraguaná and Goajira. Thus the most extensive portion of the dry zone lies over the southern Caribbean Sea rather than over the land to the south (fig. 10-3). The Oceanic portion appears to be a contiguous area in contrast to the complex intermingling of dry and wet zones found over the land mass of northern South America. This complexity over the continent is clearly related to topography.

The essential topographic picture of the area (figs. 10-1 & 10-4a) is one of long stretches of mountain-bounded shorelines oriented east-west, interrupted by shorter coastlines aligned north-south. Interspersed between these mountains are pocket-shaped valleys, some opening to the east to the Caribbean Sea, others to the west. From the southwest the towering Sierra Nevada de Merida complex intersects these east-west coastal ranges and forms the western border of the extensive plains known as the Llanos. To the westward lies Lago Maracaibo and its adjacent lowlands, which in turn are bounded to the west and north by the Sierra de Perija, and the low-lying Goajira Peninsula. Still farther west, adjacent to the coast, is the Sierra Nevada de Santa Marta node.

Some of the driest areas, with approximately 500 millimeters average annual rainfall (figs. 10-1 & 10-4b) lie along the east-west oriented stretches of mountainous coastline and extend inland for varying distances. Other dry areas are farther inland and are separated from the sea by zones of much higher rainfall of obvious orographic origin. Dry area precipitation deficiency is due both to the infrequent number of days with rain, fifty days per year, and the limited falls of rain on rainy days.

Constricted valleys, opening eastward to the sea, record large annual totals of precipitation, the average amounts varying from 2,000 to 2,500 millimeters.

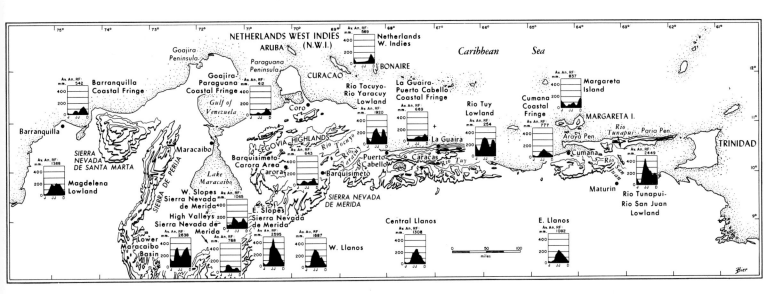

Fig. 10-1. The northern coastal margin of South America, and the southern Caribbean Sea.

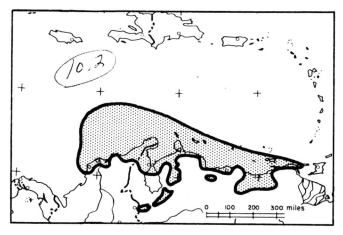

Fig. 10-2. The dry area along the northern coast of South America.

The contrasting rainy and arid areas are clearly distinguishable during all seasons, as can be seen from the precipitation pattern for January, April, July, and October (figs. 10-5–10-8). Large portions of northern South America (fig. 10-6) receive more rain in April than in preceding months, but the largest rainfall totals are centered in the southern Lago Maracaibo Basin.

By summer the center of maximum rainfalls has shifted. During June the area of greatest rainfall totals spreads progressively across the eastern Andean slopes and western Llanos. By July (fig. 10-7) and August the area of greatest rainfalls has shifted to the central and eastern Llanos, and some of the weather systems which bring ample summer rainfalls to this area also cause smaller increases in rainfall amounts in the dry area to the north. Also the narrow *east-west* oriented coasts receive less rainfall, have smaller probabilities of rain falling, and have higher evaporation than the *north-south* oriented coastlines.

In autumn (fig. 10-8), the area of maximum rainfall is again centered over the southern Lago Maracaibo Basin. Rainfall amounts falling over the Llanos decrease, while storms more frequently, though erratically, cause heavier rains over the dry area.

Winter (fig. 10-5) brings general aridity to most of northern South America. Exceptions are the coastal ranges, southern Venezuelan Andes, and southern Maracaibo Basin. Also, the normally drier coastal and offshore regions of extreme northeastern South America receive greater winter rainfalls. The abrupt change from moderate winter rainfalls over northern and northeastern coastal areas, to pronounced aridity over the Llanos, suggests that the rain-producing storm-systems are centered to the northeast of the continent and that their passage is blocked by mountainous coastal terrain. It is noted on the accompanying maps that the dry area is relatively drier than its surroundings during all months, even during its rainier season.

In discussing temporal precipitation characteristics of the dry area, attention will be focused on the Netherlands West Indies, which are near the heart of the dry area, have a relatively dense precipitation observation network, and have no high mountains to complicate storm rainfall analysis.

A daily precipitation analysis was undertaken for the entire available Netherlands West Indies (N.W.I.) station network. Precipitation classes based on frequency, "widespreadness," and daily amounts were established by extensive scrutiny of the data. The following precipitation classes were defined (fig. 10-9):

A. Dry days (no station in the N.W.I. received rainfall)
  1. Short dry spells
  2. Extended dry spells of five or more days duration

B. Wet days (at least one station in the N.W.I. receives measurable rainfall during the day)
  1. Widely scattered light rains
  2. Widespread rainfalls
    a. Widespread moderate rains
    b. Widespread, locally heavy rains
    c. Widespread, locally very heavy rains
  3. Scattered rainfalls of variable intensities

Fig. 10-3. Frequency of rain: percent of total observations during which rain was occurring.

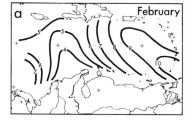
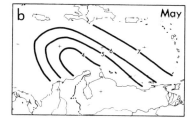
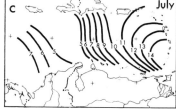
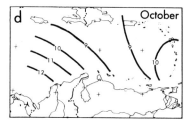
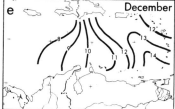

Details of the classification are found in Appendix A of this chapter.

To smooth the curves of daily rainfall frequency, five-day running averages and percentages were computed for each of the frequency categories and plotted on the graph. On this graph six precipitation seasons are discerned for the dry area.

The dominant season is that of the *extended dry spells of late winter and spring,* extending from the middle of February to June 10. From June 10 through August 20 there is a *summer period of light, scattered rainfalls* with a few widespread rains interspersed. After August 20 through September 6 there is a *late summer period of increasingly frequent extended droughts.* From September 6 through October 15 there is a *season of increased*

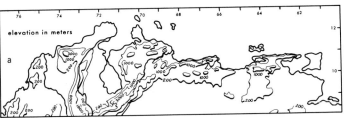
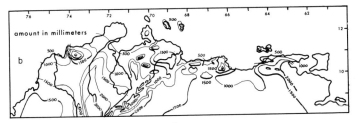

Fig. 10-4. Topography (a) and average annual rainfall (b) along the northern coast of South America.

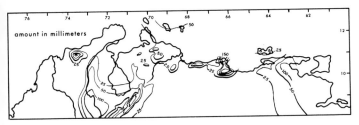

Fig. 10-5. Average January rainfall along the northern coast of South America.

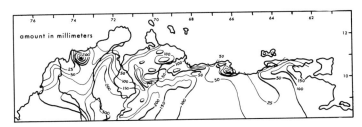

Fig. 10-6. Average April rainfall along the northern coast of South America.

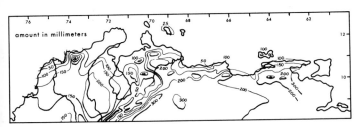

Fig. 10-7. Average July rainfall along the northern coast of South America.

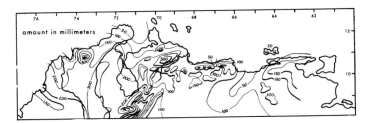

Fig. 10-8. Average October rainfall along the northern coast of South America.

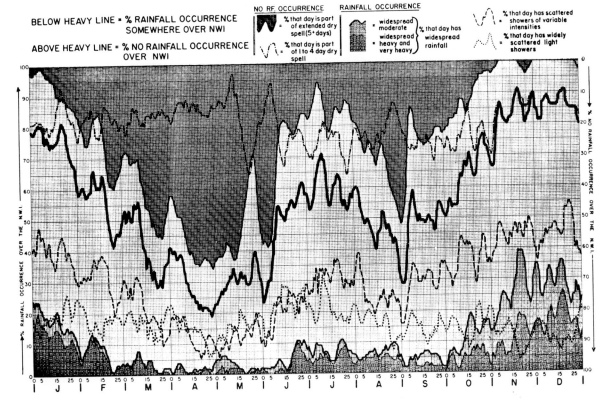

Fig. 10-9. Running five-day rainfall frequencies for the Netherlands West Indies.

frequency of scattered, light rainfalls of early fall. By October 15 the *primary rainy season* of the dry area has begun and continues to about January 23, with peak frequencies of widespread rains occurring between November 14 and December 23. From January 23 through mid-February there is a *season of decreasing widespread rains* and rains of variable intensities.

The foregoing analysis of precipitation over northern South America focuses attention on a number of problems:

1. Why is the dry area relatively drier than its surroundings during all months?
2. What meteorological conditions are associated with the extended droughts of spring?
3. Why are the frequent summer rainfalls over the N.W.I. so scattered and light?
4. Do some synoptic systems in fall and early winter differ from those occurring in other seasons so as to produce more frequent and heavier rains over the dry area?
5. What characteristics of winter storm systems cause less rainfall westward across the dry area and inland over the Llanos?

Since clouds and rainfall are basically a result of upward motions of air, it is reasonable to find answers to these questions in the windfield.

Seasonal Characteristics of the Dry-Area Windfield

Surface winds flow into the dry area from the east-northeast during all seasons (figs. 10-10 to 10-13). Over the Caribbean Sea surface winds increase in velocity downstream, then decrease in velocity in the immediate vicinity of the north coast of South America. Over the northern or Caribbean Sea portion of the dry area, seasonal streamlines are anticyclonically curved or straight. These streamlines assume cyclonic curvature nearer to the coast. Hence, surface trade flow along the coast is northeasterly, while out over the Caribbean it is east-

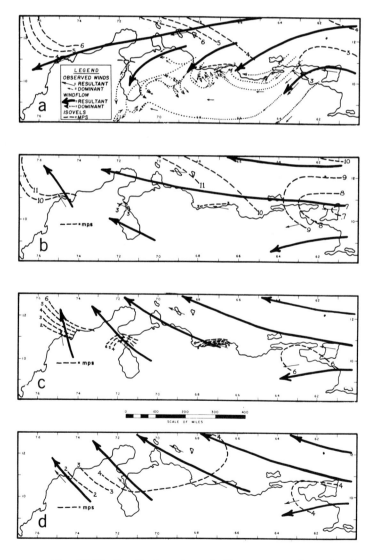

Fig. 10-10. Resultant spring windflow over northern South America, at (A) the surface, (B) at 3,000 feet, (C) at 6,000 feet, and (D) at 10,000 feet.

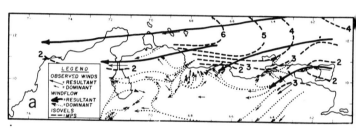
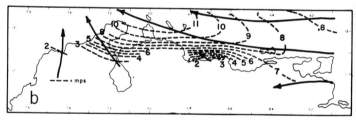
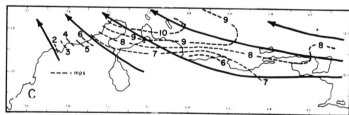
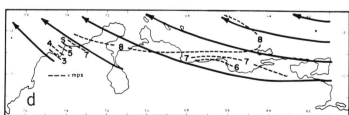

Fig. 10-11. Resultant summer windflow over northern South America, at (A) the surface, (B) at 3,000 feet, (C) at 6,000 feet, and (D) at 10,000 feet.

Origin of Dry Climate

northeasterly. With these flow patterns there is surface directional divergence in the windfield near the coast and speed divergence offshore. Over mountainous sections of northern South America the trade flow is reduced to a complex series of eddies.

At 3,000 feet and 6,000 feet the resultant windflow is from land to sea over northern South America throughout all seasons (figs. 10-10 to 10-13). Associated with this offshore flow is an increase in resultant windfield velocities from the coast to the N.W.I. This increased speed is pronounced during all seasons at 3,000 feet and is less evident at 6,000 feet only during fall.

At 10,000 feet the offshore component of the wind persists throughout the year (figs. 10-10 to 10-13). However, resultant wind velocities increase from land to sea only during the drier seasons of spring and summer.

The 14,000-foot resultant windfield is more nearly easterly than those between 3,000 and 10,000 feet. At this level there is a general decrease in velocity downstream during the drier seasons of spring and summer (figs. 10-14 & 10-15). Fall and winter, the wetter seasons, have complex velocity patterns (figs. 10-16 & 10-17).

Each season has distinctive resultant flow patterns at 20,000 feet (figs. 10-14 to 10-17). These differences, reflecting the seasonal changes in positioning of major hemispheric anticyclones and troughs, will be analyzed in terms of the contrasting significance of each to the climate of the dry area.

An inquiry can now be made into the manner in which these seasonal windfields are related to the basic physical geography of northern South America and to the greater aridity of the dry area than its surroundings.

Seasonal Resultant Windfields Over Northern South America

Next, the seasonal resultant windfields up through the 20,000-foot level were analyzed by the objective Bellamy-Bennett method of computing divergence and verti-

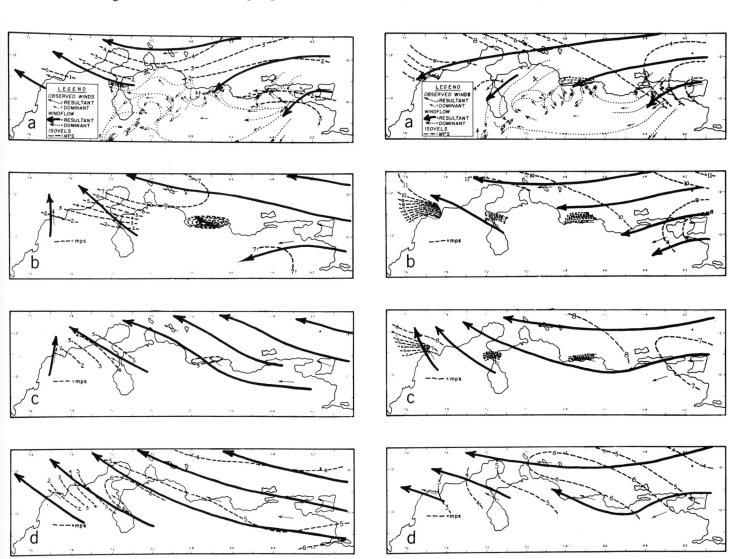

Fig. 10-12. Resultant autumn windflow over northern South America, at (A) the surface, (B) at 3,000 feet, (C) at 6,000 feet, and (D) at 10,000 feet.

Fig. 10-13. Resultant winter windflow over northern South America, at (A) the surface, (B) at 3,000 feet, (C) at 6,000 feet, and (D) at 10,000 feet.

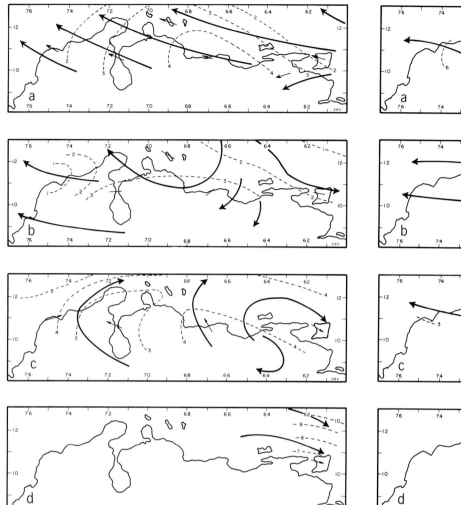

Fig. 10-14. Resultant spring windflow over northern South America, at (A) 14,000 feet, (B) 20,000 feet, (C) 25,000 feet, and (D) 30,000 feet.

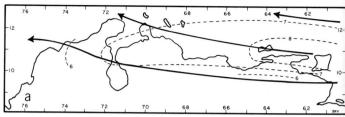
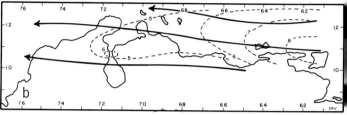
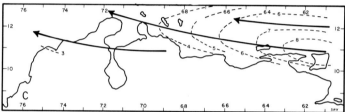
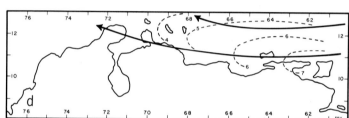

Fig. 10-15. Resultant summer windflow over northern South America, at (A) 14,000 feet, (B) 20,000 feet, (C) 25,000 feet, and (D) 30,000 feet.

cal velocities. It must be realized that gross distances between upper wind observing stations (fig. 10-18), plus the basic assumption of linearity in the gradient of the wind velocity field leads to problematic results, particularly in those seasons of diverse synoptic conditions. The Bellamy-Bennett method of computing divergence and vertical velocities employs triangles through which the wind flows, and brings either a net inflow of air or a net outflow of air from the triangle. The formula for this computation is found in Appendix B. If the result of the calculation is positive the resultant air motion is downward; if negative, upward.

In this analysis, triangle A has as its vertices Maracaibo, Venezuela; Hato Field, Curaçao; and La Guaira, Venezuela. All of these stations are within the dry area, but several land areas within the triangle are humid. Triangle B, which fits the eastern half of the dry area well, has as its vertices Hato Field; La Guaira; and Port of Spain, Trinidad. It will be noted that the largest portion of triangle C — with vertices Hato Field; Port of Spain; and Beane Field, Santa Lucia — is northeast of the dry zone. This triangle thus encompasses an area

TABLE 10-1

Seasonal Vertical Velocities of Windfields Over Northern South America

Units are in centimeters per second; + indicates downward motion, − indicates upward motion

Season & Location*		Altitude of Windfield (in feet)				
		(3,000)	(6,000)	(10,000)	(14,000)	(20,000)
Spring	A	+0.35	+0.81	+1.10	+1.13	+2.16
	B	+0.08	+0.28	+0.64	+1.25	+2.69
	C	+0.006	−0.10	−0.09	+0.37	+1.23
Summer	A	+0.41	+1.16	+1.67	+2.27	+3.17
	B	+0.36	+0.20	+0.34	+1.25	+1.23
	C	+0.08	+0.009	+0.009	−0.06	−0.32
Fall	A	−0.03	−0.11	−0.39	−0.64	+0.24
	B	+0.20	−0.24	−0.65	−0.40	+1.06
	C	−0.21	−0.60	−0.83	−1.20	−1.55
Winter	A	−0.29	−1.27	−2.33	−3.81	−5.44
	B	−0.36	−1.08	−1.71	−2.61	−3.41
	C	−0.12	−0.36	−0.34	−0.03	+0.32

*Triangle A: Vertices are Maracaibo, Venezuela; Hato Field, Curaçao; and La Guaira, Venezuela.

Triangle B: Vertices are Hato Field; La Guaira and Port of Spain, Trinidad.

Triangle C: Vertices are Hato Field; Port of Spain; and Beane Field, Santa Lucia.

Origin of Dry Climate

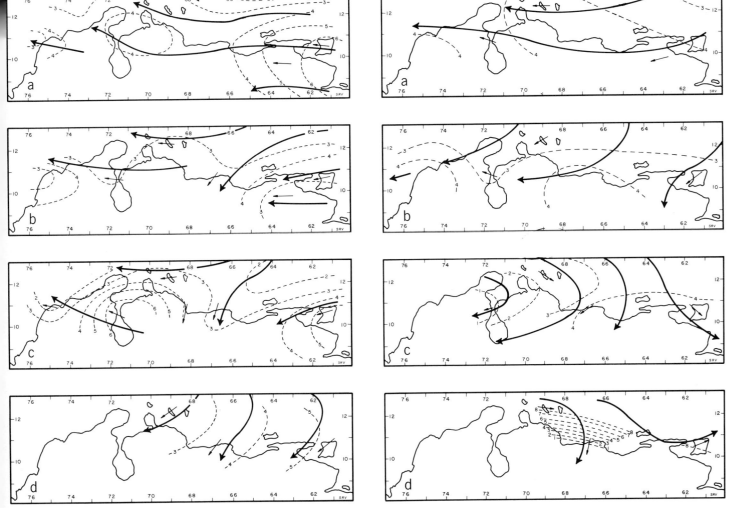

Fig. 10-16. Resultant autumn windflow over northern South America, at (A) 14,000 feet, (B) 20,000 feet, (C) 25,000 feet, and (D) 30,000 feet.

Fig. 10-17. Resultant winter windflow over northern South America, at (A) 14,000 feet, (B) 20,000 feet, (C) 25,000 feet, and (D) 30,000 feet.

where there is a pronounced increase in rainfall frequency during both spring and summer. The western portion of the dry area was not analyzed, owing to a lack of properly positioned stations.

The spring and summertime values of vertical motion in triangles A and B (table 10-1) indicate that deep easterlies over the dry area are characterized by strong downward movement from the lower layers to at least 20,000 feet. This subsidence is in accord with the aridity of east-west oriented coasts and offshore islands during this season. However, portions of the land area under triangle A, over the states of Falcon and Lara, are demonstrably rainier than they should be with mean subsidence over the triangle. These rainier areas are characterized by mountainous topography, north-south oriented coasts, and valleys that open eastward to the Caribbean Sea.

Over triangle A during autumn, a rainier season over the dry area, there is upward vertical velocity to 14,000 feet and a slight downward vertical velocity at the 20,000 foot level. This mean upward vertical velocity agrees well with the generally rainier conditions of the area but, again as in spring and summer, does not explain the rela-

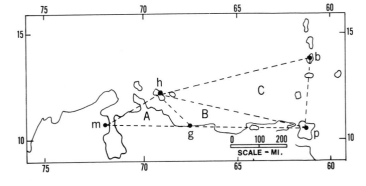

Fig. 10-18. Bellamy triangle network for computing divergence and vertical windfield velocities.

tive contrasts between the drier east-west coasts and the wetter north-south coasts. The area within triangle B also is generally rainier during the autumn season. However, it is less rainy than the areas within triangles A and C, and in this triangle B upward motion is found only above 3,000 feet and below 20,000 feet.

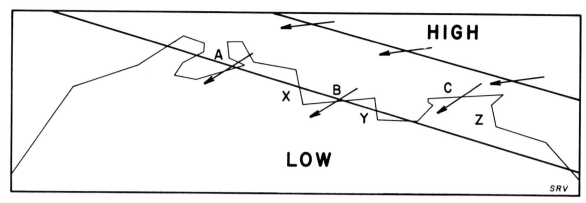

Fig. 10-19. Windflow over the northern coast of South America. See text for significance of areas A, B, C, X, Y, and Z.

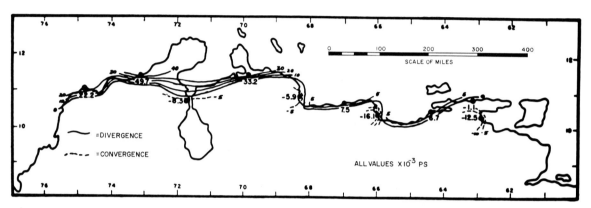

Fig. 10-20. Stress-induced spring windflow divergence over northern South America.

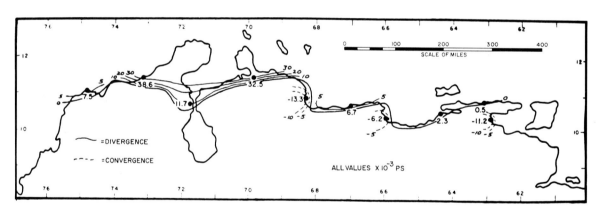

Fig. 10-21. Stress-induced summer windflow divergence over northern South America.

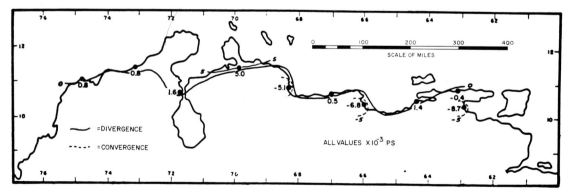

Fig. 10-22. Stress-induced autumn windflow divergence over northern South America.

During winter the resultant vertical motion over all triangles, in accord with rainier conditions of this season, is upward. However, as in the other seasons, east-west coastal fringes are much drier than north-south oriented coasts, and this fact is not indicated by the divergence and vertical motions values. The following computation of low-level divergence, which is induced by differential frictional stresses, is addressed to the solution of this problem.

Over northern South America and the southern Caribbean, the flow is from the east-northeast toward a low-pressure area to the south (fig. 10-19). Air flowing onto the rough land surface from the less rough sea surface may be expected to be increasingly turned toward lower pressure, that is, toward the south. At the same time the trade wind flow, well out over the Caribbean Sea, will not be subjected to as great a frictionally-induced turning effect and, as a result, should remain more nearly easterly. These patterns of flow are observed to occur. Such conditions should give rise to zones of surface and wind directional divergence along east-west oriented coasts (areas A, B, and C in fig. 10-19), and air should subside into these areas to replace the diverging flow. Stabilization of the lower layers of the atmosphere and a reduction in rainfall would be the logical consequence. In areas where the northern South American coast is oriented north-south (areas X, Y and Z in fig. 10-19), the easterly winds cross the coast almost at right angles. Increased frictional drag on this surface flow, as it moves from sea to land, should slow down the air and cause a convergent or "stowing" effect. It is in areas X, Y, and Z that the highest annual rainfall is observed.

Seasonal values of divergence and convergence due to differential land-sea stress contrasts may be calculated by a method used by Bryson and Kuhn. The equation for this is found in Appendix C. Computed values of divergence for the spring season (fig. 10-20) show divergence along all east-west shores in the dry area. Only along the north-south oriented coasts, where the windflow is on-shore at almost right angles to the coast, are values of convergence found. These figures then agree with east-west coast aridity and north-south raininess. Computed values for the summer season show similar results (fig. 10-21). Fall values (fig. 10-22) show lesser divergences along east-west coasts, in accord with the generally greater precipitation found along such coasts during this rainy season. The relatively smaller rainfall totals along such coasts are still evidenced, however, by the positive values of divergence found there. Winter divergence values (fig. 10-23) are positive along east-west coasts and larger to the west.

These calculated values match well with the actual precipitation distribution during this season.

Thermal Characteristics of the Earth and Their Effects on Airflow and Precipitation

Since mountains fringe the northern edge of the South American continent, the retarding effects of Earth surface friction may reasonably be assumed to extend to higher elevations as "form" stresses. Further, since the continent is heated daily, daytime thermally-created turbulent eddies are set up in the atmosphere. These turbulent eddies or "thermals" add heat to, and supplement the retarding effect of the land surface on the middle troposphere. Over the southern Caribbean Sea these form and turbulent stresses cannot be as pronounced because of the relative smoothness of the sea surface and the fact that thermally-induced turbulent eddies must be damped out by sea temperatures which are cool relative to the air above (figs. 10-24 to 10-26). This damping effect would be particularly in evidence with the cool Caribbean waters during winter, spring, and summer, but less so in fall (fig. 10-27) when water temperatures are higher — a situation possibly related to the conductivity capacity of water and to the slower autumn flow of the trade winds, with consequent smaller upwelling then.

Since in all seasons the resultant windflow from 3,000 to 10,000 feet is from land to sea over the dry area, it is hypothesized that the constraint imposed on this mid-tropospheric flow by terrain roughness and thermally-induced turbulence over the continent suddenly diminishes as the air flows out over the Caribbean. This must result in an acceleration of the flow — a condition actually observed at these levels. The additional mass trans-

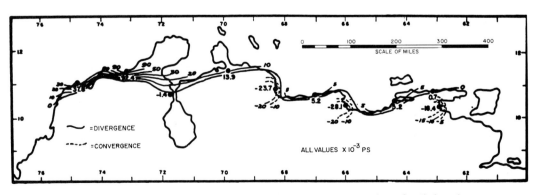

Fig. 10-23. Stress-induced winter windflow divergence over northern South America.

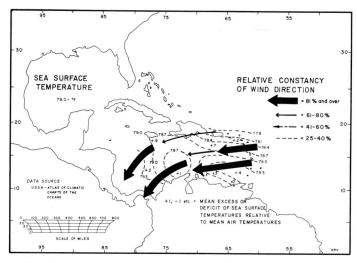

Fig. 10.24. Mean January sea surface temperature, and direction and constancy of surface winds over the southern Caribbean Sea.

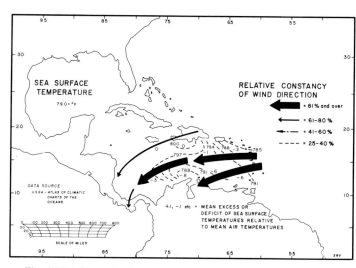

Fig. 10-25. Mean April sea surface temperature, and direction and constancy of surface winds over the southern Caribbean Sea.

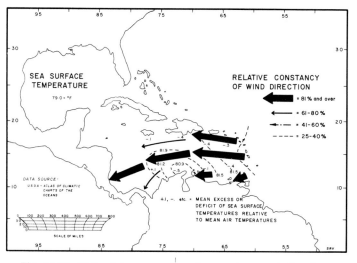

Fig. 10-26. Mean July sea surface temperature, and direction and constancy of surface winds over the southern Caribbean Sea.

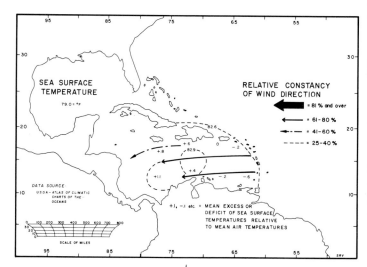

Fig. 10-27. Mean October sea surface temperature, and direction and constancy of surface winds over the southern Caribbean Sea.

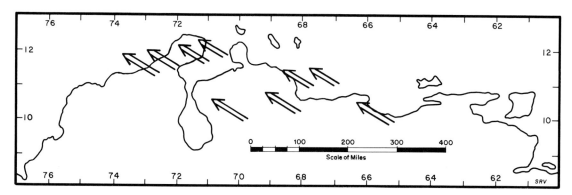

Fig. 10-28. Direction of air-mass divergence from land over the Caribbean Sea at altitudes of 3,000 to 10,000 feet.

Origin of Dry Climate

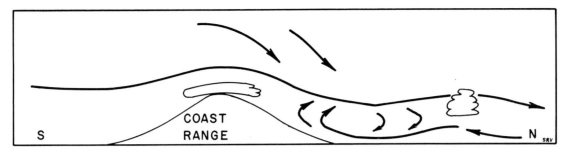

Fig. 10-29. Direction of replacement of air masses moving out over the Caribbean Sea.

ported seaward (fig. 10-28) by this increased air movement must be replaced from above or below, and the previous calculations of subsidence indicate that air is subsiding from higher levels and must replace an appreciable amount of this mass with "banked off" diverging surface air providing the remainder (fig. 10-29). The accompanying stabilization of the air over the coastal area and offshore islands should serve to diminish rainfall potential. This effect can be seen in the photo (fig. 10-30) which shows clouds evaporating and dissipating as they move from south to north across the coastal Venezuelan Andes northwest of Caracas during July of 1950.

Lettau (1967) has formulated a generalized physical theory in which he relates diurnal heating contrasts on mountain slopes, the terrain contours of these mountains, the Coriolis forces, latitude, and large-scale barometric patterns, to the character of the atmospheric circulation along these mountains. This motion system is called a "thermotidal wind" by Lettau. This new physical theory should have application to the arid zones of northern South America.

The combination of divergence and subsidence induced by the warmed mountainous slopes and colder sea could account for the lack of "weather" associated with easterly

Fig. 10-30. View toward the east-southeast from near the Caribbean Sea to the west of La Guiara, Venezuela, showing clouds evaporating and dissipating during July.

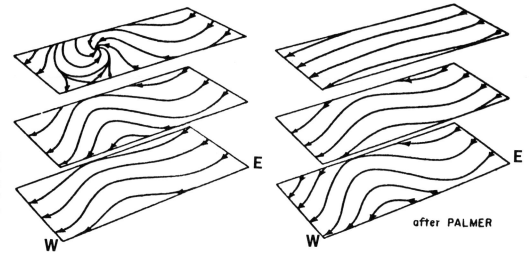

Fig. 10-31. Three-dimensional structure of waves. They may increase in amplitude with height (left) or decrease in amplitude (right).

after PALMER

waves passing over this region, as observed by E. C. Kindle (1945). Palmer and others (1955) and Riehl (1945) point out that many easterly waves have their greatest north-south amplitude and weather activity between 5 and 15,000 feet above the surface (fig. 10-31, right) and on the eastern sides of such waves. Such easterly waves are probably partially "damped out" by top-

ographic controls. Further, the subsidence and good weather typically found to the west of the crest of an easterly wave are normally associated with winds from the north-easterly quadrant. Hence, when easterly waves approach northern South America from the east, the winds most directly onshore should be part of a subsident system ahead of the wave trough. With such onshore

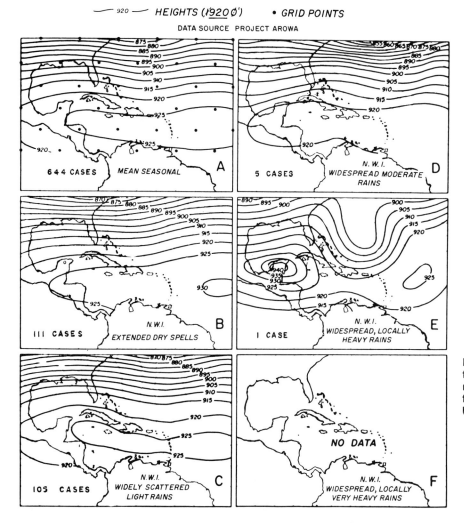

Fig. 10-32. Mean 500-millibar heights for various Netherlands West Indies rainfall types and their deviations from mean spring 500-millibar heights.

Origin of Dry Climate

winds, pronounced orographic lifting or funneling is required if rain is to be caused along the coast — a condition met only along north-south oriented, and wetter, shorelines. Only when easterly waves intensify aloft (fig. 10-31, left) should one expect orographic effects, which diminish rain on east-west coasts, to be overcome.

Broadscale Airflow Patterns and Their Effects on the Dry-Area Climate

If "weather creating" atmospheric disturbances in the low-level easterlies can be neutralized over the dry area by contrasting frictional stresses exerted on air by land and sea, then only those disturbances which extend through greater depths of the troposphere or actually become more intense aloft should be associated with heavy and widespread rainfalls over the dry area. Such deep storm systems are usually part of broadscale atmospheric disturbances in the middle latitudes.

To demonstrate the relationships between upper flow systems and the climate of the dry area, mean 500-millibar height maps were constructed for the various classes of Netherlands West Indies precipitation.

Spring (fig. 10-32)

During the frequent extended dry spells of spring over the N.W.I., the mean 500-millibar map B shows (a) an elongate subtropical high cell, (b) higher than normal easterly index, and (c) a blocking tendency over Bermuda. These factors indicate that easterly troughs over the Caribbean are either absent or fast moving and weak during extended dry spells over the N.W.I., and thus the topographic controls are effective in creating and controlling the dry climate. During the occasional widespread and intense springtime rainfalls over the N.W.I., the mean 500-millibar maps D and E show an appearance of upper troughs that extend from the western Atlantic across the central Caribbean Sea. The convergence associated with such deep extended upper trough systems counteracts the frictionally-induced low-level divergence and subsidence over the southern Caribbean dry area.

Summer (fig. 10-33)

The increasingly frequent scattered rainfalls over the dry area in summer are associated with the northward shift of the subtropical high (see maps A and C) and an increase in transitory easterly troughs, eddies, and low-

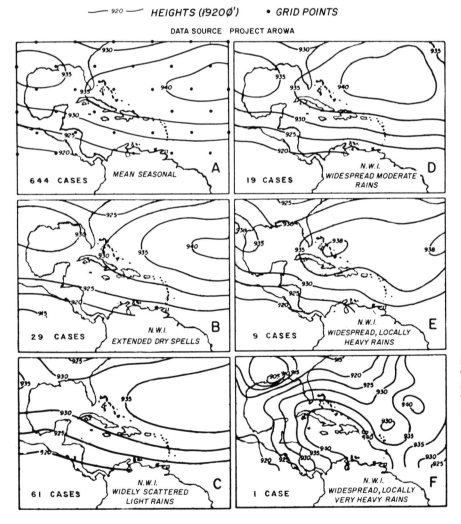

Fig. 10-33. Mean 500-millibar heights for various Netherlands West Indies rainfall types and their deviations from mean summer 500-millibar heights.

level asymptotes of convergence over northern South America. These bring heavy rains to the Llanos and to valleys opening eastward onto the Caribbean. However, these same easterly troughs are not deep, as shown by the fact that there is no evidence of them on the 500-millibar map C. Hence, stress-induced divergence dominates, and only scattered rains fall.

Extended dry spells are associated with a normal latitudinal position of the summer subtropical high, but an easterly trough is located over the Lesser Antilles, with a consequence that subsidence on the west side of the wave trough is found over northern South America on such occasions.

Widespread moderate, locally heavy, and locally very heavy rains (maps D, E and F) are all associated with deep waves in the easterlies over the central Caribbean, and these nullify the effect of topographic stress induced divergence.

These contrasts in broadscale flow are obvious when mean upper-flow maps for the months of June 1949 and 1950 are compared (fig. 10-34). June 1949, a drier than normal month over the N.W.I., showed no evidence of an extended trough over the central Caribbean, while during June 1950, an unusually rainy month over the N.W.I., a deep extended trough from the western Atlantic across the central Caribbean clearly controls the weather there.

Autumn (fig. 10-35)

Early autumn dry spells seem related to upper flow conditions (see B) similar to those of extended dry spells of summer. The infrequent late autumn extended spells have upper flow patterns characteristic of winter extended dry spells, that is, a high cell just to the northwest of South America with a subsiding northerly upper flow over the dry area and the Llanos.

Widespread moderate rainfalls seem to occur over the dry area when the normal seasonal highs aloft are shifted to the northwest, and low-level asymptotes of convergence lie entirely over the Caribbean Sea and the dry area. When deep troughs aloft extend from the western Atlantic across the central Caribbean, widespread heavy to locally very heavy showers can be expected to occur over the dry area.

Winter (fig. 10-36)

The more humid climate of early winter is associated with extended upper troughs similar to those which cause widespread, locally heavy, and very heavy rains over the area during autumn (maps E and F), and low-level east-west oriented asymptotes of convergence that cause widespread moderate rains on the south-southeast side of the subtropical high cell (map D). This high cell is located far to the north and west of its normal winter location under such conditions (compare map D with map A). As the position of the high cell shifts progressively farther southeastward, rains over the dry area diminish in frequency and intensity. Finally, when the high cell is centered over the northwestern portion of South America, extended dry spells dominate the weather of northern South America (map B).

It may be tentatively concluded that the dry climates of northern South America are created and controlled by topography and the cool Caribbean Sea which together create stress contrasts on the more typical windfield patterns in such a manner as to force the air to diverge and subside over the dry area. Only during the late autumn and winter are deep cyclonic systems prevalent enough to break this arid regime.

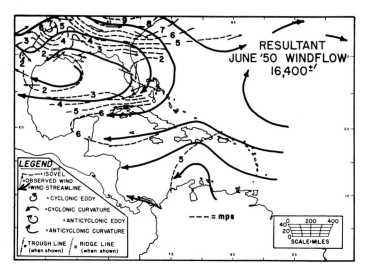

Fig. 10-34. *Above:* The resultant broadscale windflow over the Caribbean during June 1950, an abnormally wet month over the Netherlands West Indies during a season when that area is usually dry.

Below: The resultant broadscale windflow over the Caribbean during June 1949, a month during which the Netherlands West Indies received little rainfall during a normally dry season.

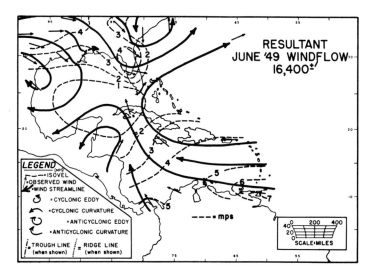

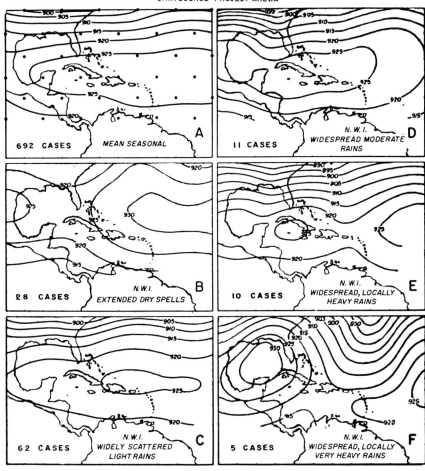

Fig. 10-35. Mean 500-millibar heights for various Netherlands West Indies rainfall types and their deviations from mean autumn 500-millibar heights.

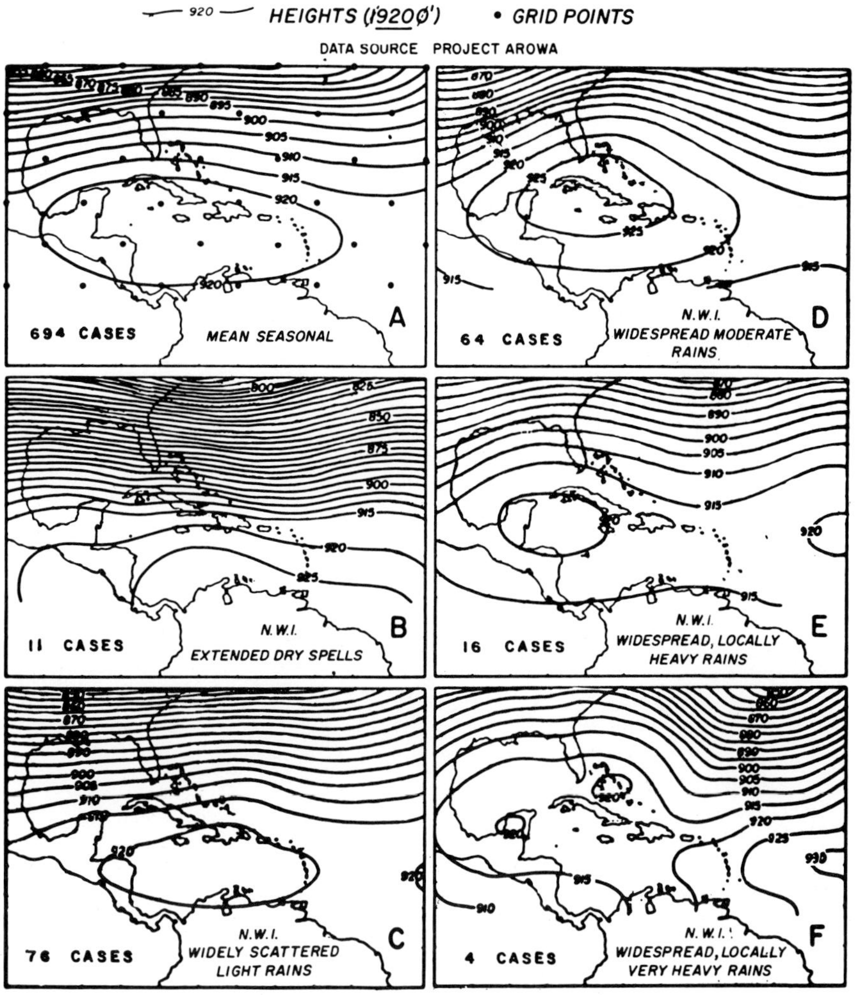

Fig. 10-36. Mean 500-millibar heights for various Netherlands West Indies rainfall types and their deviations from mean winter 500-millibar heights.

Chapter 10 Appendixes

A

Definition of Precipitation Classes

A. Dry days (no station in the N.W.I. receives rainfall).
  1. *Dry spells* — 1 to 4 days duration.
  2. *Extended dry spells* — dry day is part of dry spell of 5 or more days duration.

B. Wet days (at least one station in the N.W.I. receives measurable rainfall during the day).
  1. *Widely scattered light rains* — 20% or less of the stations receive rainfall, and no station receives more than 0.29 inch of precipitation.
  2. *Widespread rainfalls* — at least 60% of all N.W.I. stations receive rainfall during the day.
    a. *Widespread moderate rains* — no N.W.I. station receives more than 0.99 inch per day.
    b. *Widespread, locally heavy rains* — at least one N.W.I. station receives 1–1.99 inches of rainfall.
    c. *Widespread, locally very heavy rains* — at least one N.W.I. station receives rainfall of over 2 inches in the day.
  3. *Scattered rainfalls of variable intensities* — 30–59% of N.W.I. stations receive rainfall of variable intensities. A few cases of widely scattered but heavier rainfalls (that is, over 0.3) inch of rainfall) are also included in this category.

B

Formulas for Computation of Wind Divergence and Vertical Velocity

A. The formula for the computation of divergence is:

$$D = \frac{V_m \sin(\beta_{hg} - \alpha_m)}{H_m} + \frac{V_h \sin(\beta_{mg} - \alpha_h)}{H_h} + \frac{V_g \sin(\beta_{mh} - \alpha_g)}{H_g}$$

where:

m, h, g, etc. = station designations,
α = wind direction in degrees from north,
V = windspeed at m, h, or g, etc.,
H = height of triangle from point m, h, or g, etc., in meters,
β = azimuth of opposite side of triangle,
D = horizontal divergence in $(\sec)^{-1}$.

B. The formula for the computation of vertical velocity is:

$$W_2 = W_1 \left[\frac{e_1}{e_2}\right] + \tfrac{1}{2} \left(\left[\frac{e_1}{e_2}\right] D_1 + D_2\right) \triangle Z$$

where:

W_2 = vertical velocity at upper level — cm sec^{-1},
W_1 = vertical velocity at lower level — cm sec^{-1},
$\left[\frac{e_1}{e_2}\right]$ = density ratio $\frac{e_1 + e_2}{(g/cm^3)}$,
D_1 = divergence/sec at lower level,
D_2 = divergence/sec at upper level,
$\triangle Z$ = layer thickness in meters.

if W_2 is +, the air is moving downward in the mean,
if W_2 is −, the air is moving upward in the mean.

C

Formula for the Computation of Low-Level Divergence Induced by Differential Frictional Stresses

A. This formula is:

$$\text{Horizontal divergence of the wind} = \frac{\triangle C_D}{f \triangle y} \left[(V \sin\beta)^2 - (V \cos\beta)^2\right]$$

and:

$\triangle C_D$ = the difference in the drag coefficient across the coast,[5]
f = coriolis parameter,
$\triangle y$ = width of strip across coast (21 km),
V = surface wind velocity vector (mps),
β = angle between coastal axis and wind vector.

(Negative values indicate convergence while positive indicate divergence.)

Bibliographic References

KINDLE, E. C.
 1945 An application of kinematic analysis to tropical weather. U.S.A.A.F. Weather Wing, Regional Control Office, Ninth Weather Region, Morrison Field, Florida. (mimeo)

LETTAU, H. H.
 1967 Small to large-scale features of boundary layer structure over mountain slopes. *In* Symposium on Mountain Meteorology, 1967, Ft. Collins, Colorado, Proceedings. Colorado State University, Department of Atmospheric Sciences, Atmospheric Science Paper 122. 74 p.

PALMER, C. E., C. W. WISE, AND OTHERS
 1955 The practical aspect of tropical meteorology. U.S. Air Force, Cambridge Research Laboratory, Air Force Surveys in Geophysics 76. 195 p.

RIEHL, H.
 1945 Waves in the Easterlies and the polar front in the tropics. University of Chicago, Department of Meteorology, Miscellaneous Report 17. 79 p.

CHAPTER 11

NEW EVIDENCE ON THE CLIMATIC CONTROLS ALONG THE PERUVIAN COAST

Federico J. Prohaska *
Department of Geography, University of Wisconsin, Milwaukee, U.S.A.

The subtropical-tropical west coast deserts have always drawn a great deal of attention because of their apparently anomalous characteristics. Among such coasts, the coast of Peru attracted the most attention due to its population concentration and unique bioclimate (Knoch, 1930; Meigs, 1966; Rudloff, 1959; Schweigger, 1964; Unanue, 1940). The Peruvian coast is noteworthy also for other climatic peculiarities such as anomalous equatorward extension of desertic conditions, great uniformity and stability in the weather, high degree of atmospheric humidity during the entire year, and, last but not least, a nearly uninterrupted stratus layer that hovers over a great part of the coast during the winter all the way from northern Chile northward to about 8°S. Other striking features are the large number of hours with precipitation that totals only a negligible amount, the large annual and small daily ranges of temperature (unusual for a tropical latitude), the increase of thermal oceanity inland (decrease of annual temperature range), and the high persistency and shallowness of the trade winds.

One wonders why it was just in this atypical tropical coastal climate that the most important place of worship in pre-Columbian time (Pachacamac) developed, a worship of the God of Sun in a place where the sun is not visible nearly half of the year, while only a few miles inland it is shining practically all the time. By the same token, one wonders why Pizarro founded Lima just in this spot, to the surprise of the Incas, rather than on the coast or farther inland, for certainly he was aware of its rather dreary climatic characteristics. However, as it has turned out, the presence of the cloud cover and continuous advection of marine air, which regulate the temperature, probably have been the most important factors conducive to the fast growth of the Peruvian capital. Metropolitan Lima, with far more than two million inhabitants, has become the largest city of the Pacific coast of South America and is outstanding among tropical coastal cities for its high concentration of nonnative population.

Discussions of the Peruvian coastal climate in geographical literature have always mentioned the obvious controls, such as cool ocean temperatures and temperature inversion; but such discussions were limited by lack of aerological information and had to be based on what few surface data were available and what comparisons could be made with similar coastal deserts in other parts of the world (Lydolph, 1957; Trewartha, 1961). Such studies could pretty well explain temperature and humidity conditions at the coast but could not explain fully the almost complete lack of precipitation, which is the most outstanding climatic feature of the area. Such explanations can be adequately based only on upper air data, which indicates the degree of stability in combination with the available water vapor.

In tropical coastal regions no lack of precipitable water ever exists (Tuller, 1968). Therefore, the key to aridity lies in stability considerations, which at best can be deduced only in general terms from surface pressure fields. Even these are not well known in this area. Radiosonde data, however, are available from over Lima for the decade of the 1960s at least, and much more is known about the circulation and thermal structure of the eastern Pacific Ocean (Wyrtki, K., 1964; Wyrtki, L., 1967). This new information allows a more genetic approach to understanding the special features of the Peruvian coastal climate. This chapter will concentrate on an analysis of the more recent data of the central Peruvian coast. Only the pertinent facts will be treated, and no climatography of the general area will be attempted.

Sources of Data†

For Lima about 2,750 radiosonde records have been summarized for the period 1957–65 in the form of averages of height, temperature, and relative humidity for the different pressure levels at 50-millibar intervals. Wind statistics exist only for the mandatory isobaric levels (surface, 850 mb, 700 mb, and 500 mb) over the period 1957–1962.‡ In addition, the author made use of daily radiosonde observations of the entire year 1967.

---

* Deceased.

†The author gratefully acknowledges the help of authorities and staff of the following agencies in preparing and copying data: Agrarian University, La Molina; Servicio Agrometeorologico e Hidrometeorologico (SAH); Corporacion Peruana de Aeropuertos y Aviacion Comercial (CORPAC); Dirección General de Meteorología.

‡These data are available from Environmental Science Services Administration (ESSA), Weather Bureau, Washington, D.C., in microfilm or print-out form.

The launching hours were at 0700 Lima time (1200 GMT) from October 1957 through June 1961, and 1900 Lima time from October 1957 through September 1960 and since July 1961.

The mean monthly surface pressure values of the American west coast between 30°N and 30°S were taken from World Weather Records, 1951–1960, and from unpublished records of the Peruvian coastal stations at Talara, Chiclayo, Lima, and Tacna. Moreover, for the year 1967 the following data have been analyzed: Hourly surface observations at Lima Airport, 12°01′S, 77°07′W, 13 meters elevation, 3 kilometers inland in a straight line or 6 kilometers in an upwind direction from the coast; climatic statistics of the meteorological observatory "Alexander von Humboldt," La Molina, 12°05′S, 76°57′W, 238 meters elevation, 11 kilometers from the coast; three daily observations at Matucana, 11°50′S, 76°24′W, 2,374 meters elevation, 80 kilometers inland from the coast. Personal observations of cloudiness and other phenomena during the year 1967 added insight to these data. A tropical location such as this has the distinct advantage of consistent weather patterns so that the results of a thorough study of even a short period of time, such as one complete annual cycle, can be instructive and quite significant.

Southeast Pacific Trade-Wind Circulation

The trade winds have a manifold effect on the coastal climate. They cause cool ocean currents and upwelling, especially at the coast, and hence keep the marine layer cool and moist. By contrast the air above the marine layer is dynamically heated by subsidence. The result is a well-expressed *trade-wind inversion* and with it a high degree of stability that counteracts any development of convective air currents (which might bring on precipitation) and thereby produces "dry" oceans and arid or desert coasts.

The average pressure distribution exhibits in winter a high-pressure ridge over South America connecting the South Pacific and South Atlantic anticyclones; in summer a low-pressure trough extends from the equator south over the tropical west coast. Over the southeast Pacific at 90°W the pressure difference between the subtropical high and the equatorial trough, which activates the trades, is approximately 10 millibars. During the year this difference is nearly constant but shows a latitudinal displacement. During January it extends between 25°S and 5°N, while in July it shifts to a position between 20°S and 10°N. As a consequence of this latitudinal shifting by seasons, the atmospheric pressure along the Peruvian coast during the southern hemispheric winter is 3 to 4 millibars higher than during the summer. This range decreases toward the equator to around one millibar. This seasonal variation in pressure pattern is only another expression of a more intense subsidence in winter than in summer that is also evidenced by the higher temperatures above the trade-wind inversion in winter than in summer.

The marine layer below the inversion has just the inverse annual temperature variation. As a consequence, the thermal differences between the two layers, separated by the inversion, is greatest in winter; therefore the stability is at its maximum. Such a seasonal variation of the intensity of the inversion is observed along the coastal area from Ecuador to northern Chile. The main purpose of this chapter is an analysis of the structure of the inversion at the central Peruvian coast and its climatic implications.

Figure 11-1 exhibits the annual variation of the atmospheric pressure field along the tropical Pacific coast in detail. It shows that (1) in no season does the meteorological equator coincide with the geographical equator but lies between 10°N and 15°N latitude; (2) the intertropical trough varies only slightly in position and intensity during the year; (3) the pressure gradient continues across the geographical equator during the entire year along the South American coast and is stronger in winter than in summer because the high-pressure influence extends nearly to the equator during the winter season; (4) a distribution of absolute values and annual variations of pressure along the Pacific coasts of South and North America that is different and asymmetric with not only the geographical equator but also the intertropical convergence zone (ITC), or however this region may be defined.

Figure 11-1 also shows the known inverse relationship between atmospheric surface pressure and oceanic surface temperature. Furthermore, it reveals that genuine tropical temperatures (above 26°C) are found only in the Northern Hemisphere. The transition to cooler subtropical waters is concentrated in each hemisphere in a relatively small area (between 20°N and 25°N and between 0° and 5°S), and so are the pressure gradients. Even though the pressure data for figure 11-1 are based on only five stations in the Southern Hemisphere and seven in the Northern Hemisphere, the pressure pattern derived can be considered reliable since it is confirmed by the independently observed ocean temperatures.

The climatic effects of the trade winds and the concomitant phenomena along the coast are determined to a great degree by the orientation of the coastline in relation to the direction of the trade-wind flow and by the topography of the hinterland and its influence on the circulation pattern (Lettau, 1967; Lydolph, herein). Such features are different from continent to continent which exhibit low-latitude dry coasts. The Pacific coast, except for the area between 30°S and 40°S, has a NNW-SSE direction to about 14°S and then changes to a NW-SE direction southward to the Chilean border. Such an orientation coincides closely with the prevalent trade-wind direction. In addition, the Andes rise almost immediately from the coast to a mean crest height of 4,000 to 5,000 meters poleward of 8°S. The coast, therefore, is well protected from tropical air masses in the Amazon basin, and the orientation of the mountains parallel to the coast must have a channeling effect on the trade winds.

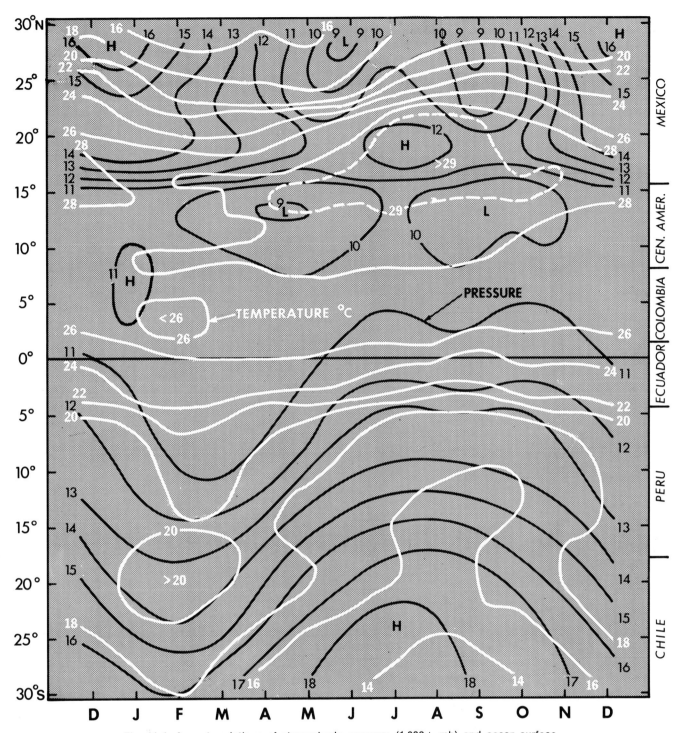

Fig. 11-1. Annual variations of atmospheric pressure (1,000+ mb) and ocean surface temperature (°C) along the west coast of the Americas between 30°N and 30°S.

All these facts taken into consideration, there is no reason to expect any similitudes of climates at corresponding latitudes in the two hemispheres along the tropical west coast of the Americas. Thus, the equatorward extension of the Peruvian coastal desert may be considered a climatic abnormality only as long as it is referred to the planetary setting, that is, the geographical equator. If one realizes that the low-latitude coastal deserts are not functions of astronomical parameters but are dependent on the trade-wind circulation and its intensity to determine their geographic location, then the Peruvian coastal desert starts at a similar latitudinal distance from the intertropical trough as the coastal desert in Baja California does, and the Baja California desert extends even further toward the intertropical convergence zone than the Peruvian desert does.

Winds Along the Peruvian Coast

Throughout the year the trade winds blow along the Peruvian coast and extend northward to and even across the geographical equator. The nearly constant direction of the pressure gradient implies an extraordinary persistency of the winds, which is corroborated by figure 11-2 and table 11-1. At Lima at 7 AM and 7 PM (launching hours of the radiosondes) 74 percent of all *observed* surface winds (excluding calms) are from the south (62 percent) and south-southeast (12 percent) and have, therefore, a small onshore component. The 24-hour frequency distribution is similar. The only difference is a higher frequency of the direction parallel to the coast (SSE), which is observed above all during the night and in winter, and from the west, which is characteristic of the morning hours in summer. It is noteworthy that no offshore component is manifest at all.

Similar conditions are found in Puerto Chicama at 7°42′S, where at 7 AM wind directions are from SE (37 percent), SSE (43 percent), and S (15 percent), accounting for 95 percent of all winds. The change of the prevalent wind direction from S in Lima to SSE in Puerto Chicama is due to change in direction of the pressure gradient. The afternoon winds (4 PM) at Puerto Chicama are similar to those in the morning and show the same total of 95 percent of the winds from a southerly direction if SSW winds are included. Thus, the trades here also show a slight onshore component, which is enhanced in the afternoon when the SSW direction increases from 2 percent at 7 AM to 20 percent at 4 PM. The small offshore component (SE) decreases during the same time interval from 37 percent to 26 percent. Neither at Puerto Chicama nor at Lima is it possible to speak of a land-sea breeze circulation, since the ocean-coast temperature contrast, even though rather large in some months, is able to shift the trade-wind direction just a few degrees. Rather incomplete wind statistics for Tacna at 18°S (Peru, Dirección General Meteorológica, 1956 & 1957) and nine guano islands (Valdivia Ponce, 1957) that lie between 7°S and 18°S latitude show

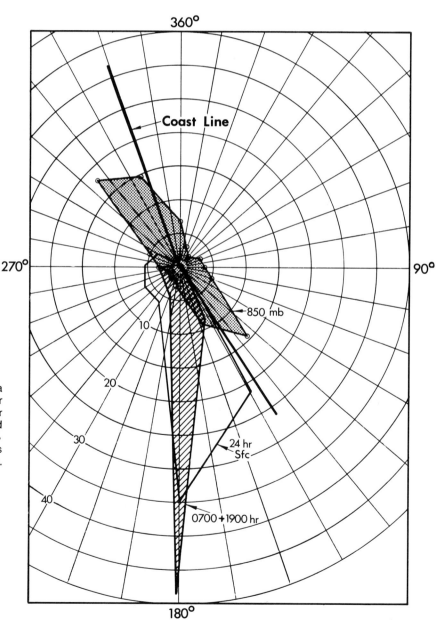

Fig. 11-2. Annual wind roses at Lima (airport) in percent of all observations for surface in 8 directions for 24-hour observations (1967) with 20% calms, and 0700 + 1900 hours (1957–1962) with 24% calms, and 850 millibars in 16 directions (1957–1962) with 1% calms.

TABLE 11-1
Monthly Averages of Pressure Differences and Winds Along the Peruvian Coast

	J	F	M	A	M	J	J	A	S	O	N	D
Pressure Differences in millibars between Tacna, 18°14'S and Talara, 4°04'S (1957–1966)												
	2.3	2.1	2.2	2.8	3.4	4.0	4.0	4.0	4.3	3.7	2.9	2.7
Prevailing wind direction (D), frequency (%), and speed (Sp) in knots												
Puerto Chicama, 7°42'S, 4 PM (1941–1953)												
D	SE	SE	S	S	S	S	S	S	SE	SE	S	SE
%	43	40	42	45	43	40	45	48	45	48	43	46
Sp	8.2	7.8	8.2	9.4	9.0	8.2	7.8	8.2	9.4	9.4	9.0	8.6
Lima, 12°01'S, 7 PM (1957–1962)												
D	S	S	S	S	S	S	S	S	S	S	S	S
%	53	61	65	66	53	53	69	56	65	72	69	66
	6.4	4.8	4.7	5.3	3.5	4.3	4.6	6.5	4.8	6.0	5.3	6.2
Resultant wind direction (azimuths)												
	175	184	181	178	183	186	177	184	182	181	183	177
Resultant wind speed (knots)												
	5.7	3.9	4.1	4.7	2.2	2.8	3.9	5.4	4.1	5.4	4.7	5.6
Steadiness of resultant wind vector (%)												
	89	81	88	88	63	64	84	84	87	89	89	89
Tacna, 18°14'S, 3 daily observations (1956)												
D	S	S	S	S	S	S	S	S	S	SW	S	S
Sp	3	5	2	4	3	3	–	6	6	6	5	5

exactly the same predominance of S and SE directions during the entire year.

Similarly uniform is the wind speed along the coast and across the islands offshore, where it is generally stronger (table 11-1). The average velocities are limited to between four and nine knots, with a tendency toward the lower limit during the first half of the year and toward the higher limit in the second half. In Lima, the resultant daily wind direction is from the south during the entire year with a persistency in its vector between 60 percent and 90 percent. Therefore, the resultant wind speed is only a little less than the actual speeds.

The trade winds have a vertical extent of no more than 1,000 to 1,500 meters at the central Peruvian coast, as can be judged from radiosonde data taken at Lima, the only aerological station along the Pacific coast between Panama (9°N) and Antofagasta (23°S). Figure 11-2 and table 11-2 reveal that at 850 millibars, or about 1,500 meters above sea level, NW is the most frequent wind direction and is the direction of the resultant wind. This direction is most predominant during December through April. Also during the period of the strongest development of the trades, June through September, the NW winds are still more frequent than the SE winds. The steadiness of the resultant wind vectors at this altitude is very low, particularly in winter (around 10 percent), and the resultant wind velocity is only a fraction of the actual wind speed. Further aloft, between 850 and 500 millibars, the wind speed is stronger in winter than in summer. Thus, in winter there is a rapid decrease in wind speeds downward toward and within the friction (marine) layer. This change is no doubt enhanced by the strong temperature inversion that exists during this time of year. The result is a nearly opposite annual variation of wind speed in the marine layer below the inversion where a summer maximum prevails and in the subsiding air above the inversion where the winter maximum prevails. These aerological conditions over Lima will be treated in more detail elsewhere in connection with the comparison with the climate of the Peruvian highlands.

Temperature Inversion

The equatorward extension of the trade winds is accompanied by a continuance of the trade-wind inversion. Salinas, at 2°S (U.S. Navy, 1959), shows inversion conditions similar to those at Lima 12°S. The inversion is a function of both subsidence-induced adiabatic heating from above and cooling from below, and hence its day-to-day and season-to-season variations depend upon surface and upper air conditions, both of which are a direct consequence of trade-wind intensities and pressure fields. To obtain an insight into frequencies of occurrence and annual variations of height, intensity, and thickness of the inversion, it was necessary to evaluate daily radiosonde data according to the significant levels for one entire year at least, since ESSA data averaged for each 50-millibar level smoothed out the characteristic inversion features. The following discussion is based on the daily 7 PM radiosonde observations for the year 1967 at Lima, a year which experienced no major climatic abnormality on the coast.

During 1967, of all soundings (320 total), 95 percent showed a surface inversion, 2 percent showed isothermal structures, and 3 percent, normal lapse rate conditions, up to 850 millibars at least. The seven days with isothermal structures were scattered between December and April, and the ten days with average lapse conditions were scattered between January and March. Hence, from May through November, the inversion was present without interruption. Even though radiosonde observations were skipped a few times within this period, it seems safe

TABLE 11-2
Resultant Wind Direction (D) in Azimuths, and Average Wind Speed (Sp) in Knots, Over Lima in 1957–1962

	J	F	M	A	M	J	J	A	S	O	N	D
850 mb (1500 m approx.)												
D	357	328	342	325	360	13	307	331	301	244	334	332
Sp	5.0	5.4	5.3	5.4	6.6	7.7	7.2	7.2	7.1	6.4	5.9	5.9
Resultant Speed												
	1.3	1.3	1.6	1.4	0.6	0.9	0.9	1.1	1.9	0.4	0.8	1.3
Steadiness of resultant wind vector (%)												
	26	25	31	21	10	12	12	15	3	7	13	22
700 mb (3140 m approx.)												
D	300	327	342	113	148	137	148	131	140	120	18	337
Sp	7.4	7.0	5.9	6.8	8.8	8.5	7.9	7.7	7.6	7.3	7.3	6.3
500 mb (5860 m approx.)												
D	130	109	99	13	341	10	356	10	16	63	64	147
Sp	9.1	8.4	8.3	9.2	12.3	12.7	13.6	13.0	10.8	10.1	9.8	10.3

to assume that the inversion was present throughout the period.

Since an inversion constitutes a convectional "lid" between the marine layer and the layers above; it becomes a controlling factor on the coastal climate throughout the year. Such a persistent inversion is extraordinary anywhere on earth, let alone in tropical latitudes. The comparisons of 10-year averages for Lima with the U.S. Standard Atmosphere at 15°N (U.S. Standard Atmosphere, Supplements, 1966) demonstrates that up to 800 millibars (2 kilometers) the troposphere over Lima is markedly cooler than the Standard Atmosphere, which reveals the strong oceanic influences, but between 700 and 270 millibars (10 kilometers) it becomes as much as 4°C warmer than the corresponding averages for 15°N, which indicates the great importance of subsidence over Lima. The two effects together are responsible for the extraordinary stability at the Peruvian coast.

In spite of the persistency of the inversion, its structure and variability differ markedly from summer to winter in conjunction with the change from low-pressure to high-pressure influence along the Peruvian coast. As an example, the mean monthly structures of the vertical temperature distribution for the two seasons are represented in juxtaposition in figure 11-3. It becomes immediately evident that the winter type represents the typical trade-wind (subsidence) inversion, and the summer type represents the combined effects of heating by subsidence from above and cooling by oceanic influences from below. The change of the structure of the inversion is in relation to and only a symptom of the change in stability that takes place not only at the coast and along the west side of the Andes but over the entire western part of tropical-subtropical South America south of the equator. The trade-wind inversion type, which is manifest June through September or October, coincides with the dry season in the Andes and the upper Amazonian watershed and signifies the high-pressure influence over this entire region at this time of the year, while the summer-type inversion coincides with the rainy season, December through March or April, and signifies the influence of low pressure.

The different controls and structures of the inversion in different seasons have different influences on the marine layer and hence on the coastal climate. Therefore, the analysis of the local climate will be broken down by seasons.

Winter or Trade-Wind Inversion and Climatic Consequences (May 16 through September 30, 1967)

Along the coast of central Peru, winter conditions prevail as long as the primary cause of the inversion is heating by subsidence, normally from June through September or October. It is apparent that winter in 1967 started on May 16 when this type of inversion was first observed. The end of the winter was arbitrarily taken as September 30 in order to have climatic statistics averaged over a period approximately equal in length to that considered to be the summer season, as discussed later. This

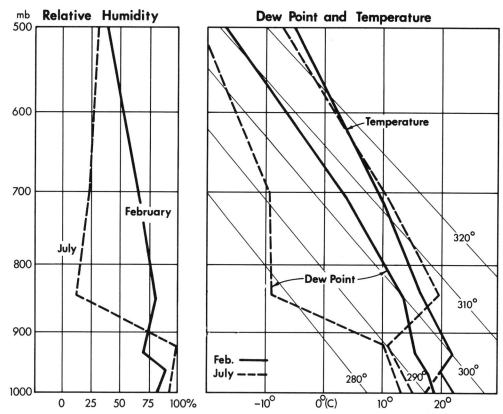

Fig. 11-3. Average vertical cross-sections of temperature, dew point, and relative humidity for February and July 1967 over Lima.

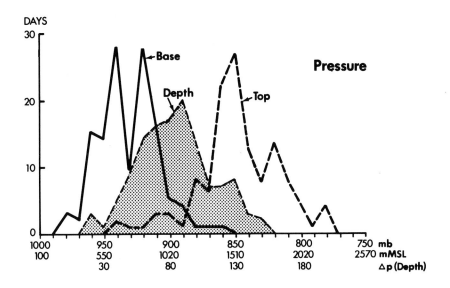
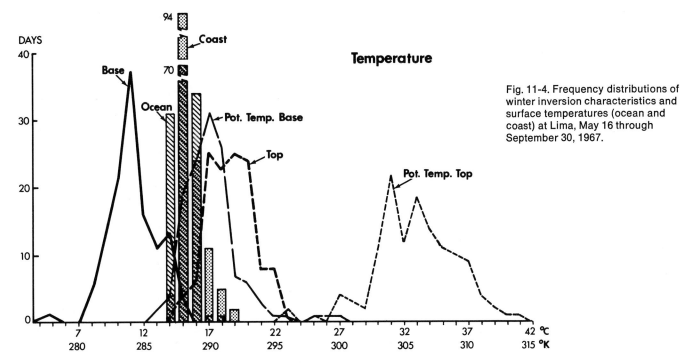

Fig. 11-4. Frequency distributions of winter inversion characteristics and surface temperatures (ocean and coast) at Lima, May 16 through September 30, 1967.

was justified somewhat by the fact that other parameters such as day length, solar radiation, and surface temperatures started to show seasonal changes around the end of September in spite of the fact that the trade-wind inversion type, and with it the stratus layer, persisted into October.

Characteristics of the Inversion

The main characteristics of the inversion during the winter season are illustrated in figure 11-4. The inversion base fluctuated between 300 and 1,400 meters above sea level, with most frequent occurrences between 600 and 800 meters. The temperatures of the inversion base were most frequently between 10° and 11°C. The inversion top fluctuated between 650 and 2,250 meters, with the mode at 1,500. The temperature at the inversion top had a mode between 17° and 20°C. The inversion depth averaged about 830 meters, had its mode at 870, and varied from 200 to 1,500. The temperature increase through the inversion was generally between 6° and 10°C but amounted to as much as 15°C. The corresponding increase of potential temperature varied generally between 14° and 18°K but reached a maximum of 23°K. The average lapse rate through the inversion, thus, amounted to approximately −11°C per kilometer and showed a frequency mode of −15°C per kilometer.

The increase of the variability upward was common to all elements represented in figure 11-4. The interdiurnal variability increased also for both pressure and temperature from the surface up to the top of the inversion. At the top of the inversion, 35 percent of the interdiurnal variabilities fluctuated within the limits of ±10 millibars (90 m), while at the inversion base 43 percent of the interdiurnal variabilities lay within these limits. The maximum interdiurnal variability at the top was 100 millibars or 900 meters, while at the base it was only 60 millibars or 540 meters. Therefore, the interdiurnal variability of the depth of the inversion resulted primarily from fluctuations in the height of the inversion top. Such a condition is an obvious consequence of subsidence as the primary cause of the trade-wind inversion. Associated temperature values exhibited interdiurnal variabilities within the limits of $\pm 1°C$ in 66 percent of the cases at the base of the inversion and in 56 percent of the cases at the top. At the inversion top, day-to-day changes of as much as 4° to 6°C were not infrequent. The normally inverse relationship between depth and intensity of the inversion was only weakly expressed in 1967. During the same year the data indicated no relationship between the height of the base of the inversion and the height of the top of the inversion, between temperatures at the base and the top, or between the inversion structure and vertical changes in wind direction and/or speed.

Atmospheric Humidity

The steady onshore trade winds, together with the inversion, have a strong influence on the humidity content of the marine layer. The winds increase the evaporation, but the inversion inhibits the transport of water vapor upward, and the coastal mountains intercept its lateral diffusion inland. Therefore, the humidity content in the marine layer is very high and contrasts sharply with drier layers above the inversion, a contrast that increases with the strength of the inversion. The result is a marine layer saturated nearly throughout its depth, with a low and stable stratus cover produced by continuous mixing below the inversion. At Lima Airport the 24-hour average of the relative humidity was 88 percent during the four winter months of 1967, and the hourly values never fell below 60 percent. Even at 2 PM, when the daily temperature is usually at a maximum, the mean value of the relative humidity was above 80 percent. The strong contrast between the humid marine layer below the inversion and the dry air above the inversion can best be illustrated by the mixing ratio. The uniform values of 9 grams per kilogram in the marine layer decrease to 3 grams per kilogram at the top of the inversion, which corresponds to a decrease in relative humidity across the inversion from 100 percent at the base to 20 percent at the top.

Clouds

The stratus in the marine layer is extraordinary in its persistency. During the winter of 1967, 90 percent of all days were overcast (cloud cover of more than 95 percent of each hourly observation). Only five days had less than six oktas (eighths of sky cover), and the clearest day during the entire period had an average of 3.4 oktas. The longest consecutive overcast period was 44 days, during which just 10 hours, recorded during six different days, had a cloudiness of only 5 to 7 oktas. During the entire winter period only 53 hours or 1.6 percent of the time had clear sky, generally around sunset. Previous years had shown short sunny periods but with an irregular frequency.

In spite of this stable cloudiness with ceilings generally between 150 and 300 meters, fog is rather infrequent at the coast at levels below 100 meters (airport and downtown Lima). An earlier study had shown that during the period 1943–1950, low-ceiling frequencies below 120 meters had an inverse relationship to wind speed (Graves, 1944). Therefore, it appears that nightly calms might produce fog, but the data for 1967 indicate that usually calms do not last long enough to allow the ceiling to descend to the surface.

The stratus is only a few hundred meters thick and extends upward approximately to the inversion base. Its thickness is a function of the strength of the mixing process in the marine layer (mixing condensation level) and the height of the inversion base, which depends on the simultaneous intensity of subsidence from aloft among other factors. This can be deduced from daytime cloud observations at La Molina, at 235 meters elevation, 22 kilometers east of the airport. During periods when both stations had overcast sky, the marine layer was deeper and inversion stronger than normal. That is, the inversion base was higher with a correspondingly lower temperature, and the inversion top had a higher temperature but only a slightly higher altitude. Hence, the temperature gradient across the inversion was nearly twice as great as it was when the stratus was rather thin and the sun or sky was partially visible at La Molina. Above the inversion, the western slopes and the highlands have very small amounts of cloud, 2 to 4 oktas during winter, which consist primarily of cirrus, the typical dry season cloud type.

Precipitation

The continuous advection of marine air inland and the resultant concentration of water vapor below the inversion cause frequent saturation in the surface layer during nighttime. This factor, together with strong nocturnal cooling as a result of intense net long wave radiation from the cloud top into the extremely dry atmosphere above, increases the instability in the marine layer and leads to frequent drizzlers, called *garúa*. This is a typical night and early morning phenomenon all along the Peruvian coast from about 8°S latitude southwards.

During the 1967 winter, *garúa* was recorded 877 hours at Lima Airport, which produced a total amount of precipitation of only 6.1 millimeters. It occurred most frequently between 11 PM and 9 AM, during which slightly more than half of all days recorded garúa. It was least evident in the afternoons between 12 PM and 5 PM. This pronounced daily variation applies only to the coastal strip itself. On the slopes and hills (lomas) between 100 and 200 meters and about 700 to 800 meters, where the inland advance of the marine air is lifted orographically, almost continuous drizzle is experienced in the contact zone of the stratus with the ground (fog precipitation). Here the total amount increases to between 100 and 200 millimeters during the winter, which, with the very reduced evaporation loss, is enough to produce the so-called lomas vegetation (Roessl, 1967).

As vegetation itself is a fog catcher, once its growth is started, the soil beneath the vegetation receives amounts of water several times greater than would have been produced on a barren slope by *garúa* processes alone. This in turn fosters further vegetational development. Here then is a case of vegetation acting not primarily as a water-consuming factor but as a water-producing factor in a self-increasing process. All this is plainly visible in the landscape as a zone of intensively green pasture with grazing cattle between completely desertic landscapes both below and above, which indicate respectively the lower and upper limits of the stratus. The botanical aspect of this singular and seasonal phenomenon has been widely investigated (Weberbauer, 1945), and it is not necessary to enlarge upon it here, even though it is a direct consequence of the trade winds and their intrinsic inversion. From October through December as the stratus cover and its *garúa* precipitation fades away, the vegetation slowly disappears, and the landscape changes back once more to the uniform, almost complete desert so characteristic of the lower part of the western slope of the Andes.

Temperature

Another direct effect of the stratus cloud cover, or indirect effect of the inversion, is the influence on the insolation and ultimately on the temperature in the marine layer. Since during the winter about 80 percent of the incoming solar and sky radiation is reflected or absorbed by the clouds, it has little effect on temperatures. Thus, surface temperatures along the coast depend almost entirely on the temperature of the marine air that is being advected inland by the onshore component of the trade winds.

During the winter of 1967, the daily water temperatures at La Punta (Callao) were between 14°C and 16°C except on two days (fig. 11-4). Equally consistent were the air temperatures at Lima Airport. The 7 PM values, which practically coincide with the daily averages, remained almost the same from day to day (a true tropical behavior), despite the low absolute values. On 84 percent of all days the temperatures stood at 15° or 16°C, and the extremes during the entire winter were 14° and 19°C (fig. 11-4).

The cloud cover also strongly reduces the effective outgoing surface radiation. The result is a very small diurnal range of temperature at the coast (3°C), which is completely atypical of a tropical station (table 11-3). Table 11-3 shows also the very insignificant temperature variations experienced from one ten-day period to another during the winter of 1967. As a matter of fact, during the months July through September, the interdiurnal variability never exceeded $\pm 0.9°C$, and 70 percent of the time it was $0.0°C$; only in June was it once as much as $-2.0°C$.

By comparing sea and air temperatures in table 11-3, it becomes evident that in winter the air temperature is merely a function of the sea temperature. This continues to hold even as late as the spring equinox when the noon solar zenith distance is only 15°. Obviously, as long as the trade winds and, hence, the cloud cover persist, the direct influence of incoming and outgoing radiation on surface temperatures is practically nil. This strong control by the cloud cover can be exemplified by its absence during the only clear night, 11/12 of June, when the minimum temperature dropped 4.5°C lower than it was either the day before or the day after.

During the winter, temperatures at the base of the inversion are 4°C to 5°C lower than at the surface, which results in an average lapse rate of 6.7°C per kilometer through the marine layer, if one defines the upper limit of the marine layer as the base of the inversion. The inversion top has temperatures 2° to 5°C higher than those at the surface. The mode of the potential temperature at the inversion top is about 16°K warmer than that at the surface (304°K = 31°C, a temperature which is seldom observed at the coast even in midsummer). During the entire winter the free atmosphere at an altitude of 1,500

TABLE 11-3

Ten-Day Average Air and Sea Temperatures at Lima (Callao) During Winter 1967

	June			July			August			September			Winter June 1 to Sept. 30
	1–10	11–20	21–30	1–10	11–20	21–31	1–10	11–20	21–31	1–10	11–20	21–30	
Lima Airport (Callao), Air Temperature (°C)													
Mean maxima	18.0	17.5	16.9	16.6	17.1	16.4	16.3	17.2	16.7	17.0	16.8	17.4	17.0
Mean minima	15.1	14.3	14.6	14.6	14.5	14.1	13.6	14.3	13.2	13.8	13.8	14.0	14.2
Range	2.9	3.2	2.3	2.0	2.6	2.3	2.7	2.9	3.5	3.2	3.0	3.4	2.8
Average	16.6	15.9	15.7	15.6	15.8	15.3	14.9	15.7	15.0	15.4	15.3	15.7	15.6
Callao (La Punta), Ocean Surface Temperature (°C)													
Average	16	16	15	15	15	15	15	15	14	14	14	14	15

meters was 3.1°C warmer than Lima Airport. At this elevation the western Andean slopes experience, too, the highest monthly temperature averages. One has to go to an elevation of 2,300 meters to find daily mean temperatures as low as those at sea level, and to find diurnal maximum temperatures as low as those observed along the coast one has to ascend to 3,500 to 4,000 meters. Such high daily maximum temperatures are the consequence not only of subsidence but also of intense solar radiation and dry soils. In combination with intense heat loss during nighttime, the daily ranges of temperature increase with elevation even at steep slope exposures and show the typical behavior of tropical arid zones.

At the climatic station at Matucana, 2,374 meters, 90 kilometers east of Lima Airport, the average diurnal range of temperature is more than 10°C in winter. On less-inclined slopes, valleys, or rolling plains above the inversion, as in southern Peru and northern Chile, daily ranges of more than 20°C are regionally observed even at elevations above 3,000 meters, and certain places may have the highest daily ranges of temperatures anywhere at the earth surface (Weischet, 1966). Hence, in regard to temperature, the inversion produces a stronger contrast over a smaller distance than the Andes do, and most of the atypical climatic features in this oft-mentioned area are products of the extremely narrow coastal zone below the inversion layer.

In Lima the winter temperature is 9°C below the latitudinal average, a temperature observed in Salvador (Bahia) at the Atlantic coast of South America at the same latitude. In the northern hemisphere the same latitude and exposure (west coast of Nicaragua) has 10° higher temperatures than Lima. To find temperatures similar to those in Lima, one has to go to 24°S along the west coast of Africa and to 28°S along the west coast of Australia. The Australian coast best represents the latitudinal mean (cf. Chapt. 24).

The Peruvian coastal weather in winter, and its stability in space and time, is, thus, the consequence of the persistence of the inversion together with the onshore component of the equally persistent trade winds. The mixing processes in the marine layer produce the quasi-permanent stratus cover belt below the inversion, which largely inhibits water-vapor diffusion into the higher atmospheric layers.

It should be pointed out, however, that the inversion and the associated stratus are not produced by or limited to the area of the Peruvian coastal current, or to the coast itself and its hinterland, as a result of friction and onshore-upslope movements, since both the inversion and the stratus extend far into the Pacific, well beyond the western fringes of the Peruvian current. This can be documented by occasional radiosonde observations taken by oceanographic vessels and can be seen very clearly on ESSA 3, ESSA 5, and ATS 3 satellite pictures, which show an immense and uniform cloud shield over the subtropical southeast Pacific that starts at the South American coast and extends west-northwestward to be dissolved eventually in the typical trade wind cumuli (fig. 11-5).

Summer Inversion
(January 1 through May 15, 1967)

Typical summer conditions prevail from January through April. In this period, too, an inversion is almost continuously manifest, the trade winds blow with the same persistency as in winter, and occasional rains give only scarcely measurable amounts. But with the exception of the unidirectional steadiness of the wind, these similarities are only apparent; subsidence is restricted to higher levels (Foehn effect), and precipitation falls not in the form of drizzle but as big rain drops. Other climatic parameters, such as cloud cover, radiation, and temperature, are also quite different from what they are

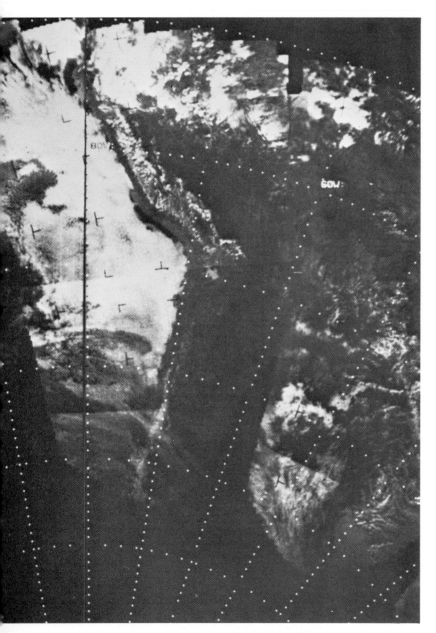

Fig. 11-5. West coast of South America as seen by Satellite ESSA 5, PASS 1381–1393, August 7, 1967.

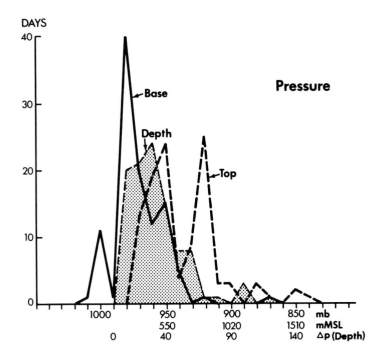

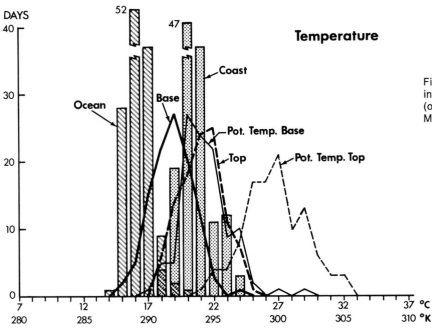

Fig. 11-6. Frequency distributions of summer inversion characteristics and surface temperatures (ocean and coast) at Lima, January 1 through May 15, 1967.

in winter. Consequently, a much greater seasonal contrast is observed at the coastal area than farther inland where the cloud and rain regimes are the primary climatic determinant. This relatively large contrast adds another atypical aspect to the Peruvian coastal climate.

Characteristics of the Inversion

During summer, the surface inversion is primarily the consequence of the cool ocean. In contrast to the winter inversion, it starts nearer to the surface, is less pronounced, and shows a greater day-to-day change in intensity, depth, and altitude. In fact, during summer the inversion top is normally at or below the altitude of the inversion base in winter. With a few exceptions the top lies between 400 and 900 meters (fig. 11-6). The inversion base most frequently lies between 200 and 500 meters, and the inversion depth is generally only 100 to 300 meters (dp 10 mb to 35 mb), in contrast to the winter inversion when the thickness is around 800 meters.

During summer the temperature increases through the inversion normally by only 1° to 4°C. The mode of the lapse rate through the inversion is the same as in winter, −15°C per kilometer. These inversion characteristics are representative only for the coastal strip itself, where a marine layer exists. Inland, the fast-rising slopes soon penetrate the inversion, and over the ocean the inversion probably starts nearly at sea level. This can be assumed because the temperature of the inversion base over Lima is higher than the ocean temperature just offshore, which ipso facto excludes a genuine marine layer. Furthermore, the extrapolation of the inversion slope downward to sea level gives approximately the actual ocean temperatures (fig. 11-3). The thickness of the inversion over the sea is, thus, between 600 and 800 meters, a depth similar to that found in winter. The mode of the temperature increase through the inversion over the sea has to be between 5° and 6°C, or an increase in potential temperature of about 10°K.

Atmospheric Humidity

The mixing ratio decreases slowly from 13 grams per kilogram at sea level to 12 at the inversion base and to 11 at the inversion top. The relative humidity decreases accordingly from 80 percent to 90 percent in the marine layer to a little less than 70 percent at the inversion top. Saturation is generally not observed at or below the inversion base. These are typical features of a "cold water type" inversion. Subsidence becomes increasingly important only at higher altitudes. A vertical cross-section of relative humidity makes this evident in spite of the fact that the isohumes are based on the average values for 50-millibar surfaces (fig. 11-7). Figure 11-7 also demonstrates the great seasonal contrast in the vertical distribution of humidity over the dry coastal areas as high as data are available (up to 10 km).

The temperature inversion is strong enough to restrict the diffusion of water vapor upwards. Along the coast the result is a high atmospheric humidity, notwithstanding desert soils. At Lima Airport the average dewpoint is 2° to 4°C higher than the ocean temperature offshore and coincides with the mean minimum temperature at the airport, which signifies frequent dew formation during the early morning hours. By noon, the time of strongest advection and insolation, the dewpoint has generally increased slightly by about ½ °C. The similar daily variation of air and dewpoint temperatures yields a continuously high relative humidity throughout the day, which even at noon is seldom below 65 percent.

If one accepts Scharlau's limit of sultriness of 18.8 millibars vapor pressure (Scharlau, 1952), or 16.6°C dewpoint, one can see that from a bioclimatic standpoint the Lima area is a continuously sultry region, since the hourly values of humidity are above the limit of sultriness throughout the entire summer and they are considerably above this limit during midsummer. Some days the dewpoint is more than 21°C during the early afternoon, which is equivalent to a vapor pressure of more than 25 millibars. In comparison, the vapor pressure of the upper Amazonian basin gives values of 30 millibars if extrapolated to sea level.

Clouds

The inversion controls the upper limit of the maritime summer fog. The fog is formed far offshore near the western border of upwelling and is driven onshore by west winds. This kind of sea breeze is observed only in the morning and only on those days when the preceding night has experienced a lull in the winds. It lasts only until the south winds take over again, generally around noon. The fast advance of a fog or cloud bank toward the coast is a typical sight in the first hours after sunrise. At the coast the fog is quickly transformed into low fracto-stratus or fracto-cumulus clouds, which are dissolved inland. The inversion prevents the upward diffusion of water vapor and causes the high atmospheric humidity in the thin marine layer. Moreover, the inversion, together with the increasing subsidence aloft, inhibits the formation of the typical tropical cumulus clouds.

The small amount of low clouds is more than balanced by the sharply increasing amount of middle and high clouds towards the Andes. During summer the Peruvian part of this mountain massif is embedded in the deep tropical easterlies, which is evidenced by the wind distribution over Lima from the 500-millibar level upward. These moist easterlies, rising to great heights, wrap the Andean peaks in towering clouds which are visible from the Pacific side sometimes even as an impressive foehn-

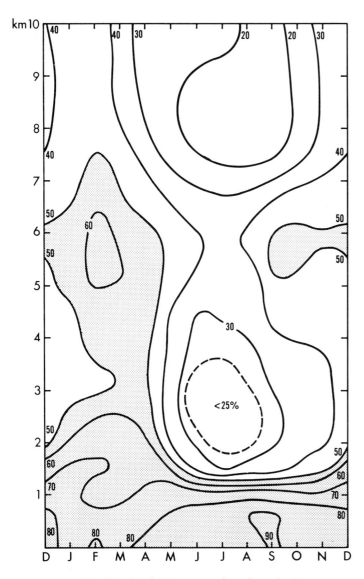

Fig. 11-7. Vertical cross-section of relative humidity over Lima, 1957–1965.

Fig. 11-8. West coast of South America as seen by Satellite ESSA 3, PASS 1531–1543, February 1, 1967.

wall or "Föhnmauer." Descending the westward, leeward slope, the foehnwall dissolves into alto-cumulus lenticularis, the most frequently observed cloud form at the coast. These ground observations are corroborated by satellite pictures which consistently show, at this time of year, extensive cloudiness over the upper Amazon basin which extends across the Andes westward to the Pacific coast where generally it cuts off rather sharply, indicating the stabilizing influence by subsiding movements together with the cool ocean (fig. 11-8). Clouds in so many different forms and levels in toto produce a rather high amount of total sky cover in summer as well as in winter. Although a completely overcast sky seldom occurs in summer, neither does a cloudless sky throughout a full 24-hour period. During the summer of 1967, the monthly means varied around 6 to 7 oktas. The lowest daily mean was 0.5 okta, and the most frequent value observed was 7.5 oktas.

Precipitation

The characteristic abnormality of the summer climate along the Peruvian coast is the almost complete lack of precipitation. It is obvious that the main obstacle to the extension to the Pacific coast of the rainy season, which is so well expressed throughout tropical South America, is the high rising Andes which cause a great loss of water vapor through intensive orographic precipitation on the windward, east side of the mountains and a great drying of this same air by forced downward movement (Foehn effect) on the leeward, west side. In addition, even the weakened trade-wind circulation is strong enough to enhance the forced subsidence and, together with the cool-water inversion, inhibits the normally local precipitation processes in Pacific air masses. The overall stability is far less strong and uniform than it was in winter, but instead of the one intensive trade-wind inversion that is experienced in winter, multiple smaller inversions typically occur one above another during the summer. Unfortunately, the monthly mean data averaged for each 50-millibar surface smooth out these multiple inversions and produce a lapse rate which shows an overall absolute stability.

The occasional light rainfalls that do occur in summer, consisting of large raindrops and lasting a maximum of only a few hours during nighttime, are mostly not of Pacific origin but are a spillover of the intensive precipitation in the Andes and the Amazonian watershed. This is logically deduced from the observations of cumulonimbus clouds advancing westward over the Andes and is corroborated by the humidity cross-section shown in figure 11-7, which clearly indicates a continuous moist layer between 6,000 and 4,000 meters, which is the altitude of the base of the summer clouds. On days with light coastal rains, the surface inversion may disappear and nearly moist adiabatic conditions characterize the vertical atmospheric structure. The rapid decrease of precipitation down the western slope is well illustrated by data from February 1967 (table 11-4). During this month the exceptionally heavy precipitation in the Andes produced flooding and avalanches in the valleys of the Pacific watershed, yet the coast had only three days with measurable rain that totaled just 2.4 millimeters. The sharp decrease in precipitation as the western Andes slope downward is depicted clearly by the changes in vegetation. This becomes more and more xerophytic at the lower elevations and finally disappears completely at around 2,000 to 1,500 meters above sea level.

Temperature

Coastal temperatures during the summer are the result of the interaction of radiative and oceanic influences, as well as the inversion. The inversion has to be taken into consideration as a third factor because it is responsible for the moist marine layer which inhibits normal nighttime cooling that is generally intense in desert regions. The resulting relatively high minimum temperatures in this area are not, therefore, the consequence of direct oceanic influences, as they generally are in other coastal climates, but rather as mentioned above they are caused by the higher humidity so that they are on the average 3°C higher than the water temperatures offshore. The daily range of temperature along the Peruvian coast is twice as large in summer as in winter, but even so it is only 6° to 7°C. Mean maximum temperatures during the summer along the central Peruvian coast are high and rise to as much as 10°C above the ocean surface temperature and even higher where the coastal plains are wider. They are only 3° to 4°C lower than in "normal" tropical coastal situations at similar latitudes as, for instance, at Salvador (Bahia). A few miles inland, in downtown Lima, the mean maxima are similar to those experienced at Salvador (28° to 29°C in Lima-Campo Marte; 30°C in Salvador). Since the atmospheric humidity is quite similar to that in Salvador, at Lima the

TABLE 11-4

Precipitation Profile of the Western Andean Slope During Summer 1967

Station	Latitude	Longitude	Elevation (meters)	Precipitation (millimeters)	
				February	January–April
Cerro de Pasco	10°40′S	76°20′W	4360	220	530
Jauja	11°47′S	75°30′W	3387	191	473
Matucana	11°50′S	76°24′W	2374	148	334
La Molina	12°05′S	76°57′W	238	5	9
Lima Airport	12°01′S	77°07′W	13	2	6

"effect" of the summer temperature is just as tropical as at the same latitude on the Brazilian coast where Salvador shows almost exactly the same effective temperatures as Lima.

Seasonal Changes (Transition From Winter to Summer-type Inversion) and the Inversion as Climatic Divide

The transition from the trade-wind to the cold-water inversion type and vice versa occurs in a different pattern during the astronomical spring than during the autumn. In spring, periods of winter and summer type alternate until the latter become more frequent and extended and eventually prevail at about the end of the year or the beginning of January. In autumn, on the other hand, the transition is a matter of only a few days, or at the most a few weeks. As a matter of fact, in 1967 the trade-wind inversion was established between one day and the next, and, taking cloudiness as the criterion, the winter stratus layer appeared between one hour and the next. The sunny warm summer weather changed to the cool damp winter conditions on May 16 between 3 AM and 4 AM. This change of inversion type and concomitant weather conditions is represented in figure 11-9. During the first half of May the most frequent hourly cloud observation was 0 oktas, while during the second half of May it was 8 oktas. The mode of the maximum temperature dropped from 23° to 20°C, and the average relative humidity rose from 85 to 89 percent. At the same time the summer inversion characteristics changed to the trade-wind type.

The sudden start of winter conditions was conspicuous in previous years as well, as can be documented by aerological data. In other years the change was not as dramatic as it was in 1967, but it never extended over a period of many weeks or even several months, as is regularly the case of the transition in the opposite direction. This fact is expressed in the rapid decrease in the mean temperature from the summer to the winter level (by 8°C) and the slow increase during the reverse process at the opposite season. Neither transition in temperature is due to oceanic influences alone, since the sea temperature has an annual range of less than 3°C. Actually, the sudden establishment of the trade-wind inversion at the beginning of winter is an expression of a change in the general circulation pattern which brings on both the beginning of a wet season onto the Pacific coast and the dry season to the tropical Andes and the Amazonian basin.

Consequently, the seasonal change in the intensity of the trade-wind circulation and, thus, of the inversion type is the main control of the Peruvian coastal climate and of the atypical features which it possesses in a tropical location: large annual and small daily ranges of temperature, higher relative humidity, more extensive cloudiness, and more frequent precipitation during the dry season (hemispheric winter) than during the tropical rainy season. These abnormal conditions are restricted to the narrow coastal area. A few miles inland, along the Andean slopes, the climate becomes nearly "normal." Humidity, cloudiness, and precipitation are higher during the rainy season (hemispheric summer) than during the dry season, although not very high owing to the lee effect of the Andes; the daily range of temperature increases while the annual range decreases. In fact, the annual range of temperature decreases from the coast inland at the same rate as the mean annual temperature rises and becomes nil at an elevation of about 1,500 meters, the same altitude at which the highest mean annual temperature is observed. Above this elevation the annual variation even becomes inverted (the monthly averages become slightly higher during the dry season than during the rainy season). This is due to the fact that the monthly mean temperature at the west-facing slopes below the mean elevations of the highlands depends more on the annual variation of the maxima than of the minima (as it does in the highlands) because of good air drainage. Furthermore, the maxima are primarily a function of insolation, and along the desertic western slopes this is more effective at raising sensible temperatures in the winter months when there is less cloudiness and a drier atmosphere than in the summer. For instance, in Matucana, the mean daily maxima during the period January through March 1967, were 17.3°C and during July through September, 20.8°C. During the same periods the minima were 11.1°C and 9.5°C respectively. Hence, the minimum temperatures varied in the opposite direction from the maximum temperatures, as is normally the case, but to a lesser degree. As a result, the mean temperatures were 14.3°C for the astronomical summer and 15.3°C for the astronomical winter.

The stronger increase of the maximum temperatures, in contrast to the lesser decrease of the minimum temperatures, from summer to winter results from the stronger heating by subsidence in winter, which acts in favor of high maxima and counteracts low minima. In fact, at similar elevations on the east side of the Andes the minima are significantly lower than they are on the west side at this time of the year, in spite of the higher atmospheric humidity in the east. The aerological data prove that subsidence plays a decisive role in this sense on the

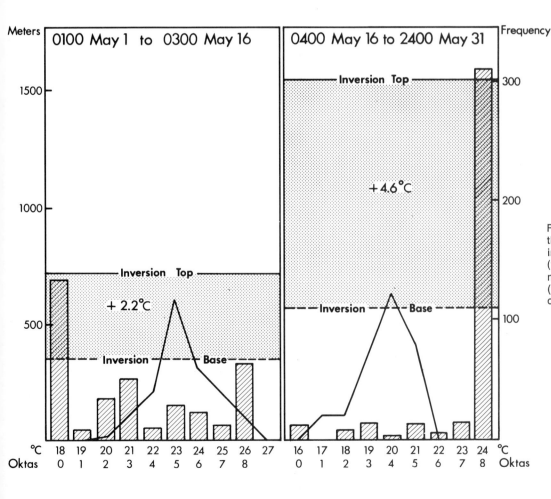

Fig. 11-9. Inversion (altitude, thickness, and temperature increase), surface temperature (frequency curve of daily maxima), and cloud cover (frequency histogram of 24 hours observations) at Lima in May 1967.

west side of the Andes, since temperatures are higher in winter than in summer between 1,500 and 4,500 meters altitude. Therefore the free atmosphere has the same seasonal temperature variation as the western slopes and hence a seasonal variation that is inverse to that experienced at the coast itself. In the free atmosphere above 4,500 meters, and in the highlands as well, the annual temperature varies in accordance with the corresponding change of seasons.

Remarks on the El Niño Phenomenon

No study on the Peruvian coastal climate seems to be complete without at least a mention of the El Niño problem. A large bibliography exists on this rather infrequent phenomenon (Schuette, 1968) in contrast to a few existing studies on the coastal climate in general. Nevertheless, the El Niño investigations have not produced a complete understanding of the phenomenon. There are obvious reasons for this. First of all, the occurrences are rare, supposedly only five times during the past century, and they have mostly happened at times of no systematic and regular meteorological and oceanographic observations. In addition, different criteria have been used to identify the phenomenon. For oceanographers, abnormally warm surface water is a sufficient criterion to speak of the Niño "current." Often the destruction of oceanic fauna or the variable salinity content is considered as additional criterion, since it may identify the origin of the water. Meteorologists, on the other hand, consider high coastal air temperatures and/or abnormal precipitation at the coast and/or in the hinterland to be essential features.

Nearly all of these phenomena may occur simultaneously and may have a common cause, but they may also be observed independently or in different combinations. Whatever the phenomena associated with the El Niño abnormality, threshold values of the meteorological and/or oceanological parameters and their extension in space and time have never been objectively defined. For all these reasons the results and conclusions depend on the criteria used and differ accordingly. In this dilemma, instead of seeking a generally acceptable definition, the best solution would be to rescind the El Niño concept completely in scholarly papers, as already suggested by Schweigger (1959). In addition, no "Niño-current" exists as such; at best it could be equated with the Peru Countercurrent, identified by Wyrtki (1963). The only disadvantage would be that a poetic and emotional term would disappear from the scientific literature, since El Niño means "the Christ child."

For a long time it has been known that an inverse relationship exists between pressure and temperature at the Peruvian coast. This relationship holds for the regular annual variation and for its irregular positive and negative deviations in winter and summer as well. The seasons are evident in view of the preceding discussion. A more intense or more equatorward position of the subtropical high causes a stronger than normal trade-wind circulation, since, in comparison, the changes of the pressure and location of the Inter-tropical Convergence zone (ITC) are relatively small. This, in turn, induces more intensive upwelling and cooler ocean surface temperatures, a stronger subsidence inversion with more cloudiness and less incoming radiation, and, as a result, a cooler marine layer and greater atmospheric stability. Years with extreme high pressure at the Peruvian coast are abnormally cold (Graves, Schweigger, and Valdivia, 1955). During the southern hemispheric winter when the subtropical high is in an equatorward position, the pressure gradient always seems strong enough to maintain a certain degree of the trade-wind circulation. But during the opposite season when the high shifts poleward, the pressure gradient may occasionally become so small that the predominance of the trades ceases, and other factors may become decisive, which leads eventually to the warming of the northern Peruvian coastal region and its consequences — El Niño.

Combining this deduction with the present knowledge on the complicated structure of the different currents off the Peruvian coast, J. Bjerknes (1966a) states that: "in weak trade-wind regimes the heat transfer to the atmosphere, both in sensible and latent form, is below normal, and more than normal amounts of heat remain stored in the ocean, thus raising its surface temperature. . . . Weakness of the trade winds has more sudden and spectacular ocean effects in terms of temperature induced by cessation of upwelling. . . ." In this context it is understood that by "cessation of upwelling" the Peru Countercurrent surfaces and/or extends toward the coast. Also in normal summers the side-by-side existing cold and warm waters are directly visible by their different albedo to observers flying at low altitudes over the coastal waters.

A decrease in the strength of the trades is also accompanied by a decrease in subsidence, and hence a decrease in stability or increase in instability, which is enhanced further by the increase of surface temperatures. The result can be intensive precipitation, sometimes even before the warming process at the ocean surface takes place. In general, it can be assumed that warm surface water is not a necessary or sufficient condition to produce coastal rains and that the El Niño phenomenon, whatever its definition may be, is neither produced by a shifting of the ITC to the south of the equator nor by an atmospheric circulation-induced reversal of the Peru Current as such. As Bjerknes (1966b and 1969) and Doberitz (1968) have shown, the El Niño phenomenon is just one of various side-effects of disturbances of the global tropical circulation. In this case, by and large, the normal "anomalies" of the equatorial eastern Pacific, that is, the coolness and dryness of the Ecuadorian and northern Peruvian coast, become anomalously "normal," and record temperatures and precipitation that are in conformity with tropical latitudes.

Bibliographic References

BJERKNES, J.
1966a Survey of El Niño 1957–58 in its relation to tropical Pacific meteorology. Inter-American Tropical Tuna Commission, Bulletin 12(2):25–86.
1966b A possible response of the atmospheric Hadley circulation to equatorial anomalies of ocean temperature. Tellus 18(14):820–829.
1969 Atmospheric teleconnections from equatorial Pacific. Monthly Weather Review 97(3):163–172.

DOBERITZ, R.
1967 Zum Kuestenklima von Peru (Coastal climate of Peru). Seewetteramt, Einzelveroeffentlichungen 59. 115 p.
1968 Cross spectrum analysis of rainfall and sea temperature at the equatorial Pacific Ocean: A contribution to the "El Niño" phenomenon. Bonner Meteorologische Abhandlungen 8. 53 p.

GRAVES, M. E.
1944 An objective method for forecasting ceiling heights. Panagra, Lima, Peru. 22 p.

GRAVES, M. E., E. SCHWEIGGER, AND V. PONCE J.
1955 1954, un año anormal. La Compañía Administradora del Guano, Lima, Boletín Científico 2:49–71.

KNOCH, K.
1930 Klimakunde von Suedamerika. In Koeppen and R. Geiger (eds), Handbuch der Klimatologie, bd. 2/G. Gebrueder Borntraeger, Berlin. 249 p.

LETTAU, H. H.
1967 Small to large-scale features of boundary layer structure over mountain slopes. In Symposium on Mountain Meteorology, 1967, Ft. Collins, Colorado, Proceedings. Colorado State University, Department of Atmospheric Sciences, Atmospheric Science Paper 122. 74 p.

LYDOLPH, P. E.
1957 A comparative analysis of the dry western littorals. Association of American Geographers, Annals 47(3):213–230.

MEIGS, P.
1966 Geography of coastal deserts. Unesco, Paris. Arid Zone Research 28. 140 p.

PERÚ, DIRECCIÓN GENERAL DE METEOROLOGÍA
1956– Promedio mensual meteorológico del Perú, Septiembre 1955–Diciembre 1956. Revista Meteorológica del Perú 2(3):37–43; 3(4–5):47–55.

ROESSL, J. C.
1967 Las lomas de la costa y los beneficios que pueden prestar a la economía nacional. Perú, Servicio Forestal y de Caza, Informe 29. 50 p.

RUDLOFF, W.
1959 Meteorology in Peru. United Nations Technical Assistance Programs, Report TAA/PER/8. 46 p.

SCHARLAU, K.
1952 Die Schwuelezonen der Erde. Deutscher Wetterdienst in der U.S.-Zone, Berichte 42:246–249.

SCHUETTE, K.
1968 Untersuchungen zur Meteorologie und Klimatologie des El Niño — Phaenomens in Equador und Nordperu. Bonner Meteorologische Abhandlungen 9. 152 p.

SCHWEIGGER, E.
1959 Die Westkueste Suedamerikas im Bereich des Perú-Stroms. Keysersche Verlagsbuchhandlung, Heidelberg-Muenchen. (Geographische Handbuecher) 513 p.
1964 El litoral peruano. 2d ed. Universidad Nacional "Federico Villarreal," Lima.

TREWARTHA, G.
1961 The earth's problem climates. University of Wisconsin Press, Madison, Wis. 334 p.

TULLER, W. E.
1968 World distribution of mean monthly and annual precipitable water. Monthly Weather Review 96(11):785–797.

UNANUE, J. H.
1940 Observaciones sobre el clima de Lima y su influencia en los seres organizados, en especial el hombre. 5th popular edition. Comisión Nacional Peruana de Cooperación Intelectual, Lima. 250 p.

U.S. ENVIRONMENTAL SCIENCE SERVICES ADMINISTRATION (ESSA)
1966 World Weather Records, 1951–1960. Vol. 3: South America, Central America, West Indies, The Caribbean and Bermuda. U.S. Government Printing Office, Washington, D.C. 355 p.

U.S. ENVIRONMENTAL SCIENCE SERVICES ADMINISTRATION (ESSA)/NATIONAL AERONAUTICS AND SPACE ADMINISTRATION (NASA)/UNITED STATES AIR FORCE (USAF)
1967 U.S. Standard Atmosphere supplements, 1966. U.S. Government Printing Office, Washington, D.C. 289 p.

U.S. OFFICE OF NAVAL OPERATIONS
1959 U.S. Navy Marine Climatic Atlas of the World. Vol. 5: South Pacific Ocean. Direction of the Chief of Naval Operations NAVAER 50–1 C-532. Washington, D.C. xv, 267 charts.

U.S. WEATHER BUREAU
1965 World weather records, 1951–60. Vol. 1: North America. U.S. Government Printing Office, Washington, D.C. 535 p.

VALDIVIA PONCE, J.
1957 El clima y las aves guaneras en 1955. Compañía Administradora del Guano, Boletín 33(6):7–19.

WEBERBAUER, A.
1945 El mundo vegetal de los Andes peruanos, estudio fitogeográfico. Estación Experimental Agrícola de "La Molina," Edit. Lumen S.A., Lima, Peru. 776 p.

WEISCHET, W.
1966 Zur Klimatologie der nordchilenischen Wueste. Meteorologische Rundschau 19(1):1–7.

WYRTKI, K.
1963 The horizontal and vertical field of motion in the Peru current. University of California, Scripps Institution of Oceanography, La Jolla, Bulletin 8(4):313–346.
1964 The thermal structure of the eastern Pacific Ocean. Deutsche Hydrographische Zeitschrift, Ergänzungsheft, Reihe A, 6. 84 p.
1967 Circulation and water masses in the eastern equatorial Pacific Ocean. International Journal of Oceanology and Limnology 1(2):117–147.

CHAPTER 12

DISTINCTIVE HYDROGEOLOGICAL CHARACTERISTICS OF SOME PAMPAS OF THE PERUVIAN COASTAL REGION

Pierre Taltasse
Department of Geology, University of São Paulo, Brazil

The Pacific littoral of Peru, from the desert of Sechura in the north to the valley of the Rio Yauca in the south, is bordered by sedimentary plains known as *pampas*. For the most part these are sandy, more or less inclined toward the sea, and divided by valleys that often are moderately entrenched.

These plains, or plateaus, which cover both crystalline and sedimentary rocks, are composed of fluviatile, lacustrine, or palustrine materials. River deposits *(llapanas)* or eolic formations (dunes, barkhans) often are associated with them.

From the Sechura desert to the valley of the Pisco River, these pampas open up directly to the coast; from the valley of the Pisco to that of the Yauca, they cover the pre-Andean trench, which separates the Western Andean chain from the Coastal Cordillera (the pampas of Villacuri, Huayuri, Nazca, and Sacaco, to mention but the major ones).

The valleys of the coastal area, most of which have seasonal flow only, delimit the individual pampas. Until recently these were the only areas of agricultural development in the coastal region of Peru.

The pampas, with their very fertile soil, represent excellent areas for agricultural expansion. But the necessary irrigation networks would pose complex problems of a technical character (for example, siting of diversion dams, length and alignment of feeder canals), and would involve investment beyond all proportion to the amount of water transported and to the areas brought under cultivation.

Because it is not possible to rely on irrigation by gravity, under economically reasonable conditions, wells have been dug or drilled here by private initiative. The pampa of Lanchas, southeast of Pisco, was one of the first to be surveyed to this end.

Contrary to all expectations, aquifers, sometimes carrying groundwater, often were discovered beneath the pampas, furthering their rapid agricultural development. In order to give an idea of the importance of the quantities that such groundwater horizons might hold, we will quote as an example the pampas of Lanchas and Villacuri, between Pisco and Ica, where in 1964 about 2.5 cubic meters per second (88.29 cu. ft. per second) were withdrawn.

From the geological and hydrological point of view, these groundwater horizons, whose existence is just being recognized but whose importance is not yet fully appreciated, present an initial problem: how did they originate and by what mechanism were they supplied?

Using as a basis the observations on the groundwater horizon at Lima, which is simply the underflow of the Rimac River, one must assume as a working hypothesis that the groundwater horizons, gradually being discovered in the pampas, were similarly the underflows of littoral water courses; or at the most they represent fossil groundwater drains, augmented by water from the valleys above the pampas.

Even if some such cases should prove to be correct, in particular as regards the pampas of Chilca, Lurin, and Huacho, the systematic studies undertaken in 1962–64 by the Ministry of Public Works (directorate of Irrigation and the National Institute of Mining Research and Development), by the Agricultural University La Molina and by the Hydrogeological Mission of UNESCO, lead to the conclusion that this explanation cannot be applied to all the pampas under which aquifers have been found.

The pampas of Lanchas and Villacuri, situated between Pisco and Ica, were chosen for a pilot study, as being most representative of the pampas of the pre-Andean trench, and having many wells and drill sites permitting observations.

The study included a detailed geological survey, records of water discharge, hydrochemical analyses, the preparation of relevant maps (iso-piestic, iso-thermic, and so on), and of limnigraphic data, all over a period of three years.

The main fact to note is that this groundwater horizon discharges into a fault complex of Cretaceous schists, sandy strata of the Miocene, and into the lower levels of the pampa series (not yet stratigraphically divided), and that the Rio Seco follows the axis of this fault complex. This groundwater horizon drains westward to the valley of the Rio Pisco. Its maximum temperature of 32°C (hacienda Cabo Blanco) distinguishes it clearly from the water of the underflow of the Rio Pisco (18°C). So also does its hydrochemical composition: the groundwater being of the sulphatic-chloridic-sodic-calcic water type, as compared to the chloridic-sodic water of the underflow of the Rio Pisco.

The fault zone mentioned above is replaced to the northeast by the Huamany fault, directly on the right bank of the Rio Pisco; it feeds the springs of "Agua Santa."

Similar studies were later undertaken in the pampas

of Sacaco, between Lomas and Acary, that is, in the southern part of the pre-Andean trench; in the pampa of Ventanilla* in the vicinity of Lima, and in the pampa of Saña† south of Chiclayo. They provided comparable conclusions: charged groundwater horizons, always thermal (28–32°C, 82–91°F being the maximum temperatures recorded), connected with special tectonic events, but without any apparent relation to the lithology, with slight variations only in water level (contrary to the surface water), and drained directly or indirectly by the underflow of neighboring water courses, with a hydrochemical composition distinct from them.

The existence of such groundwater horizons constitutes an extremely unique hydrological phenomenon. Their conditions of sedimentation have a number of features in common: the presence of a local structural event or of a structure complex particularly favorable for the charging of groundwater. Still, the question of the necessarily remote origin of these groundwater horizons is far from being solved. Is this groundwater charged by rainfall, originating in the Andes, discharging at great depth at a variable rate in accordance with local structure or favorable conditions? Or is this a hydrogenetic phenomenon, as might be assumed from the occurrence of these waters in dry locations at sometimes extreme altitudes, and their relation to the peri-Pacific situation of the Andes, already responsible for magmatic phenomena less well known?

At the present state of knowledge, hydrochronological studies appear particularly appropriate to appraise the art played by each of these hypotheses.

From the economic point of view, the existence of systems of aquifers of this type, which we only have just begun to discover, renders urgent a classification of the pampas of the Peruvian coastal region as regards the nature and the origin of their subterranean water resources. There are pampas with an "autonomous" water supply (for example, Lanchas and Villacuri), pampas fed by fossil alluvial beds (for example, Chica and Huacho), and possibly also pampas of a mixed water supply.

This classification, through the detailed studies on which it is dependent, will throw new light on the geological structure and recent geomorphological development of the Peruvian littoral. Furthermore, it will make it possible to adapt the development programs, and particularly the exploitation of subterranean water (pumping and more particularly artificial recharge), to the specific hydrological conditions of each individual pampa.

In conclusion, it appears that this rather special hydrological phenomenon considerably modifies the conception previously held of the natural conditions in these desert areas, the coastal pampas of Peru. It also accounts for some of the beautiful oases of greenery and life, still at their dawn, which man, by his faith, courage, and determination, has been patiently able to create here.

---

*The study was carried out by A. Pérez, geological engineer, Ministry of Public Works.

†The study was carried out by A. Flores, geological engineer, Ministry of Public Works.

Bibliography

Dollfus, O.
  1965 Les Andes centrales du Pérou et leurs piémonts (entre Lima et le Péréné), étude géomorphologique. Institut Français d'Etudes Andines, Lima, Travaux 10. 404 p.

Solignac, J. L. M.
  1961 Investigación de las aguas subterráneas de las zonas de la Costa y de la Sierra. FAO, Rome, EPTA Report 1268. 162 p.

Taltasse, P.
  1963 A les eaux souterraines de la zone aride côtière du Pérou. Arid Zone, UNESCO 20:8–13.
  1965 Mouvements récents dans le sillon pré-andin au Pérou. Académie des Sciences, Paris, Comptes Rendus 261:4464–4465.

THE COASTAL DESERT OF CHILE

Reynaldo Börgel O.
Institute of Geography, Santiago, Chile

The coastal desert of Chile is not limited to the coastal area itself but includes some of its continental hinterland. From north to south we find not one but many deserts, as the geomorphological and climatic characteristics, as well as those of the vegetation, to be found in the northern area of Chile, are quite varied.

Almost all of the coast of Chile, comprising the area between Arica in the north and the Valley of Elqui in the south, is arid. Not included is the semiarid transition zone and the area of mediterranean-type climate between the Elqui Valley at latitude 30° South and the coastal region of Maule at latitude 36°.

The coastal desert of Chile is part of the great global belt of deserts of the southern hemisphere. However, an important factor in the formation of this particular coastal desert is the Humboldt current, which, owing to its thermal characteristics, subjects the northern part of Chile to a thermal inversion, which obstructs the approach of the mass of warm and humid air from the South Pacific. This thermal inversion creates a cloak of cold air of low humidity, which establishes a wedge of high pressure between 1,500 meters altitude and sea level. Above this height high pressure dominates, caused by the descent of tropospheric air, from which results the total rejection of the humid air masses from the Pacific. This situation prevails up to a height of 3,200 meters; above this height the pluviometric system, unlike that of the winds, arrives from the Amazon basin covering all the frontier region which comprises the north of Chile, southwest Bolivia, and northwest Argentina, as far as latitude 27° South. These rains, known as those of the "Bolivian Winter," reach a height of between 3,500 and 4,000 meters and occur between the months of November and February. But these rains do not extend westward from the Andes to the coastal desert. According to all available records and to the reports of explorers who surveyed the coastal desert of Chile, rain in this desert occurs only once every 30 to 40 years.

If we consider as a coastal desert the whole territory subjected to the thermal inversion of the Humboldt current, the coastal desert of Chile reaches as far as 100 kilometers inland. While the Humboldt current is the essential factor explaining the location of this desert, one must refer to additional factors which determine its particular regional characteristics. These factors can be divided into geomorphology, drainage, and vegetation.

Geomorphology

Chile rises from the Pacific Ocean along a fault cliff. This feature limits the transition from ocean to continent to a narrow continental shelf and to rather small coastal plains, the most extensive of which reaches an average width of scarcely 2 kilometers. On these small plains the principal ports of northern Chile have been built: Arica, Iquique, Tocopilla, Antofagasta, and others. In the majority of cases the coastal fault forms a rectilinear wall which is, at the same time, the windward slope of the coastal mountains.

In these conditions, aridity manifests itself at the very border between the ocean and the continent. The influence of the coastal morphology makes itself felt to varying degrees in the interior. North of Antofagasta, after traversing the coastal escarpment which is from 200 to 400 meters high, one enters a complex area of gentle hills, linked by extensive coalescing piedmont plains. This topography known as *pampas* corresponds to a high plain which forms part of the longitudinal depression interposed between the Coastal and the Andean Cordilleras along the whole distance from the north to the south of Chile.

From Antofagasta to the south the steep coast rises to 600 meters; beyond it the Sierra Vicuña Mackenna culminates at 3,000 meters. From this and lesser heights the leeward hills of the Coastal Cordillera descend toward the interior, forming glacis (gently sloping banks) and bolsons (interior basins) of extreme aridity. In these hills lie the main saline deposits which comprised the wealth of Chile at the end of the past century through the exploitation of nitrate deposits. Crossing the central pampas one ascends over inclined plains from 1,200 to 1,500 meters. These are composed of systems of coalescing gently sloping piedmont plains, separated from the foothills of the Andes. In these one can observe the climatic events of the Middle and Late Quaternary, noting phases of humidity which can be placed in chronological order by means of archeological remains which testify to an ancient local population.

The foothills constitute a denuded or truncated mountain complex, divided into separate parallel chains, with long and narrow intermontane basins of indubitable lacustrine origin. While the mountain chains consist of Cretaceous and Jurassic sediments with intrusions of

Tertiary plutonic rocks, the lacustrine sandstone belongs to the end of the Tertiary and to the Early Quaternary periods, corresponding with palaeoclimatic periods of greater humidity.

To the east of the foothills the Tertiary rises in wide steps, composed of a series of alternating ridges of relatively low altitude and of saline basins, presumably of hydrothermal origin. This entire system culminates on the frontier of Argentina in a volcanic alignment with manifestations of early and late effusions of basalt and andesite.

Following this regional synthesis described in an east-west cross-section, it is worthwhile to consider the geomorphological development of this area, but in a north-to-south direction across the intermediate depression or the great central pampa. This is an area of active sedimentation, owing to the combined action of the coalescent piedmont plains which create a veritable *playa* of arid sedimentation, situated between the eastern foot of the Coastal Cordillera and the great sloping fans which descend from the western foothills of the Andes.

The arid zone of the pampas descends slightly from north to south over the approximately one thousand kilometers from Arica to Taltal. The pampas are an undulating terrain, the topography of which results from the coalescence of glacis descending from various directions. The development of these forms of accumulation on the east side of the pampas is much more pronounced because they are derived from the higher Andes. Therefore, the bolson bottoms occur nearer to lower Coastal Cordillera, which was not able to supply as much sedimentary material. We believe that one of the causes that gave rise to the existence of deposits of nitrate of sodium and potassium in the eastern strip of the Coastal Cordillera is the result of cutting off drainage by the accumulation of glacis and fans. The northern pampa of Chile is carpeted by a surface cover of varied materials produced by weathering. In general, there occurs a type of pavement, homogeneous from the point of view of its granulometric and morphographic character, as determined by the petrographic nature of the different materials. In general, the liparites form pavements of slabs of material which is very flat but sharp, cutting like a knife; the basalts form a cover of thick debris and scattered blocks in a heavy matrix of geometric arrangement; while the calcareous tufa of sandy consistency are easily moved by the wind, creating eolic sediments, usually found at the western foot of the piedmont plains.

There are pronounced variations in temperature with depth in this cover of weathered material. Thus, when air temperature at some centimeters above the surface is 25°C (77°F), in contact with the rocks the temperature is 33°C (92°F), yet at a depth of 15 centimeters (0.6 inch) beneath the surface it drops to 8°C (46°F).

Despite the fact that these measurements were carried out on piedmont plains in the salar (salt flat) region at 3,800 meters, the thermal conditions have a similar relationship on the piedmont plains of the central pampas situated at heights which range between 1,200 and 2,100 meters altitude.

In its geomorphological aspects the coastal desert of Chile presents many zonal variations due to its great extent, north to south. The factors we mentioned are representative for all the area situated between Arica in the north and the river Huasco in the south, that is extending some 1,750 kilometers from north to south.

To the south of the Huasco River the coastal desert continues, but a greater humidity of the hinterland is clearly indicated by a more vigorous development of vegetation and influences the coastal area as well. The climate here is of a cyclic semiarid character. This semiarid area in which the resources of subterranean water are sufficient to generate the appearance of a spontaneous springtime vegetation every five or six years, reaches fairly southern latitudes. This southern limit of the semi-arid area is difficult to establish precisely, since it continually fluctuates between latitudes 30° and 32° South.

The coastal desert of Chile can be subdivided for the purpose of a detailed study into the following sections: the coastal desert of the small plains *(pampitas)*, the Pampa del Tamarugal, the Pampa de Atacama, the coastal desert of the undulating or southern pampas, and the coastal desert of the transitional pampa. These sections together extend from Arica in the north to the Valley of Elqui in the south, that is, from latitudes 18.5° to 30° South.

Drainage

Within the area determined above as the coastal desert of Chile, the drainage presents some distinct phenomena, which it is convenient to indicate in order to explain the features of vegetation and the geomorphological evolution, both zonal and regional.

The geological history since the middle of the Tertiary and during the Quaternary period is an indispensable aid in explaining the maintenance of water resources and in distinguishing between forms originating in palaeoclimates of greater humidity, and forms due to aridity even more marked than the present one.

The dominant formations are the Cretaceous and Jurassic, consisting of marine and terrestrial sediments, pyroclastic and volcanic ones, with intercalations of layers of gypsum. In limited sectors of the littoral, in the Moreno range to the north of Antofagasta, metamorphic rocks appear, the parent material of which consists of Paleozoic and partly Precambric rocks. All this has been intruded by plutonic rocks, principally tonalites and granodiorites. These plutonites do not correspond to one another chronologically: they might date to any period between the late Paleozoic and the Tertiary.

In the hinterland, the transition zone to the Andean Cordillera and on the altiplano Tertiary and Quaternary formations dominate. In both cases they consist of andesitic-basaltic volcanites at the highest altitudes of the mountains and rhyolites on the flanks descending to the central pampas. Whereas the basalts are essentially Quaternary, the rhyolites are Tertiary. These last form the vast arid pediments which ascend from the great central pampas to the Western Andean Cordillera.

The Pliocene was characterized by vast deposits of andestic lava in the Western Cordillera. During the Quaternary this material was redistributed over a wide area, extending from the frontier of Bolivia and Argentina to the eastern border of the great central pampas. In the redistribution of the volcanic material we can observe numerous surfaces, some of them more or less humid by origin, and others more arid. It is interesting to note that on the inclined arid plains two fundamental forms or piedmonts can be found: arid pediments and glacis. The first are formed on liparites and are most frequently found as a transitional belt between the volcanic area which serves as Chile's frontier with Bolivia and Argentina and the salars of the Western Cordillera. In these pediments, consisting of only a little consolidated lava, erosion by meltwater of ice and snow has created deep gorges (quebradas). At the lower end of these are hardly any fluvial deposits as the water acts upon the liparites, dissolving the ashes.

Much more important from the point of view of drainage is the distribution pattern of the piedmont. In these powerful detritic accumulations important movements of water took place until recent times. Even more interesting is the relationship which these piedmonts show to other features of associated forms, especially the salars of the Pre-Cordillera. In effect, the ascent from the great central pampas to the Cordillera does not imply a palaeomorphology of closed compartments. If at present this is the impression which the geomorphology of the Chilean desert implies, in the Lower Quaternary there existed rivers connecting these geomorphic units by means of a well-developed exoreic drainage system. Fluvial deposits with scattered fragments of decomposed basalt and andesite certify to the existence of connecting rivers between the lacustrine systems of the altiplano, with a clearly developed drainage proceeding from the great altiplano of Bolivia and northwest Argentina toward the Pacific coast. Although it is certain that the piedmont bears witness to a movement of material by means of groundwater, it seems likely that prior to this process of a semiarid nature there occurred processes of violent outflow of the water from the lakes of the altiplano and the Western Cordillera toward the Pacific or at least into the great central pampas of the north of Chile. To this extent the coastal desert of Chile did partake during the Quaternary in hydrological resources, the present reserves of which should be found beneath the present salars, and should percolate through the detritic fans.

From the point of view of geology and archeology, the changes in surface drainage can be followed in the alternate changes which the prehistoric population in this region underwent, especially insofar as the river Loa is concerned. This river is important, since, apart from being the only one to maintain its exoreic flow right to the sea, its tributaries are found in the very area of the Western Cordilleras. These two facts convert the area irrigated by the waters of the Loa into a morphological unit which is ideal for a chronological study of the palaeoclimatic sequences which occurred in the region.

The coastal desert of Chile is a geological desert, by which definition we wish to emphasize that the development of its forms, as also of the climatic history can be traced to the Tertiary, or in the majority of cases, at least to the Early Quaternary.

The superaridity of the coastal desert of Chile is a consequence of natural and human factors. Among the former we have already mentioned global reasons which are linked to the existence of the Humboldt current, and, in consequence, the thermal inversion which affects an extensive part of the Peruvian and Chilean littoral. In the second place there is the devastation which man has wrought on the natural vegetation which existed in the central pampas, especially concerning the *tamarugo* tree *(Prosopis tamarugo),* which was cut in great numbers for use as fuel in the processing of nitrate at the end of the last century.

With the exception of the river Loa there exists no surface drainage in the coastal desert of Chile. On the other hand the phreatic groundwater horizon existing in some areas has recently been lowered to such a degree that the cattle farms situated to the south of Arica, which had made great economic progress and which appeared to be quite flourishing, have begun to fail in the last ten years, and at present have been partly abandoned due to the drop in water level.

Nevertheless, experiments undertaken at Canchones hold promise for the possibility for reforestation with *tamarugos* and for the development of a sheep farming industry in the area.

Vegetation

Between Arica and Caldera (latitude 18.5° to latitude 27° South) no vegetation exists. The descriptions of Charles Darwin and the classical accounts of the explorers who traveled in the desert in the last century, men like A. Bertrand, E. Barros, R. Philippi, F. San Roman, and others, refer only to *tamarugos* in the central pampas and to some mosses on the coastal cliffs. The vegetation only reappears above an altitude of 2,800 meters, where *llaretales, tolares,** and the growths of wild grasses which reach as far as 4,000 meters constitute veritable carpets of red, yellow and green.

---

*Llareta is a densely cushioned resinous Andean herb of the umbilliferae family. Irrespective of their very slow growth rate they are cut for fuel, sometimes by the truckload. Tolar is any of several South American plants of the genera *Baccharis* and *Hepidophyllum*.

If one compares the windward and leeward sides of the Coastal Cordillera, one observes a great variety of plant and cactus life on the slopes facing the ocean. By contrast, all traces of vegetation are missing on the leeward slopes. Since the small green strips of shadow afforded by the *tamarugos* of the central pampa have disappeared, due to destruction by man, the coastal desert of Chile offers an overall picture of desolation and forlornness.

Conclusion

Due to the geologic characteristics of the coastal fault of the Chilean littoral, where the windward side of the coast range coincides with the coastal escarpment, the coastal desert of Chile comprises not only the Coastal Cordillera but also the central pampas for a distance of 100 kilometers to the east. This area is affected by the climatic features which the thermal inversion imposes, a direct result of the Humboldt current.

The Chilean desert is superarid, one of the few extreme deserts of the world where, as a result of its great extent and its topography, one cannot even expect the occurrence of the great torrential floods, which, as features of the "Bolivian Winter," do affect the altiplano of the Cordillera de los Andes between November and February.

The vegetation has suffered intensive destruction in the recent past. Today it is nearly nonexistent, although potentially renewable on experimental farms. At the end of the nineteenth century the destruction of the *tamarugo* for use as fuel in the nitrate industry created a desolated countryside which accentuated the superaridity of this desert. Furthermore, there exists evidence of the gradual disappearance of groundwater resources which, until a few years ago, created an atmosphere of prosperity in a number of coastal farms devoted to cattle and horse raising.

Finally, geomorphological studies in this region provide evidence of ancient connections between basins which function today as unconnected geomorphic units. This augurs well for a scientifically improved use of groundwater in this area.

Bibliography

ALMEYDA ARROYO, E.
 1946 Biografía de Chile. 14th ed. Editorial San Francisco, Padre las Casas. 494 p.

BOERGEL O., R.
 1966 Mapa geomorfológico de Chile. *In* Atlas de la República de Chile, p. 108. Instituto Geográfico Militar, Santiago.

MUÑOZ CRISTI, J., AND O. GONZALEZ F.
 1966 Mapa geológico de Chile. *In* Atlas de la República de Chile, p. 107. Instituto Geográfico Militar, Santiago.

MUÑOZ PIZARRO, C.
 1965 El desierto florido. Museo Nacional de Historia Natural, Santiago. 31 p.
 1966 Sinopsis de la flora chilena. Claves para la identificación de las familias y géneros. Ediciones de la Universidad de Chile, Santiago. 500 p.

TRICART, J.
 1966 Un chott dans le désert chilien: La Pampa del Tamarugal. Revue de Géomorphologie Dynamique 16(1):12–22.

CHAPTER 14

A CLIMATIC PROFILE OF THE NORTH CHILEAN DESERT AT LATITUDE 20° SOUTH

César Caviedes L.
Department of Geography, University of Saskatchewan, Regina Campus, Canada

It is the purpose of this chapter to evaluate the climatic factors affecting the North Chilean desert and their various combinations, and to analyze the relationship between the atmospheric dynamics and the different types of arid climate in the great desert of northern Chile.

The Landscape

The relief of North Chile near the twentieth parallel is composed of three well-defined units in a north-south alignment: the Coastal Cordillera, the central plateau, and the high Cordillera of the Andes.

The Coastal Cordillera as far as 20° South appears as a compact range whose massive rounded summits reach an altitude of 1,200 to 1,600 meters above sea level. Near the Pacific the range terminates in an almost vertical escarpment, reaching 500 to 700 meters above sea level. At the foot of the escarpment are narrow coastal plains on which isolated centers of settlement have arisen — Pisagua, Iquique, Tocopilla.

Toward the interior there extends a long and wide high plain that rises gently toward the Andes in a moderately inclined piedmont. This is the nitrate desert or the "Pampa del Tamarugal," the name of which stems from the scattered stands of the *tamarugo* tree, *Prosopis tamarugo* (Mimosaceae). Inland from Iquique, in an area a little before the Rio Loa, the high plain at about 1,000 meters above sea level splits up into a series of barely defined basins with flat bottoms occupied by *salars* (salt flats). On the other hand, north of Iquique, the continuation of the high plain is interrupted by deep canyons forming the region of the "quebradas" (gorges), which contain the few places where an oasis agriculture is possible.

Over the accumulation of volcanic deposits that form the Andean piedmont, one ascends to the high peaks of the Andes and reaches the great Bolivian altiplano beyond them. The piedmont is strongly dissected by "quebradas," which at the latitude of Iquique do not affect an interruption of the continuity of the pampas.

This complex region has impressive relief features, most of which are no longer in an active stage of formation. Only through the quebradas rush violent floods, locally known as *avenidas,* after the rainstorms of the summer months.

Climatic Regions of North Chile

The analysis of the climatic characteristics in a profile traced from the coast to the Andean Cordillera, at the latitude of Iquique, provides a clear picture of the variety of characteristics which, within this short transverse stretch, constitutes the arid climate.

The observations are based on the following stations: Iquique, at sea level; Los Condores, at an altitude of 500 meters (1,640 feet), at the top of the coastal escarpment; Canchones, 960 meters (3,150 feet), in the center of the nitrate pampa; Pica, 1,280 meters (4,200 feet), on the Andean piedmont; and also on data from stations in the Cordillera: Chusmiza, 3,500 meters (11,480 feet), Collahuasi, 4,805 meters (15,764 feet), and Ollague, 3,695 meters (12,123 feet). The most complete observations were carried out at Iquique (1916–45). Los Condores has data from 1949–50, 1956–57, 1960–64; Canchones from 1942–48; Pica 1964–65; Chusmiza 1929–36 (precipitation only); Collahuasi 1915; and Ollague 1964–65.

Examination of the data at our disposal confirms the climatic regions suggested by Fuenzalida Ponce (1950), for which Weischet (1966a & b) later presented the dynamic foundations. By adding new stations which were not considered by these authors and by making use of data on insolation obtained recently, additional support is supplied for the definition of the climatic areas proposed by them.

The data presented in the profile demonstrate the existence of the following climatic regions (fig. 14-1): (1) the coastal desert, (2) the interior desert or pampa, (3) the arid Andean piedmont, and (4) the arid high Andes.

The differentiation between the desert and the arid climate is intentional. The former is typified by the almost complete absence of precipitation; the latter receives precipitation but less than the evaporation potential for this region.

The Coastal Desert

The coastal desert climate is characterized by small annual and daily variations in temperature (fig. 14-2). Absolute temperatures here are the most moderate of

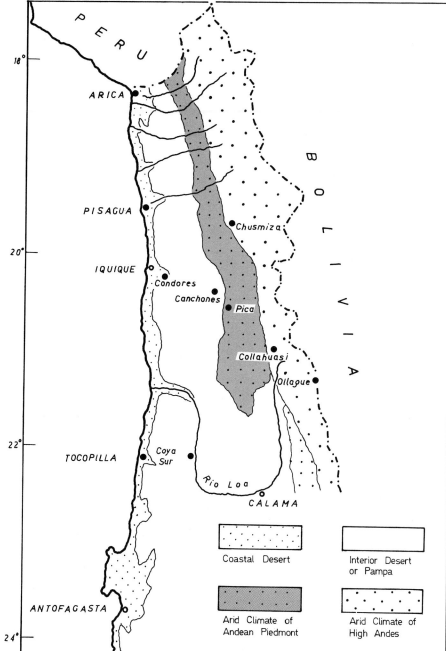

Fig. 14-1. Climatic regions of the north Chilean desert.

the whole profile and become less extreme as one approaches the coast. This is illustrated by a comparison of the extreme values at Iquique and Los Condores shown in table 14-1.

It is thus evident that the occurrence of the extreme temperatures varies: August and February show the most extreme values. But as regards the average temperatures (fig. 14-2), the coldest part of the winter occurs at Pica and Los Condores in the middle of July, at Iquique toward the end of that month, at Canchones and at Collahuasi in June. The warmest part of summer occurs in February at Iquique, Los Condores, Canchones, and Ollague, and in January at Pica and Collahuasi. That is, there is no single reason for the different dates of temperature extremes for stations as close to each other as Iquique and Los Condores, Canchones and Pica.

Relative humidity at the coastal stations is remarkably high, as may be seen in table 14-2, wherein the relative humidity at Valparaiso (33°S) is added for comparison with that at Iquique. At the other stations in the profile, humidity decreases with increasing altitude.

At the coastal stations the winter months have the

TABLE 14-1

Extreme Values of Temperature at Several Chilean Stations

Station	Maximum °C	°F	Month	Minimum °C	°F	Month
Iquique	31.3	88.3	3	8.0	46.4	8
Los Condores	28.0	82.3	2	3.2	37.8	6
Canchones	35.8	96.4	1-2	0.7	33.3	8
Pica	34.5	94.1	1	2.5	36.5	7
Ollague	23.5	74.3	1	−32.3	−26.1	6

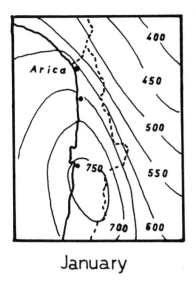
January

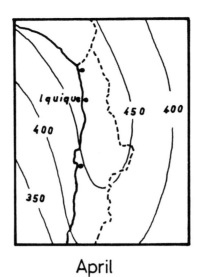
April

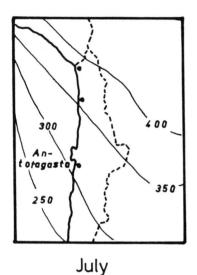
July

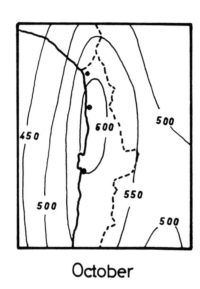
October

Fig. 14-2. Isopleths of insolation for the north Chilean desert, according to J. Hirschmann, 1962 (gram-centimeters per square centimeter per day).

highest atmospheric humidity. Weischet (1966 a) calculated that the saturation deficit in the coastal area varies from 7 grams per cubic meter during the summer months to 3 or 4 grams during the winter. It may thus be seen that one of the hygrometric characteristics of the coastal desert is a relatively high humidity of the air, maintained throughout the year with great regularity, while at the same time there is a continued saturation deficit preventing precipitation. Only under optimal conditions can condensation produce the dense mist known as *garúa* or the *camancháca,* an often dense fog layer.

Coastal cloudiness is equally high; for 7 months of the year it is over 50 percent, and during the other 5 months (December to April) it is never less than 30 percent (table 14-3). This cloudiness, generally at a low altitude, is due to the almost permanent occurrence of a condensation threshold that indicates the start of an inversion at altitudes varying from 500 to 1,000 meters above sea level. Mist is frequent during the winter and spring, but mitigates slightly every day at about midday.

TABLE 14-2

Average Annual Relative Humidity at Several Chilean Stations

Station	0800 hours	1400 hours	1900 hours
Iquique	78	68	76
Los Condores	77	64	68
Canchones	72	23	34
Pica	38	16	26
Valparaiso	85	72	74

TABLE 14-3

Cloudiness in Northern Chile

	Iquique	Los Condores	Canchones	Pica
Jan	4.0	3.5	2.6	3.0
Feb	3.5	2.6	2.5	4.5
Mar	3.1	3.1	0.9	5.3
Apr	4.3	3.9	1.2	0.9
May	6.0	4.5	2.5	2.5
Jun	7.0	6.1	1.3	1.9
Jul	7.6	6.5	1.0	0.9
Aug	7.5	6.3	1.1	1.7
Sep	7.7	6.7	1.3	1.9
Oct	6.6	6.1	1.1	2.1
Nov	5.5	5.3	1.2	2.8
Dec	4.2	3.9	2.4	6.0

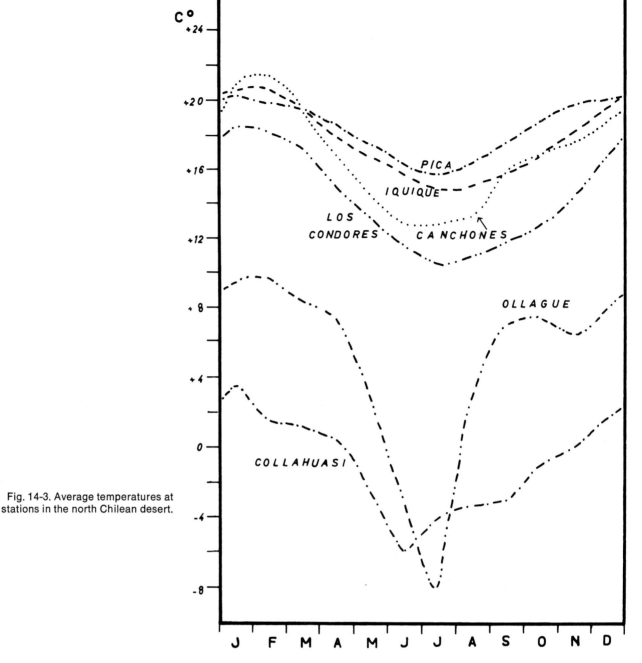

Fig. 14-3. Average temperatures at stations in the north Chilean desert.

While the atmospheric humidity is sufficient to cause this cloudiness, it does not suffice to cause precipitation. Iquique, the records of which extend over 30 years, barely has an average annual rainfall of 2.6 millimeters (0.1 inch); and Los Condores, with over 9 years observations, has had in only one year an annual precipitation of 16.3 millimeters (0.64 inch). At these two stations the rare precipitation has occurred during June, July, and August, the winter months, and is linked to frontal rains that commonly affect Middle Chile.

These climatic characteristics are limited to a narrow coastal fringe enclosed and defined by the upper limit of the condensation threshold of the inversion: between 1,000 and 1,200 meters (Weischet, 1966a). The climate of the coastal desert also penetrates inland along canyons and valleys that dissect the Coastal Cordillera, at least in the form of low mist bearing inland the humidity of the coast.

The Interior Desert, or Pampa

The climate of the desert interior is totally different from the coastal desert climate. Here the largest annual variations of temperature are recorded at the non-Andean stations of the profile (see for example the curve for Canchones in fig. 14-3). The daily amplitudes are also large, as Weischet (1966a) indicates for the Coya Sur station, situated at the southern end of the Pampa del Tamarugal. The absolute maximum and minimum values are fairly extreme for tropical deserts despite their altitude of 1,000 meters. Humidity values for the morning hours are slightly high for an inland desert (table 14-2), but during the rest of the day, they are in every case lower than the average values found in inland deserts on the west coasts in the southern hemisphere, according to the data quoted by Száva-Kováts (1938). The extreme dryness of the air is represented by so high a saturation deficit — 14.3 grams per cubic meter in July and 18 grams in November (Weischet, 1966a) — that the evaporation at Canchones reaches the maximum values of the Pampa del Tamarugal, that is, 34 millimeters maximum and 11.2 millimeters minimum per day (Galli and Dingman, 1962).

Cloudiness is rare but increases slightly toward the summer. In winter the air is very clear, and only small wisps of cirrus cloud are to be seen. Precipitation is almost unknown; even the occurrence of a few drops of rain is quite exceptional.

Given these thermic and hygric conditions, the region of the Pampa del Tamarugal represents the central core of the North Chilean desert which extends between the peaks of the Coastal Cordillera and the foot of the Andean piedmont.

The Arid Andean Piedmont

In this climate temperatures throughout the year show the smallest variation of all the stations in the profile. Pica is known for its almost perpetual summer, where the heat never surpasses that of Iquique (compare the curves in fig. 14-3), despite the mitigating influence of the Pacific at the latter station. The absolute temperatures are nevertheless extreme (table 14-1) due to the fact that the Andean piedmont is already under a certain influence from the high mountains. In other oases of the zone, frost sometimes occurs, which destroys the subtropical crops: mangoes, citrus, and tomatoes.

Cloudiness is low (table 14-3) but still higher than at Canchones, and its annual maximum coincides with the rise in humidity of the air and in precipitation during the summer months (table 14-4).

The occurrence of summer precipitation, the moderate temperatures throughout the year, and the gradual increase in rainfall with altitude, are factors that characterize the climate of the Andean piedmont. This climatic region begins on the gentle slopes rising from the Pampa del Tamarugal; in this section of the profile, the upper limit is indicated by the 3,000-meter contour which corresponds roughly to the 100-millimeter isohyet (compare rainfall amounts at Poroma and Chusmiza, tables 14-4 & 14-5).

The Arid High Andes

This climate of the High Andes is distinguished by the large annual variation in its monthly average temperatures (table 14-1). The absolute temperatures are also quite extreme, resulting in a large daily range: 25°C (45°F) in January and 35.5°C (64°F) in August. The values as well as the times of their occurrence show clearly that they result from insolation conditions. The air has low humidity, and cloudiness is rare, apart from the increase in summer. This climate is most distinct in the high stations of the Andes where annual rainfall is about 100 millimeters, concentrated between January and March and decreasing toward the south.

The climatic characteristics of the Andean stations are closely related to the meteorological conditions that dominate on the Bolivian altiplano, and differ markedly from those of the other stations considered in this profile.

Explanation of the Climatic Regions

The pronounced desert climate of North Chile, with its different climatic subtypes, is explained by the interaction of the following factors on a regional scale:

1. The location of the region on the eastern margin of the South Pacific anticyclone.
2. The presence of a cold offshore current flowing toward low latitudes which, because of oceanic and atmospheric dynamics, has cold water upwelling at the surface.
3. The existence of an atmosphere constantly undergoing massive subsidence with the resultant deficit of humidity.
4. The effect of the relief on the various combinations of climatic elements.

The South Pacific anticyclone is distinguished by a great stability which is reinforced in winter by the relative cooling of the continental block, causing an isobaric bridge to the South Atlantic anticyclone. Thus the predominance of an anticyclonic condition over the whole region, preventing the invasion of humid air from the open Pacific, is assured in winter. Nevertheless, occasionally this stability is liable to be disturbed, and with the rapid advance of subtropical maritime air masses, light winter rain of little importance may occur along the coast. As a result of the anticyclonic condition, the offshore air is subjected to a constant downward movement, is dynamically warmed, and causes a corresponding deficiency in humidity inland. The heating of the air mass creates an inversion layer at a certain level above the sea, expressed as a layer of cloud at an altitude of about 1,000 meters. Underneath this layer, the air cooled by the cold current, assures the stability of the lowest

TABLE 14-4
Rainfall in the Piedmont Oases

Station	Altitude (m)	(ft)	Year (mm)	(in)	January (mm)	(in)	February (mm)	(in)	March (mm)	(in)
Pica	1,280	4,200	9	0.35	3	0.12	4	0.16	0.5	0.02
Pachica	1,650	5,415	15	0.59	4.2	0.16	7	0.28	0.0	0.0
Aroma	2,100	6,900	52	2.05	13	0.51	17	0.67	17	0.67
Chapiquilta	2,500	8,200	53	2.09	19	0.75	20	0.79	7.4	0.20
Sibaya	2,680	8,800	76	2.99	27	1.06	19.5	0.77	15.2	0.55
Poroma	2,830	9,300	92	3.62	32.8	1.26	31.6	1.24	19.3	0.76

TABLE 14-5
Rainfall at Stations in the High Andes

Station	Altitude (m)	(ft)	Year (mm)	(in)	January (mm)	(in)	February (mm)	(in)	March (mm)	(in)
Chusmiza	3,500	11,480	171	6.73	62	2.44	41	1.61	39.6	1.56
Collahuasi	4,805	15,764	105.8	4.16	10.7	0.42	75	2.95	2.7	0.10
Ollague	3,695	12,123	90	3.54	20	0.79	70	2.76	–	–

atmospheric layer and prevents all convection and formation of precipitation other than the low altitude mists — *garúa* and *neblinas*. The low temperature of the air over the sea allows only a high relative humidity, but the saturation deficit persists throughout winter and summer, and it is this which is responsible for the scant precipitation. The cold current flowing offshore, reinforced by the upwelling of cold water to the surface, contributes to the stabilization of the air by cooling the lowest layer of the atmosphere. This current is also responsible for the pronounced thermal anomaly existing at the latitude of Iquique and along the coast of central Chile. At 30° south latitude the annual deficit is −30 kilogram-calories per square centimeter and increases to −40 to −60 at 20° South (Geiger, 1964).

The influence of these factors is limited to the coastal strip, since the Coastal Cordillera acts as an effective barrier preventing the penetration of any littoral air to the interior. Therefore, in the interior, the desert climate of the pampas is determined by different climatic factors.

Equally, the influence that the more humid air masses, originating from the tropical interior of South America, could bring is canceled by the high and wide Andean barrier preventing the passage of these air masses. Thus the air, without the possibility of receiving humidity from any source, reaches its greatest saturation deficit in the intermontane depression. In the absence of any effective factor except topography — relief and altitude — the temperatures depend on insolation. The solar radiation reaches the ground through a clear atmosphere, almost totally lacking in clouds, and its variations are directly reflected in the daily, monthly, and annual temperature curves.

This relation is illustrated by a comparison of an isotherm map with the isopleths of insolation maps of North and Central Chile, compiled at the Laboratory of Solar Energy of the Technical University Federico Santa María at Valparaiso (Hirschmann, 1962). These isopleth maps demonstrate in a new way that the core area of the North Chilean desert is located inland, to the east of the Coastal Cordillera, i.e., in the Pampa del Tamarugal, as is also clearly demonstrated by Weischet (1966a & b).

During the month when temperatures are highest (January) the center of insolation is located over the continent, south of Antofagasta, since the abundant moist air and clouds from the altiplano obviously lower the transmissivity of the atmosphere in the Pampa del Tamarugal. But in spring and autumn it is the pampa which becomes the center of the highest insolation values in North Chile. Throughout the year it is the hinterland of Iquique which shows optimal conditions of insolation. This whole area is enclosed by the daily isopleths of 350 to 400 gram-calories per square centimeter in winter (July) and 600 to 700 in summer (January).

The summer precipitation originating in the east, which is characteristic of the climate of the Andean piedmont and of the high Andean mountains, is explained by the dynamics of the air masses over the Bolivian altiplano. In fact, there is a center of low pressure at the surface over this region in summer, which by convection produces a high-pressure center aloft, located over the high Andes around 15 to 20°S. The sources of this high thermal pressure are the heat rising from a channel of intense convection produced in the lower atmosphere and from additional compensation phenomena. Because of the convection, storms and cloudbursts are common in the altiplano and in the Western Andean Cordillera. As a result of this "emptying" process, northeasterly winds prevail in this area, filling up the deficit in the air mass left by the rising air, and bringing new humidity to the low layers of the troposphere. These winds, the action of which is in addition to that of convection, cause summer rain on the altiplano and over the western Andes, as is indicated by the relatively high precipitation figures at stations above 3,000 meters (table 14-5), and by their effect on the Andean piedmont (Gutman and Schwerdtfeger, 1965). A comparison between the precipitation at Chusmiza, Collahuasi, and Ollague clearly shows the influence of air from the altiplano on the Andean rainfall, and shows furthermore the decrease in precipitation which marks the transition from the arid Andes to the desert Andes (23–26° South).

When the easterly air masses descend to the intermontane depression of North Chile, dynamic heating results and increases the saturation deficit. This does not, therefore, bring humidity to the dry air which prevails over the Pampa del Tamarugal, thus excluding possibility of precipitation. In the winter the anticyclonic conditions, usually affecting the whole region to just beyond the Andes, prevent the invasion of outside air, and the dryness of the interior region is absolute.

Bibliography and References

ALMEYDA ARROYO, E.
  1948 Pluviometría de las zonas del desierto y de las Estepas Cálidas de Chile. Editorial Universitaria, Santiago. 162 p.

CAVIEDES L., C.
  1967 Radiación solar y temperatura en el núcleo del desierto norte de Chile. Asociación de Geógrafos de Chile, Boletín 1(1):5–10.

CHILE, OFICINA METEOROLOGICA
  1964 Climatología de Chile. Valores normales de 36 estaciones seleccionadas. Período 1916–1945. Proyecto Hidrometeorológico de Chile-Naciones Unidas, Santiago.

FUENZALIDA PONCE, H.
  1965 Clima. *In* CORFO, Geografía Económica de Chile (Texto refundido) 1(4):98–152. Talleres de Editorial Universitaria, Santiago. 885 p.

GALLI OLIVIER, C., AND R. J. DINGMAN
  1962 Cuadrángulos Pica, Alca, Mantilla y Chacarilla, Provincia de Tarapacá. Chile, Instituto de Investigaciones Geológicas, Carta Geológica de Chile 3(2–5). 125 p.

GEIGER, R.
  1964 Wärmetransport durch Meeresströmungen (Heat transport by ocean currents). In R. Geiger, Die Atmosphäre der Erde, 2. J. Perthes, Darmstadt. (Maps, Scale 1:30,000,000)

GIERLOFF-EMDEN, H. G.
  1959 Der Humboldtstrom und die pazifischen Landschaften seines Wirkungsbereiches. Petermanns Geographische Mitteilungen 103:2–4, 7–10.

GUTMAN, J. C., AND W. SCHWERDTFEGER
  1965 The role of latent and sensible heat for the development of a high pressure system over the subtropical Andes, in the summer. Meteorologische Rundschau 18(3):69–75.

HIRSCHMANN, J.
  1962 Estado actual de las investigaciones para evaluar la energía solar en Chile. Scientia 23(117):5–38.

KOEPCKE, H. W.
  1961 Synökologische Studien an der Westseite der peruanischen Anden. Bonner Geographische Abhandlungen 29. 320 p.

LYDOLPH, P. E.
  1957 A comparative analysis of the dry western littorals. Association of American Geographers, Annals 47(3):213–230.

PROHASKA, F. J.
  1962 Factores advectivos en el clima de la Puna Argentina. Universidad Nacional de Cuyo, Boletín de Estudios Geográficos 9(35):43–53.

SCHWEIGGER, E.
  1959 Die Westküste Südamerikas im Bereich des Peru-Stroms. Keysersche Verlagsbuchhandlung, Heidelberg-München. (Geographische Handbücher) 513 p.

SCHWERDTFERGER, W.
  1966 Strömungs-und Temperaturfeld der Freien Atmosphäre uber den Anden. Meteorologische Rundschau 14(1):1–6.

SZAVA-KOVATS, J.
  1938 Verteilung der Luftfeuchtigkeit auf der Erde. Annalen der Hydrographie u. Maritimen Meteorologie 66:373–378.

TREWARTHA, G.
  1961 The earth's problem climates. University of Wisconsin Press, Madison, Wis. 334 p.

WEISCHET, W.
  1966a Zur Klimatologie der nordchilenischen Wüste. Meteorologische Rundschau 19(1):1–7.
  1966b Die Klimatologischen Entstehungsbedingungen der Extremen Wüste der Erde. Freiburger Universisitätsblätter 12:53–67.

CHAPTER 15

NEW DIRECTIONS IN THE CHILEAN NORTH

Donald D. MacPhail and Harold E. Jackson
Department of Geography, University of Colorado, Boulder, Colorado, U.S.A.

"In northern Chile where is the driest climate in the world are villages, because even there the desert is not absolutely rainless, and where there is rain there are streams and settlement beside them" (Bowman, 1924). These words by Isaiah Bowman nearly fifty years ago are appropriate today, in the early 1970s, in the vast desert called the Atacama. Occupied for millennia, its opportunities, then as now, are limited by the scant supply of potable water. As one travels into the interior, mindful of the harsh environment, one is constantly surprised at the numbers of people who do manage to live there in those places favored by water or minerals (fig. 15-1). Even so, it is thinly settled overall. This sometimes neglected area, known in Chile as the "Norte Grande," contains approximately one-third of the national territory, and from it comes over one-half of the value of Chilean exports (Duisberg, 1959; James, 1969).

What is this region today? Is it still under a "colonial" style of economy as it was for decades, wherein the riches are withdrawn to serve the needs of the distant and traditional coreland focused on Santiago? It is not the purpose of this paper to try to up-date Bowman's comprehensive study of 1924. Rudolph (1963) has already accomplished this. Rather it is our intent to examine the direction and nature of the pulsations of the current scientific and applied research as they relate to the regional development of Atacama. These activities will provide some new perspectives and give insight into the area's immediate future.

The study area is bound on the north by the frontier with Peru, an unnatural boundary, for there is no difference in the nature of the Atacama desert on the two sides of the frontier. To the east, the altiplano commonly associated with Bolivia and Peru is in fact also shared by Chile. The border between Bolivia and Chile only occasionally follows the Andean crest. Because of the lack of conformity to natural features, frontier conflicts arise. For example, several streams have their genesis in Chile and their demise in a Bolivian salar. This presents political problems that have repeatedly entered the news.

The desert boundary within Chile to the south is not so clearcut. The transition from arid to semiarid is a gradual one, varying from year to year. Convenience, tradition, and physiognomy dictate a limit placed for this study at latitude 27 degrees south, which approximates Punta Morro near Copiapó in Atacama Province.

The climate of the area is a difficult one on which to generalize (Schneider, 1967; Fuenzalida V., 1963). The coastal section, extending perhaps 50 kilometers (30 miles) inland, is a cool desert (classified as BWkn). Small diurnal temperature ranges, frequent fogs, and generally cool temperatures are indicative of this coastal zone. Inland, away from the moderating influences of the ocean, the temperatures rise, fog is absent, and seasonal extremes are felt. Diurnal ranges of 40 degrees centigrade (104 degrees F) are frequent (Fuenzalida V., 1963). The Andes form the east flank of the desert. The zone of annual rains begins at approximately 1,524 meters (5,000 feet) elevation (Walton, 1969). Annual rains do not preclude aridity in this region, and the precipitation is seldom sufficient to find its way to the sea.

The topography of the Atacama is as varied as its climate (fig. 15-2). Adjacent to the Pacific are found marine terraces, here and there interrupted by the Coast Range descending directly into the sea. The Coast Range varies from 20 to 60 kilometers (12 to 37 miles) in width, and reaches elevations as high as 3,100 meters (approx. 10,170 feet). To the east of the Coast Range is a structural depression, the Pampa del Tamarugal. This depression has been filled with alluvium from the Andes and Coast Range. It is generally 30 kilometers (19 miles) wide (Schmieder, 1965). It is a northerly extension of the Central Valley of Chile, and the entire depression can be traced 3,000 kilometers (1,260 miles) southward of the Gulf of Reloncaví. To the east of this depression lie the Andes.

The aridity of this region is emphasized by the solitary river traversing the Pampa and winding its way to the sea. This river, the Loa, rises near the Bolivian border, descends through Calama, turns north and enters the Pacific at about 21.5 degrees south latitude. Other drainage, limited as it is, is into numerous salars or desert playas.

Beneath the veneer of aridity and isolation, we find important mineral resources some tapped and some not as yet exploited or perhaps even known. Large reserves of metallic and nonmetallic minerals await development. In addition to these are excellent marine biology prospects which are combined with the favorable weather conditions of the coast. The area's tourist potential just began to be used in the 1960s. Chile shows every intention of fully developing this rich region of the country.

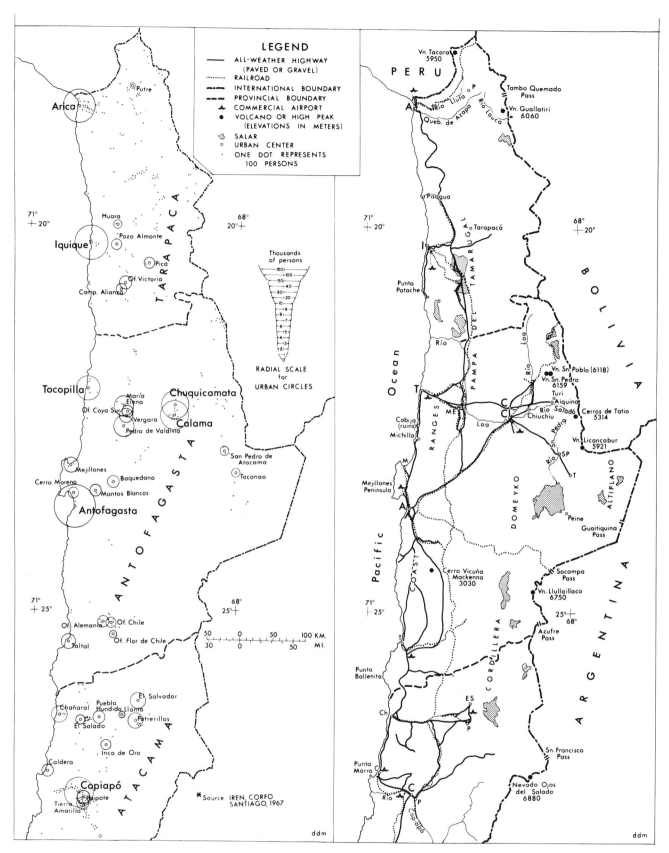

Fig. 15-1. The Norte Grande. Population distribution according to the 1960 census is shown in the first panel; major geographic regions are shown in second panel. Map is modified from that of the Instituto de Investigación de Recursos Naturales, CORFO (Saa Vidal, 1967).

Economic planning in arid regions requires long periods of time (Tixeront, 1965). What are the prospects of this northernmost region in light of the present record?

Meigs says that "the sea itself is an advantage which the coastal deserts have over inland deserts" (Meigs, 1969). Much of the Atacama is a coastal desert, sharing the advantages that these deserts have in common: abundant sea life, ease of transport, and a temperate and even climate. Other attributes, although untapped in the Atacama, are constant winds for a potential power source, solar energy, tidal power, and the vast potential of extracting moisture from the atmosphere. Aridity is, after all, the feature that commands the attention of the student of the Atacama.

The Past

Human occupance of the Atacama has existed for an extended period. Many archaeologists stand on a figure of six or seven thousand years while others, the controversial Father LePaige among them, claim that a more accurate figure would be 30 thousand to 50 thousand years (*La Mañana,* Antofagasta, October 4, 1967). Knowledge of exact periods is unnecessary, as man has had little widespread effect on the landscape of this region. Until the Spanish conquest, the occupants of the Atacama were centered at the few places where adequate water supplies might be obtained. Calama on the Rio Loa was certainly the most important of these oases. Late in the pre-Spanish period, the Incas conquered the Atacama and pushed on through Chile as far as the Rio Maule. The Incan conquest changed the human ecology of the Atacama but little. A road, or trail, was maintained through the desert using Calama as the principal resting place. This slight change occurred approximately one hundred years before Almagro made the first Spanish intrusion into Chile.

Coastal fishing and floodplain agriculture were the means of subsistence for the ancient Atacameño people. Pedro de Valdivia found 1,000 hectares under cultivation at Calama when he traversed the Atacama in 1540 (Inostrosa, 1963). The amount of irrigated land there was only slightly more in the 1960s (*Ercilla,* no. 1681, Aug. 23, 1967). Even though remnants of early Spanish occupation, such as the old sixteenth century chapel at Chiu-Chiu, still remain, the basic established patterns were as unaffected by the colonial Spaniard as by the Inca before him (fig. 15-3). Then, in the eighteenth century the tempo changed; silver was exploited at San Lorenzo de Tarapacá and was smelted and exported through the nascent port at Iquique (Antonioletti Ruiz, 1966). Silver mining declined in this region so that it had almost stopped by the end of the same century. The wars of independence, 1810–20, certainly contributed to the neglect of the mines.

No discussion of the Atacama would be complete without mentioning the nitrate exploitation beginning in the

Photo by Hans Schneider

Fig. 15-2. Interior basin and range topography of the Atacama.

nineteenth century and continuing sporadically into the 1970s. Nitrates were obtained even before the Spanish arrived on the scene, but the extraction was disorganized. New technological developments, high explosives in particular, led to a rapid expansion of nitrate mining in the first third of the 19th century. The first exportation of nitrates occurred in 1831, and, by 1860, 50,000 tons were being exported annually. At this point, the political situation was rapidly deteriorating. The boundaries of Peru, Bolivia, and Chile had never been completely

Photo by René Saa V.

Fig. 15-3. Terraces at the oasis of Aiquina on the Rio Salado, Antofagasta Province. For centuries, the Atacameño people have farmed here using spring-fed waters for irrigation.

agreed upon. Now the lack of accord became important. Temporary agreements between the three states came to nought as war erupted in 1879. The War of the Pacific, or "guerra del salitre" as Oscar Schmieder calls it, terminated in 1883 with Chile gaining the Atacama northward to Arica (Schmieder, 1965).

The nitrate period was marked by rapid influxes of people into the desert. The nitrates were mined in the Pampa del Tamarugal and were then sent by rail to the ports of Iquique and Antofagasta. The development of synthetic nitrates, perfected by Germany during World War I, led to collapse of the industry in the postwar 1920s, with the decline continuing to the present (James, 1969).

The ports, not surprisingly, have had a mixed development. Arica, at the Peruvian border, was early used to export the silver from Potosí, Bolivia. Cobija, whose sparse skeletal remnants would hardly qualify it even as a "ghost town" today, was chosen for a Bolivian port during that country's dominance of the Atacama. Potable water determined the location of Cobija (Cruz Larenas, 1965). Iquique's and Antofagasta's fortunes waxed and waned, first with nitrate production and later with copper. However, the rugged and plunging coastline of the Norte Grande possesses no favored sites for ports (figs. 15-4, 15-5, & 15-6). Each was sited, not necessarily because of natural endowments, but because of economic development in the hinterland to the east.

The coastal cities are today the outstanding foci of human occupance. Through time the emphasis of settlement has gradually shifted from the remote interior to the sea. In pre-Columbian days, the oases of the altiplano and those on the fringe of the pre-cordilleran ranges were predominant. Later, man's major activities centered on the mining camps of the pampas further west. Finally, it is the ports which now dominate the desert North (Santana Aguilar, 1970).

Contemporary Development

Development in the Atacama during the 1960s covered a wide variety of endeavors. Some of the development has been in the traditional area of mining. The most recent period is noteworthy because substantial development has occurred outside mining. The short time during which these new economic areas have been open make it difficult to ascertain their progress. Short-term success can be found in the industrial and agricultural sectors, while the results from fishing investments are positive, with some reservations.

Courtesy of El Mercurio

Fig. 15-4. The industrial district of Iquique where, in 1967, the Chilean government cut back operations of some economically submarginal fish-meal plants in order to organize an integrated economic development program for the region (*El Mercurio,* Santiago, May 26, 1967). The City of Iquique is seen in the distance.

Courtesy of the Instituto Geografico Militar de Chile

Fig. 15-5. A vertical view (April 14, 1956) of Arica and the entrenched Azapa Valley, noted for its irrigated olive and citrus orchards and vegetables. Portions of the Lluta Valley are visible on the northern margin of the photo. The small white island of Alcarán near Arica was exploited for years for its guano. Now it is connected, through landfill, with the mainland and will be developed for tourism. Just north of it, a large, L-shaped jetty has been constructed to improve the harbor of Arica. Compare with figure 15-6.

Fig. 15-6. The Port of Arica with the famous landmark, El Morro, visible at the right. Fishing trawlers, freighters, and naval vessels are anchored behind the new jetty, which extends from the base of El Morro into the Pacific. Compare with figure 15-5.

Courtesy of El Mercurio

Minerals

Copper has replaced nitrate as the benefactor of Chile. While this is not a recent change, it is emphasized because the 1960s saw the virtual end of the nitrate industry in Chile. The Chilean industry had depended on the Shanks process, and this had put it in a poor competitive position (Duisberg, 1959). Only two large processing plants, operated briefly by a consortium of Compañía Anglo-Lautaro and CORFO (Corporación de Fomento de la Producción), have continued to function, under state control (*Ercilla,* no. 1718, May 28, 1968). CORFO had spent forty million dollars to initiate new steps in nitrate production to help make it more competitive. Of the few plants still operating in the late 1960s, only two still maintained significant production — María Elena and Pedro de Valdivia. There was also a small plant in Antofagasta Province. Some Chileans see a reversal in the decline of the industry as new ways are developed to extract potassium and iodine from the nitrate beds (*Ercilla,* no. 1716, May 14, 1968). Only time will prove if their optimism has a basis.

In contrast to nitrates, copper production rose in the 1960s, and the future will see an acceleration of this trend. The proportion of Chile's income derived from copper sales has increased steadily. With the opening of new mines, combined with the Chileanization of the older ones, production of copper may rise dramatically. What effect this will have on the world market price remains to be seen. CORFO had hoped to double the 1965 copper production of 630,000 tons by 1971 (Corporación de Fomento de la Producción, 1966), but production was only a disappointing 571,224 tons then (CORFO, *Chile Economic Notes,* no. 95, 1972).

Chuquicamata is the largest open-pit copper mine in the world (fig. 15-7). Its operation was assumed in 1910 by the Anaconda Copper Corporation after many years of prior production. Chuquicamata is located near Calama and has sent most of its production out of the country through the port at Tocopilla. But an increasing amount will be shipped from Antofagasta. In addition to increased production at Chuquicamata, CORFO would like to see an increase in fabrication of finished products within Chile. More complete processing and finished product fabrication would require additional power and water, two commodities in short supply in the Atacama. A 1967 prediction indicated that the better oxide ores would be depleted at Chuquicamata in four to five years at the current rate of extraction (*El Mercurio,* Santiago, May 28, 1967). The ownership at Chuquicamata in 1970 was shared between Anaconda and the Chilean government, with Chile owning controlling interest. The following year, however, the Government of Chile expropriated this and other large mines, to complete total nationalization. In 1968 and 1969 this single mine accounted for over a third of the nation's export income.

Chuquicamata is one of several copper mining sites in the Atacama. La Exótica, a 1960 copper discovery located near Calama, was originally developed jointly by Anaconda and the Chilean government. It has since been completely nationalized under the Corporación del Cobre (CODELCO). It is estimated that there are 150 million tons of 1.35 percent copper ore at this site, which will require 20 to 30 years to deplete (*El Mercurio,* Santiago, May 28, 1967). Much of this ore is copper oxide, and the extant processing facilities at Chuquicamata will be utilized. La Exótica will produce some 100,000 metric tons of copper annually (Brown, 1968). In response to the need for housing for workers at La Exótica, CORVI (Corporación de Vivienda) is constructing a low-cost housing development in Calama, rather than at Chuquicamata nearby.

Potrerillos and El Salvador are important copper mines located east of Chañaral, which is the port through which their production is exported (James, 1969).

Photo by Hernan Gonzalez
Fig. 15-7. Blasting and shattering copper ore at the great open pit mine of Chuquicamata.

Courtesy of El Mercurio

Fig. 15-8. A geological exploration team endures difficult terrain conditions at 5,000 meters (16,404 ft.) of elevation on the Chilean altiplano near Tambo Quemado Pass, Province of Tarapacá.

Another recent copper discovery has been made at Michilla, 80 kilometers (50 miles) north of Antofagasta. Estimates put the find at Michilla at 1,000,000 tons of 2.35 percent copper (Brown, 1968).

Sulphur deposits of commercial quality represent another mineral resource found in several locations in the Norte Grande. Volcán Tacora on the Peruvian border, east of Arica, is one of the two most important locations because of its location on the railroad to La Paz (Concha M., 1966). East of Taltal lies "Plato de Sopa," another important location. Inaccessibility of most of the sites and distance to markets for the sulphur will impede development. More importantly, the extreme altitude at which the sulphur is found, 5,000 meters (approx. 17,000 feet), will hamper exploitation (James, 1969) (fig. 15-8).

Iron has been exploited extensively in Chile for some five decades. Originally, the extraction took place at El Tofo lying south of the study area. More recently a large body of ore has been found east of Antofagasta near the Argentine border. Both the distance from the coast and the elevation at which the ore is found (4,500 meters or approx. 15,000 feet) will impede development of this valuable site. The original site of El Tofo was found to be part of a rather extensive north-south trending iron ore deposit. The northern sections of this deposit lie east of Chañaral (James, 1969).

Agriculture

Students of contemporary development in the Atacama can easily overlook the obvious. Who are the permanent residents of the desert Norte Grande? The Chileans (the southerners) have traditionally played the role of transients. Much as nomads, they enter the desert to exploit the silver, nitrates, and copper, then move on to another site when the minerals are depleted. Some attention must be given to the region's permanent settlers — the descendants of the Atacameños. These people have been clustered around a large number of historic oases for centuries (fig. 15-3).

Little attention has been devoted to these oases. A joint, and very thorough, study of the oasis of San Pedro de Atacama by the Universidad de Chile and CORFO showed that a consolidation of small landholdings of widely dispersed, diminutive irrigated fields would greatly enhance the productiveness of the local residents (Aranda Baeza, 1964). The plots are organized around historic indigenous clans ("ayllus") (*Ercilla*, no. 1681, August 23, 1967). The recommendations of the researchers so far have been ignored and the oasis continues to be one of basically subsistence agriculture. Calama, another ancient oasis, has lost much of its original character. Its proximity to the copper mines of Chuquicamata and La Exótica has transformed it into a modern city. Perhaps this change has been for the best; the oasis had very poor, saline soils, a problem compounded by the application of slightly saline water from the Rio Loa.

Chiu-Chiu, located east of Calama, has a pattern of recent development that may serve as the prototype for the desert region. This is a very old settlement, and most of the contemporary inhabitants are native. Principal crops are carrots, alfalfa, and wheat, grown for the most part for the nearby urban complex of Calama-Chuquicamata. INDAP (Instituto Nacional de Desarrollo Agropecuario) tried to assist this community for several years without success. Much of the land, however, is now owned by absentee landowners who rent to the local farmers for one to two years. The general lack of organization at Chiu-Chiu has resulted in the merchants of Calama setting prices for the products of the oasis.

In contrast to the more recent development of private holdings of land, one can still find examples of pre-Hispanic land tenure systems. The traditional, indigenous, land-tenure system seems to break down in those oases where the impact and exposure to the contemporary Chilean economy is intense and prolonged. However, in Putre, the Vegas de Turi, and Peine are found vestiges of the older communal land tenure (Winnie, 1965). In general, agriculture is not accomplished communally, but livestock grazing (sheep, llama, alpaca) is. This pattern exists in areas that have little felt the effects of westernization (Duisberg, 1959).

Government technical assistance and financial aid has benefited the agricultural village of Toconao. Reconstruction of the canal systems in that village was financed by long-term credit which was fully repaid in a few years. This contributed to increased production of figs and other tropical fruit. The success was in spite of the handicap of small holdings (minifundias) inherent at Toconao.

It is evident that agriculture is the Atacama's oldest and least changed economic endeavor. Barring some radically new development in the desalinization of water, it is unlikely that irrigated agriculture will exceed 15,000 hectares in the Norte Grande (Duisberg, 1959). Except in high elevations, the crops grown on this limited acreage include tropical crops difficult to grow elsewhere in Chile. In spite of its small areal extent, however, the agriculture sector is an active one today. For example, the Rio Lauca diversion into the Azapa Valley near Arica has enlarged acreage there in olive orchards, truck crops, and alfalfa (fig. 15-5). Olives constitute half of the production of the valley (Instituto Geográfico Militar, 1966). These northern valleys have recently encountered marketing problems with olives. The traditional Peruvian market has been closed to them (*El Mercurio*, Santiago, May 26, 1967). Another sector with activity is research. There is a research station in the altiplano at Pisiga, for instance, where experiments are underway with thirteen varieties of cereals, potatoes, tamarugos, and eucalyptus (*El Mercurio*, Santiago, October 21, 1967). Another experimental station is located at La Chimba, on the periphery of Antofagasta, using water obtained from the Cordillera. Here the primary research is centered upon fruits and vegetables that would grow best in this environment. Three crops are obtained each year from the fields, but estimates still place costs at four times those of Santiago.

Agricultural scientists from the Universidades de Chile, del Norte, and Católica de Valparaiso are concentrating their efforts significantly on the traditional food-producing crops of the North. Their research focuses on plant breeding and selection, production, and disease control of maize, beans, alfalfa, and tomatoes. The Corporación de la Reforma Agraria (CORA) and INDAP (Instituto Nacional de Desarrollo Agropecuario) support this research (*Ercilla*, no. 1681, August 23, 1967).

One of the more promising experiments in agriculture deals with the "tamarugo" (*Prosopis tamarugo*). This tree grew along the whole interior from north of Iquique southward to the Rio Loa at one time. The demand for firewood during the nitrate period almost depleted the tree, and only in the 1960s were reforestation attempts begun. In 1962, CORFO began small-scale experiments in the spacing of the trees and use of the tree products as animal food. Livestock preferred the fruit of the Tamarugo to hay; they were eventually left to forage among the developed trees, seeking falling leaves and straw. CORFO found that 256 trees could be planted to the hectare, and they required special care and watering during their first three years (Ossandon Estay, 1967). Expectations are that the re-established vegetative cover will permit a density of 10 to 14 sheep per hectare (*Ercilla*, no. 1681, August 23, 1967).

Fishing

It was in 1960 that fishing was "discovered" in northern Chile. The exploitation of fish was to absorb the recently released nitrate miners and even open some additional opportunities. The Chilean fishing attempts were beset with problems from the first. The coast did not have a history of offshore fishing and therefore had no boats or trained fishermen. In addition, Chile's share of the cold-water coast differed in one important respect from the Peruvian coast to the north. The Chilean portion of the continental shelf is but 18 kilometers (11 miles) in contrast to Peru's 95 kilometers (60 miles).

Despite this, the growth of the fishing industry in the Norte Grande since 1960 has been phenomenal. The focus of fishing has been anchovetas. The scale of the growth is illustrated by the following: (1) exportation of fish meal and oil rose from 13,699 metric tons in 1959 to 299,164 tons in 1966, and (2) in 1960 the Norte Grande had a "fleet" of 24 vessels, and in 1965 this figure was 251 with ten tons of registry (Salinas Messina, 1967). As the industry grew, there were concomitant changes in the Norte Grande. The number of processing plants increased considerably. By 1965, Iquique had 23, Arica 7, and Pisagua 3, to mention only the three most important centers (Salinas Messina). These plants provided women with employment, many for the first time. As Chile only uses about 15 percent of the fish production, the balance has been exported, primarily to Western Europe (Salinas Messina, herein). The rise of fishing created a shipbuilding industry at Iquique. The majority of the north Chilean fishing fleet was constructed in Chile. These ships are modern and steel-hulled, of about 140 tons each (Salinas Messina, 1967) (fig. 15-9).

The prosperity was shattered in 1965. In that year the anchovetas disappeared from Chilean waters. Production for 1965 was less than half that of 1964 (435,000 compared to 924,000 tons). To an infant industry grown accustomed to large growth each year, this setback was critical. Two hypotheses were put forth on the disappearance of the anchovetas: (1) it is a limited resource on a rather narrow shelf, and (2) the anchovetas' disappearance is cyclical (Salinas Messina, 1967). Current research indicates that the second alternative is more likely. The anchoveta disappearance of 1965 coincided with a mild "El Niño" occurrence (Krebs, 1970). In 1966, the anchovetas returned, but the setback has had its effect. To combat future crises, attempts are being made to spread the fishing fleet out in new ports — for example, Tocopilla, Mejillones, and Antofagasta. Additional species, especially tuna and bonito, will be sought, thereby reducing the dependence on the somewhat elusive anchoveta. In the meantime, the coastal fishing industry has recovered its previous stature (Santana Aguilar, 1970).

This debate on offshore fishing limits is not reserved for Peru and the United States. Chile, Argentina, Ecuador, and Peru all claim a 200-mile jurisdiction over fishing from their coasts. At the time of this writing United States boats have not yet been detained by Chile, but it seems only a matter of time before they are. This matter is serious to Latin American states, and journals in those countries make much of the intransigence of the United States' insistence on a 12-mile limit. The Chileans claim that the United States set the precedent for the 200-mile limit (Garcia, 1967).

Urbanized Areas

Three cities, clinging to the barren coast, contain the overwhelming majority of the Atacama population; the balance of the people are centered in some few mining towns or are found threadlike along widely separated irrigable valleys (fig. 15-1). This region is an urban one. The three provinces making up the study area, Tarapacá, Antofagasta, and Atacama, are 83 percent, 92 percent, and 66 percent urban, respectively (Uribe Ortega, 1967).

It is difficult to generalize on urban growth in the Norte Grande. Each city appears to have unique reasons for growth. Antofagasta, the largest city, is the capital of its province. It is the major port in the Atacama. In 1967, Antofagasta was made a free port for Paraguayan shipments (*El Mercurio*, Santiago, July 14, 1967). The hope is that products from the Chaco of western Paraguay will be exported through Antofagasta. It is 1,580 kilometers (980 miles) from the Chaco to Antofagasta, including a long traverse through Argentina.

CORFO has built an eight-million-dollar copper-refining plant in Antofagasta, which will process 60,000 tons per year, mostly for export (Corporación de Fomento de la Producción, 1966). An addition in 1965 to an earlier sulphuric acid plant raised production to 60 tons per day. These industries underlie the heart of Antofagasta's economic nature — ancillary to mining. Much of the current construction at Antofagasta is related to the projected increase in activity as a result of the opening of the "Exótica" mine. An unusual problem in Antofagasta, and shared by other Atacama ports, is the occurrence of an occasional rain. In August of 1967, heavy rains in Antofagasta almost ruined a large amount of unprotected wheat being sent to Bolivia (*El Mercurio*, Santiago, August 27, 1967). Antofagasta's population had grown to 108,000 by 1964 (Salinas Messina, herein).

Iquique is the capital for Tarapacá Province. In 1964, the population was 90,000. There is still some mining activity remaining from the nitrate days. Emphasis is fast shifting to the burgeoning fishing industry. There were 23 fish-processing plants in Iquique in 1965; however, in 1967, this number was reduced (fig. 15-4).

The development at Arica has been directed from Santiago. Chile made Arica a free port for Chile to encourage industry in the city (fig. 15-5). This scheme was quite successful with many industries locating there. Assembly plants were begun in quonset hut factories. Manufactured goods have included car batteries, bicycles (Oxford), motor-bikes (Lambretta), textiles, autos (a

Photo by René Saa V.

Fig. 15-9. Part of the fishing fleet of the Port of Antofagasta.

wide variety including Chevrolet, Citroen, Datsun, Fiat, Ford, Opel, Peugeot, Simca, and Volvo). This city is also a Bolivian outlet (fig. 15-6). The tourist potential of Arica was augmented by the addition of a casino and many new hotels (New York Times, May 12, 1963). There were 35,000 "callampistas," or squatters, in Arica in 1960 (Hernandez Parker, 1966). Housing has continued to remain a major problem here and in the other cities throughout the Norte Grande.

The ports of the Atacama share in the disadvantages of location with respect to world markets. The products exported through these ports — for example, copper, nitrate, fish meal — are valuable enough to withstand this disadvantage but the cost is, nevertheless, borne by Chile. Imports too must be brought from distant points. Even food, most of which is obtained from central Chile and overseas, must necessarily include these high transportation costs.

Duisberg (1959) has stated that the Atacama will have no real stability until (1) sufficient potable and industrial water is available and (2) sufficient competitively priced industrial energy is available. His analysis appears to be correct. What has been passed off as "development" in the past has really been "temporary exploitation." From the silver extractive period to the era of nitrates and through the present copper exploitation, it has been much the same. However, Chile is sincere in her present attempts to develop the Atacama. Let us inspect some of the problems relating to water and energy that must be faced and the work being done to solve them.

Water

Water is the first and most critical problem. The Rio Loa is the only permanent stream in the study area, and its waters are greatly in demand. Copper mining at Chuquicamata, the agricultural oases at or near Calama, and a pipeline to Antofagasta, all compete for the limited moisture of the stream. By most standards, the Loa is already too saline for most uses when it reaches Calama. In addition to the Loa, trans-Andean sources, meager as they are, also have been tapped. The Rio Lauca, as previously mentioned, begins in Chile and flows into Bolivia. After many years of waiting upon Bolivia's cooperation, Chile began unilateral action; Chile is diverting water from the upper Lauca to the Azapa Valley in the north (Glassner, 1970). Bolivia severed diplomatic relations with Chile when the diversion order was given (New York Times, April 17, 1963). Other than limited supplies of subsurface water, these two sources exhaust the easy natural possibilities in the region.

Desalinization has been put forth as a logical solution for the Atacama's water problems. The shape of the Atacama, long and narrow, and its proximity to the Pacific Ocean, make the suggestion seem plausible. However, the various distillation techniques devised thus far have not been economically feasible except in special locations.

One method showing promise has been developed by the Soviet Union and has been used successfully in the Kara Kum. The Soviets have used cylindrical mirrors to focus the sun's rays on a glass tube, which acts as a boiler from which 75,000 tons of distilled water per year have been obtained (Walton, 1969). No cost estimates had been made at the 1969 date. Certainly the Atacama has abundant solar energy. Away from the coast, in the Pampa del Tamarugal especially, are located a number of brackish water sources.

The second promising technique of obtaining water is especially intriguing. The coastal desert portion of the Atacama is affected often with a sea-fog known in Chile as the *camanchaca* (fig. 15-10). The onshore prevailing winds drive moist air from the interior Pacific across the narrow, cold Peru Current. The drop in temperature causes stability and saturation in the lower portions of the air mass. The anomaly of aridity combined with dense fog is not unique to Chile. In Peru they call the same type of fog *garúa*. It can also be found along the west coast of Baja California, the coast of southwest Africa and parts of Morocco. During summer, the camanchaca will penetrate up to 32 to 48 kilometers (20 to 30 miles) inland through several breaks in the Coast Range (Börgel, herein).

Chilean researchers at the Universidad del Norte discovered that they could obtain moisture from these clouds. There is evidence that the indigenous people along the coast used piles of rocks to condense moisture and actually create springs (Engel, 1967). Modern scientists use a frame with nylon or saran filament stretched from side to side. Water condenses on the filaments and is directed to enclosed receptacles below the frames. The most important experiments were conducted at Mina Portezuela, 10 kilometers (6 miles) east of Antofagasta at an elevation of 850 meters (2,800 feet). The first experiment, started in November 1959, ran for 489 days. Water was obtained on 319 days, while the maximum number of consecutive days without water was ten (Muñoz Espinosa, 1967). Experiments have been on a limited basis, and it is difficult to estimate costs using this method of extraction (Muñoz Espinosa, 1969; Cornejo T., 1970). However, costs do not appear to be prohibitive. New Jersey officials read of these experiments and decided to use a modification of the technique to clear fog along major highways of that state (Time, July 14, 1967).

Moisture from sources such as the camanchaca appear to have a great deal of promise for small-scale purposes. Using water from these sources, small military posts, or isolated small mines particularly would benefit from a system using this water to replace that brought in by mule, and conserving it by a recycling process. CORFO has authorized research support for further study of the distribution of the camanchaca and developing larger more-effective devices for extraction of the moisture from the fog.

Several trees — for example, tamarugo and eucalyptus — adapt to an Atacaman microenvironment, aridity

Photo by Hans Schneider

Fig. 15-10. An oblique aerial view of coastal fog, the camanchaca, high against the flanks of the Coast Range in Antofagasta Province.

and all. Peruvian experiences with similar microenvironments have been extensive. The Peruvian garúa, with its associated lomas vegetation comprises some 150,000 hectares of coastal Peru. The experimental station at Lachay, Peru, has grown trees, oats, and potatoes in the lomas zones there (Cornejo T., 1970).

The Chilean researchers have focused on the mechanical extraction of moisture from the camanchacas; this seems to contrast somewhat from the direction of Peruvian research. Water from the Chilean fog traps, or *captanieblas* (fig. 15-11), could be used to start small trees, eventually allowing them to become their own captanieblas. Basic information must be sought about the camanchaca zone of Chile. Location, duration, direction, and intensity of flow of the fog are among the most obvious questions to which answers must be ascertained.

Research in Antofagasta has been directed toward these ends. Soils in this zone are also of primary interest, as are crops that might adjust to the soils that are found. Chilean agronomists fear that toxic sodium nitrates and other substances will limit severely even the introduction of trees in some localities.

Scientists at the University of Arizona have developed a rather complex water desalinization "system" that could provide power, water, and food on desert coasts (cf. Chapt. 5). Initial experiments were run in 1963 at Puerto Peñasco, Sonora, Mexico. Originally, the system was based on solar power, but in 1966 this was changed to diesel power. This system is a very complex one in which the diesel engine's waste heat is used to desalt seawater which in turn is piped to greenhouses where a variety of vegetables are grown (University of

Photo by Rómulo Santana A.

Fig. 15-11. A fog trap (captaniebla) located in a pass of the Coast Range near Antofagasta. This is one of several experimental designs being tested by Professor Carlos Espinoza and others at the Universidad del Norte in the zone of the camanchaca, or coastal fog.

Photo by Peter Duisberg

Fig. 15-12. The llareta *(Laretia compacta)*, a plant that grows slowly at elevations above 4,000 meters (approx. 13,000 ft.) in the cordilleras of the North. It is in danger of extinction from overuse as a fuel.

Arizona, Environmental Research Laboratory, 1969; Hodges, 1969). A system such as this could work satisfactorily on the Chilean coast. If placed at Antofagasta, Iquique, or Arica, the vegetables could be sold in those cities.

Energy

The second of Duisberg's priorities was competitively priced industrial energy. As might be expected, conventional sources of power are almost completely lacking in the Atacama. The only hydroelectric source is in the far north, where the Rio Lauca is diverted. Some 22,000 kilowatt-hours are generated as the water drops to the Azapa Valley (Millas E., 1962, *Ercilla,* no. 1402, April 4). This energy is transmitted to Arica for industrial use. Although petroleum searches have been made by ENAP (Empresa Nacional de Petroleo) there were no discoveries at the end of the 1960s. However, Chile has oil in the Punta Arenas region in the far south, and this can be shipped at a reasonable cost to the Atacama.

Lacking these two basic sources of power in the Atacama, Chile must look to more exotic methods. Nuclear power is always a possibility but requires a high capital, resource, and technological input. More concretely, geothermal possibilities exist at El Tatio in Antofagasta Province (Corporación de Fomento de la Producción, 1966).

Nevertheless, conversion of solar energy seems to offer the best hope for the Atacama. Previously mentioned in this respect were experiments in distillation conducted by the Soviet Union using solar energy for power. Measurements near Antofagasta show the solar energy to approximate 2,400 kilowatt-hours per square meter annually (Muñoz Espinosa, 1969). Galvez has been experimenting with solar energy as the power source to heat water. Using tubing of various thicknesses and made of a variety of metals — for example, copper, zinc, iron — he has made some progress toward building a pilot system (Galvez Z., 1966). The biggest need for the Norte Grande would be for industrial water heating and solar power plants (Duisberg, 1959).

On a smaller scale, very competitively priced solar power units are already on the market. Small motors powered by solar energy have been developed to 100 horsepower. Another promising development is the solar cooker. Pioneered in India, these cookers would allow the Chileans to have a viable alternative to destroying the remaining vegetation of the desert for fuel (Duisberg, 1959). The llareta, a slow-growing plant of the Atacama, has been placed on the list of plants not to be cut in Chile, but this conservation edict will have little practical effect unless the people are given some alternatives for fuel to use for cooking their food (fig. 15-12).

Conclusion

High on a barren hilltop overlooking the port city of Antofagasta stands a solitary pine. This tree was once nurtured with water obtained from the camanchaca by a Universidad del Norte fog trap. Now its leaves condense enough moisture for its needs from the atmosphere without any assistance. In many ways, this tree is a symbol of the future; it stands for radical departures from traditional concepts. The future of the Norte Grande will be increasingly tied to the renewable resources just as permanent occupance in the Atacama has been in the past. The key long-range resources for permanent occupance include water, energy, fish, and soil. Traditional patterns will be maintained in those places that have access to adequate surface supplies of water.

Water will continue to be the first priority in the North. Transmountain and transdesert aqueducts have provided some solution for the moment. But the future of the cities along the coast will be dependent on (1) desalinization technology powered by diesel or solar energy, both of which are abundant; and (2) breakthroughs in designing commercially usable units for extracting moisture from the camanchacas. The success of the scientists at the Universidad del Norte is hopeful in this regard.

The history of mining in Chile, and in other parts of the world, suggests that although copper is today "king," its importance is certain to decline. It is only a question of when the decline will occur. Copper is a nonrenewable resource, and the supply is finite. Rich, mineralized areas tend to go through cycles. One has only to look at California's gold, Potosí's silver, or Britain's coal to be reminded of this fact. The emphasis of mining changes and so do the sites. Potrerillo's copper production is today replaced by the neighboring mining camp at El Salvador. Likewise, Chuquicamata will soon be dethroned by La Exótica. In a longer view, the demand for copper in overseas industrial markets makes it likely that the Norte Grande will continue to play an important role in copper production in the immediate future. A decline in the importance of copper might be followed by more extensive development of the iron resources of the Norte Grande.

But will copper or some other mineral continue to contribute such a disproportionate amount to the total economy in Chile? If it does, then much of the country's economic future rides with the success of research programs such as those of the Instituto de Investigaciones Geológicas of CORFO searching for new mineral discoveries. If the country looks to a more balanced economic development, then in the Norte Grande at least, research and development centered on water and energy is absolutely indispensable.

A clear understanding of the Chilean North and its problems cannot always derive from economic analysis alone. One needs also to know the people. Bowman sensed this in his classic study (Bowman, 1924). So, too, did Manuel Rojas in his recent poem, "Hombre del Loa." It is the people and their concept of the place they live in that brings to fruition all of the plans, dreams, and programs for the future. Mario Bahamonde has said, "The 'Nortino' is more than a product of territory and of an inherently thirsty land . . . [and] whoever comes upon any part of this land either makes himself into the image of the Northerner or perishes" (Bahamonde, 1967). He is hardy, adventurous, perceptive, quick to act and quick to adjust to tomorrow's surprises. This applies as much to the resident man of science as to the traditional and sturdy oasis farmer.

Fortunately for Chile, the Atacama creates in those who live there an abiding love and reverence for the region. A Chilean who lived in the desert North said of the area, "It's a place where you dread to go. You weep when you arrive, but you learn to love it. You develop a kind of mystique about it. If you must leave it, then you also weep in parting." Thus, it is likely that the desert itself will help to hold fast many of those who came and now engage in research or economic development. Without them, the Chilean North will have no future.

Bibliographic References

Antonioletti Ruiz, R.
    1966   Las funciones regionales de la ciudad de Iquique. Universidad de Chile, Santiago, Instituto de Geografía, Informaciones Geográficas 16(1):133–149.

Aranda Baeza, X., R. Baraona Lagos, and R. Saa Vidal
    1964   San Pedro de Atacama. Elementos diagnósticos para un plan de desarrollo local. CORFO, Santiago. 196 p.

Bahamonde, M.
    1967   El nortino. In Ercilla, no. 1681, August 23, p. 21–23.

Bowman, I.
    1924   Desert trails of Atacama. American Geographical Society, New York, Special Publication 5. 362 p.

Brown, L. R., Jr.
    1969   Mineral industry of Chile. In U.S. Bureau of Mines, Minerals Yearbook 1967, 4:181–191. U.S. Government Printing Office, Washington. 1036 p.

Concha M., M.
    1966   Establecimientos humanos en el Altiplano Chileno. In E. Flores and others (eds), Estudios geográficos, p. 55–82. Universidad de Chile, Santiago, Facultad de Filosofía y Educación.

Cornejo T., A.
    1970   Resources of arid South America. In H. E. Dregne, ed, Arid Lands in Transition, p. 345–380. American Association for the Advancement of Science, Washington, D.C., Publication 90. 524 p.

Corporación de Fomento de la Producción (CORFO)
    1966   La CORFO y el progreso de Antofagasta. San Jorge Impresores, Santiago. 45 p.

Cruz Larenas, J.
    1965   Fundación de Antofagasta y su primera década. Editorial Universitaria, Santiago. 133 p.

Duisberg, P. C.
    1959   Arid zone possibilities in Chile. Unpublished paper prepared for U.S. Ambassador to Chile Howe. 9 p. (mimeo)

Engel, F.
    1967   Notas referentes a la adaptación de los pueblos precolombinos del Perú a la vida en tierras áridas. Paper presented at Simposio sobre Desiertos Costeros, 1967, Lima. International Geographical Union, Commission on the Arid Zone/UNESCO. 38 p. (mimeo)

Fuenzalida V., H.
    1963   Clima: Las regiones áridas de Chile. Comité Chileno para el Estudio de las Zonas Aridas, Informe Nacional sobre las Zonas Aridas de Chile, p. 13–21. Santiago.

Galvez Z., A.
    1966   Estudio en calentadores solares de agua. Universidad del Norte, Antofagasta, Revista 1:29–38.

Garcia, C.
    1967   Guerra pesquera no declarada. In Ercilla, No. 1662, April 12, p. 19.

GLASSNER, M. I.
　1970　The Rio Lauca: Dispute over an international river. Geographical Review 60(2):192–207.

HERNANDEZ PARKER, L.
　1966　Arica: La lección del optimismo. In Ercilla, No. 1315, August 3, p. 16–18.

HODGES, C. N.
　1969　A desert seacoast project and its future. In W. G. McGinnies and B. J. Goldman (eds), Arid Lands in Perspective, p. 119–126. American Association for the Advancement of Science, Washington, D.C.; University of Arizona Press, Tucson. 421 p.

INOSTROSA, J.
　1963　Las tierras de las momias en cuclillas. In Ercilla, No. 1452, March 20, p. 14–15.

INSTITUTO GEOGRAFICO MILITAR
　1966　Atlas de la República de Chile. Instituto Geográfico Militar, Santiago. 121 p.

JAMES, P. E.
　1969　Latin America. 4th ed. The Odyssey Press, New York. 947 p.

KREBS, J. S.
　1970　Land, sea, life and the 'Niño' phenomenon on the west coast of South America. University of Colorado, Boulder. (unpublished paper)

MEIGS, P.
　1969　Future use of desert seacoasts. In W. G. McGinnies and B. J. Goldman, eds, Arid Lands in Perspective, p. 103–118. American Association for the Advancement of Science, Washington, D.C.; University of Arizona Press, Tucson. 421 p.

MUÑOZ ESPINOSA, H. R.
　1969　Algunas características de las nieblas en Portezuelo, Antofagasta, Chile. Universidad del Norte, Antofagasta, Revista 2(3/4). 28 p.

OSSANDON ESTAY, O.
　1967　The reappraisal of the Pampa del Tamarugal: The Tamarugo Cattle Association — Conchones Experimental Center. Paper presented at Simposio sobre Desiertos Costeros, 1967, Lima. International Geographical Union, Commission on the Arid Zone/UNESCO. 7 p. (mimeo)

RUDOLPH, W. E.
　1963　Vanishing trails of Atacama. American Geographical Society, New York. 93 p.

SAA VIDAL, R., E. J. WEBER, AND R. ANTONIOLETTI
　1967　Mapa de distribución de la población de Chile (Escala 1:500,000, Censo 1960). Instituto de Investigación de Recursos Naturales, Santiago.

SALINAS MESSINA, R.
　1967　Un ejemplo de valoración pesquera del litoral norte de Chile: La industria de harina de pescado de Iquique. Asociación de Geógrafos de Chile, Boletín 1(2):8–15.

SANTANA AGUILAR, R.
　1970　Man and the land in the Atacama. University of Colorado, Boulder. (unpublished lecture)

SCHMIEDER, O.
　1965　Geografía de América Latina. (Traducción de P. R. Hendricks Pérez y Hildegard Schilling.) Fondo de Cultura Económica, México. 645 p.

SCHNEIDER, H.
　1967　Indices de aridez y su aplicación en el Norte Chico. Cuarto Encuentro Nacional de Geografía, Antofagasta, Chile. 8 p. (mimeographed)

TIXERONT, J.
　1956　Water resources in arid regions. In G. F. White, ed, The Future of Arid Lands, p. 85–113. American Association for the Advancement of Science, Washington, D.C., Publication 43. 453 p.

UNIVERSITY OF ARIZONA, ENVIRONMENTAL RESEARCH LABORATORY
　1969　Water, food and power for desert coasts: An integrated system for providing them. Orientation Note 1.

URIBE ORTEGA, G.
　1967　Algunas consideraciones sobre la geografía industrial del Norte. Cuarto Encuentro Nacional de Geografía, Antofagasta, Chile. 4 p. (mimeographed)

WALTON, K.
　1969　The arid zones. Aldine Publishing Co., Chicago. 175 p.

WINNIE, W. W., JR.
　1965　Communal land tenure in Chile. Association of American Geographers, Annals 55:67–86.

CHAPTER 16

THE FISH-MEAL INDUSTRY OF IQUIQUE

Rolando Salinas M.
Department of Geography, Catholic University of Valparaiso, Chile

The "Norte Grande" corresponds to the Chilean coastal desert, which is an extension of the Peruvian desert; it is a narrow stretch of land which runs along the Andes, between latitudes 17° 30′ and 26° 30′ South. Its surface extends over 222,512.7 square kilometers, and it represents 29.39 percent of the continental territory of Chile (Chile, Dirección General de Estadística y Censos, 1964). The landscape is one of unrelieved desert, with a total absence of rains and vegetation, apart from the Cordillera to the north of the Loa River.

The massive exploitation of the nitrate mines, and of the gold, silver, and copper found in the North, gave rise to weak and dispersed nuclei of population, in mining camps and processing plants (oficinas), most of which by the end of the 1960s had been abandoned. The ports of the region owed their existence to the mining activity and thus were subject to the economic instability of their hinterlands. The census of 1895 showed a population of 37,647 inhabitants in the ports of the province of Tarapaca (Iquique, 33,031; Pisagua, 3,635), and 52,103 inhabitants in the oficinas and camps (for example: Huara, 7,730; St. Catalina, 4,649). In the province of Antofagasta the coastal population reached 23,007 inhabitants, and in the inland region, 21,078 inhabitants (Espinoza, 1907).

During a more recent stage a different development of the coast came about; the ports began to depend fundamentally on local rather than regional conditions. Arica (56,000 inhabitants in 1964) has undergone a rapid growth coupled with an industrialization that lacks a regional base but depends on a national monopoly for the assembly of automobiles and television sets. Antofagasta (108,000 inhabitants in 1964) has achieved both greater regional stability and a greater functional differentiation. This city is the most stable urban center of the unstable Norte Grande. Iquique (90,000 inhabitants in 1964) represents the only attempt at utilizing the northern coast by exploitation of the fishery resources. These three examples do not, however, escape the weakness and the instability that characterize the economy of the Chilean desert.

The Fishing Boom and the Re-emergence of Iquique

Development and Governmental Initiative

In 1960 the difficult socioeconomic situation produced in Iquique by the gradual disappearance of the nitrate industry resulted in a crisis. Only one of the 99 nitrate plants that functioned in 1912 remained, and it was supported by the state in order to employ a total of 1,300 persons. This serious situation was characterized by growing unemployment, paralyzed trade, an inactive port, social unrest, and a general lack of resources for development, since the hinterland of Iquique lacks agricultural possibilities and exploitable minerals other than nitrate. A plan is under study for reforestation and for exploitation of the flocks of wool-producing sheep found in the pampas of Tamarugal.

The success of the fishing industry of the Peruvian coast, and the demand for fish meal in the world market, suggested that the solution might be found in the encouragement of this activity on the northern coast. Iquique would be the center, in order to take advantage of the resources of the sea, the labor force of unemployed men and women, and the existing infrastructure. A department of the State Planning Corporation (CORFO, or Corporación de Fomento de la Producción) drew up a plan for the development of a fishing industry based on the following points:

1. Cheap and abundant manpower existed, as a consequence of the unemployment produced both by the closure of the nitrate mines and the natural growth of Iquique.

2. Iquique lies at the geographic center of the fishing zones on the northern coast where the pelagic species which live near the sea surface are most abundant: tuna, bonito, hake, jurel, and anchovies. The oceanographic conditions of the Chilean desert coast are similar, but not identical, to those of the Peruvian coast.

3. An infrastructure already existed, inherited from the town's past as an important nitrate port and as the

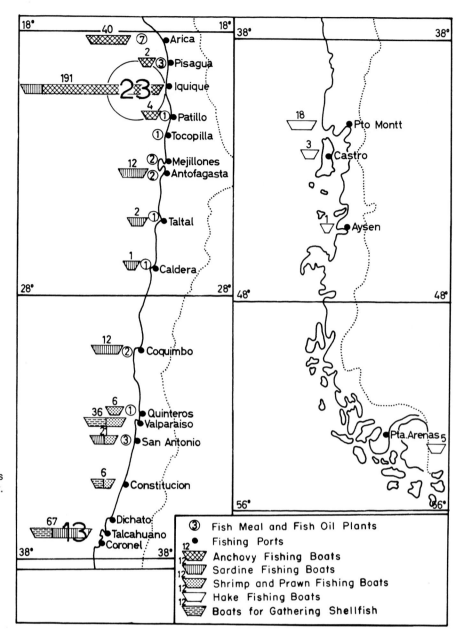

Fig. 16-1. Locations of fish-meal factories and fishing fleets along the Chilean coast.

capital of the province. The town offered a good artificial harbor, which was scarcely used; the necessary space and equipment for a port; a diesel-powered electric plant; a supply of drinking water; roads, railroad, and an airport. Furthermore, it is situated on the maritime route to Europe, North America, and Japan — the main markets for fishing products.

4. Capital was present in the region; entrepreneurs, accustomed to the risks of mining, did not hesitate to invest their capital in fishing activities. However, the largest part of the capital came from the central region of the country, as a result of the interest created by the favorable conditions provided by the state. The capital was almost entirely Chilean; only in three companies was there foreign capital associated with the national capital.

5. A policy of credit, the goal of which was to help the professional fisherman, and the fishing and industrial shipbuilders, with medium- and long-term credit and low interest loans, would make it possible for them to buy ships, fishing equipment, and industrial machinery, and to build factories. In 1963 the financial aid provided by CORFO in the form of loans, credit for contractors, and endorsements rose to 25,567,500 escudos (U.S. $9,065,200) (Nestler and Berner, 1965).

6. A tax-free industrial zone was created for Arica, Pisagua, and Iquique, permitting the free importation of ships, fishing equipment, machinery and miscellaneous other items (law 12,937). The taxes are reduced or canceled, as are, for example, purchase and sales taxes, but at the same time the businessmen who profit from these benefits must reinvest 75 percent of their profit until 1973.

7. A discount of 20 percent was allowed on the F.O.B. price of the tonnage of meal being exported, and a discount of 30 percent on the raw material of Chilean

origin included in the product (Law no. 12,937). Both discounts benefit only the manufacturers; in 1965 CORFO owed 19,422,463 escudos (approximately 4 million dollars) to the meal factories of the northern coast (Cámara de Diputados de Chile).

8. Industrial zones were created, a fundamental requirement for the northern ports, which are situated on narrow coastal plains. At Iquique an industrial zone was created with 110 lots of approximately one acre each, next to the sea, and with a supply of drinking water, a sanitary system, electricity, and roads. By 1963 this zone was filled with industries, and as the demand for spaces continued, new lots were created in the southern sector of the city.

9. The building of shipyards was encouraged, which resulted in two in Iquique and one in Antofagasta.

10. Construction began on a fishing port which would be able to serve a fleet that, according to estimates, would include 320 ships of 140 tons by 1965. The basic work on this service port was completed in 1965.

11. An industrial pilot plant was planned, which would embrace the three main aspects of the industrialization of fishing: canning, freezing, and the production of meal and oil. This plant was to be built in order to explore processes for the diversification and increasing of exports. To this end CORFO created a branch, Fishing Enterprise of Tarapaca, S.A., in which private capital could be invested.

12. The Fisheries Institute Development (Instituto de Fomento Pesquero, Santiago; IFOP) was created, in association with the Food and Agriculture Organization (FAO), to encourage the research necessary for the development of the fishing plan. Another goal of this institute was to expand technical assistance and oceanographic research by both foreign and national specialists.

Multiplication of Fish-Meal Factories

Before 1960 only three fishing companies and one plant that processed products obtained from whales, exploited the maritime resources of the northern coast. Production consisted mainly of canning for the national market, along with limited exportation of canned and frozen products, meal, and whale oil.

In 1960, in spite of the facilities offered by the fishing plan, only two factories for fish meal were installed in Iquique. Their production capacity was between 10 and 35 tons an hour. The drop in the price of fish meal on the international market (from $140 to $57 a ton), produced a crisis in the Peruvian fishing industry and interrupted the plans of the Chilean industrialists. In 1961, with the increase of the price to $100, the "fish meal fever" began. The number of boats and factories increased rapidly, as did production and export. This was a real boom, the effects of which were soon visible in Iquique, since the activity that until now had been dispersed around the central coast, became concentrated in this town. In 1966 there were 60 factories in Chile

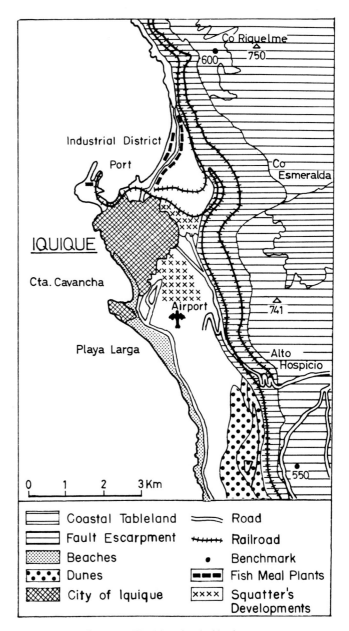

Fig. 16-2. The hinterland of Iquique.

producing fish meal, lobster, or whale products; 40 of these were on the northern coast between Arica and Taltal (with 63 percent of the productive capacity concentrated on that coast), while 23 were in Iquique (table 16-1).

Considered from the point of view of productive capacity in tons per hour, the factories of Iquique and of the Norte Grande are of medium size, as a consequence of the fragmentation of capital produced by the proliferation of enterprises (table 16-2).

In less than a decade, the production of fish meal underwent a rapid increase and became geographically concentrated in Iquique. In 1958 the northern coast produced only approximately 35 percent of the national total, and this percentage was provided by Arica; in 1959, the contribution of the northern coast to the total Chilean production increased to 50 percent, and in 1960 to 70 percent. From 1962 onward, Iquique was definitely at the head of the national production of fish meal, contributing more than 50 percent, and thus becoming a town closely associated with the prosperity or crisis of industrial fishing (table 16-3).

Because of the desert climate, no buildings are necessary, and a large part of the machinery remains in the open. The same applies to the storage of fish meal. This has facilitated the swift establishment of factories and reduced the investment costs. Every factory occupies approximately one acre of the industrial zone, and most of them are built facing the sea. This makes it possible to pump the anchovies directly from the boats to the depots. The absence of storms and the calm sea simplify these tasks, and due to these favorable natural conditions the location of the factories is not dangerous, in contrast to the rest of the Chilean coast. Mechanization is well developed, and the fact that the works were started after 1960 has allowed the use of modern machinery and facilitated the adoption of the technical innovations already proved successful in other countries.

The manufacturing process is simple and takes only half an hour, approximately. The anchovy arrives at the warehouse and must be quickly processed in order that the work be completed easily. The fish is pressed to remove all the liquid; afterward it is steamed, then it is pressed again, and the juice thus obtained is filtered; the oil and the tail water are obtained. The tail water gives a concentrate which is added to the *cake,* enriching the meal with proteins. The *cake,* which is the fish after it is cooked and pressed, is transferred to the mills where it is crumbled into little pieces, then to two driers of different temperatures, and eventually to another mill where it is pulverized. Finally it is packed in bags of paper or hemp cloth with a capacity of 100 pounds each. This normally takes place a few days afterward, to avoid the risk of fire, because it generates heat and sometimes burns spontaneously. For this reason insurance companies demand a waiting period of 21 days before shipment. One ton of fish meal is obtained from every five tons of anchovies. The meal is chemically controlled, and attempts are made to keep a standard quality.

Fishing and the Fishing Fleet

The creation of a fishing fleet was required in order to assure the supply of raw material to the factories. The quick growth of this fleet was encouraged by permission to buy ships abroad, and by the construction of dry docks, which represented a new industrial activity for Iquique. Two dry docks were established in Iquique and one in Antofagasta. The Marine Construction and Design Co. of Seattle, Washington, U.S.A., installed in Iquique a branch called Astilleros Marco Chilena S.A.I., in order to build, repair, and maintain fishing boats. This company started the mass construction of metal boats of a loading capacity of 110 tons, and in 1963 of 140 tons. The monthly production was five boats, a third of the national production of a monthly sixteen in 1963. This was the most productive year, with 52 units. They operate the biggest dry dock in the country; in 1965, 96 of the 237 boats working in Arica and Iquique came from this dry dock.

The growth of the anchovy fleet was necessarily rapid both with regard to the quantity of boats and in their capacity in tons. In 1960 it corresponded to 16.7 percent of the national industrial fishing fleet, and in 1965 it comprised 52 percent. In 1960 there were 24 boats with a capacity of 1,429 tons; in 1963 the figure was 106 boats with 8,781 tons; and in 1965, 223 boats with 20,858 gross tons. The industrial fishing fleet of the northern coast is highly specialized and shows great geographic concentration, for 223 of the 251 fishing boats (that is, 80 percent) work exclusively in anchovy fishing. The region where they fish is the coast between Arica and

TABLE 16-1

Fish-Meal Factories and Productive Capacity in Chile in Representative Years

Place	1961 Factories (no.)	1961 Prod. (tons/hr)	1962 Factories (no.)	1962 Prod. (tons/hr)	1966 Factories (no.)	1966 Prod. (tons/hr)
Arica	2	?	3	?	7	305
Pisagua	0	0	0	0	3	69
Iquique	3	?	5	?	23	785
Other ports	–	–	–	–	7	140
Norte Grande Total	5	98	8	180	40	1,301

SOURCE: Nestler and Berner, 1965; Tilic, 1966.

TABLE 16-2

Fish-Meal Factories According to Productive Capacity, in Tons per Hour, in 1966

Ports	3.5–20	25–50	51–75	90–120	Total
Arica	2	2	2	1	7
Pisagua	2	1	0	0	3
Iquique	7	13	2	1	23
Other ports	5	2	0	0	7
Total north coast	16	18	4	2	40

SOURCE: Tilic, 1966.

TABLE 16-3

Fish-Meal Production, in Tons, in Coastal Ports and Other Ports of Chile

	1958	1961	1964	1965	1966
North Coast					
Arica	6,511	31,586.0	41,308.7	22,503,4	65,405.0
Iquique	654.6	13,311.0	108,222.5	48,228.0	123,943.0
Total North coast	7,476.9	46,131	158,313.6	72,837.4	188,448.0
Total other ports	11,569.1	12,121	16,368.9	20,309.0	32,886.0
Chilean total	19,046.0	58,252.0	174,682.5	93,146.4	221,334.0

SOURCE: Nestler and Berner, 1965; Tilic, 1968; I.F.O.P.; Chile, Ministerio de Agricultura, 1965.

Fig. 16-3. The industrial section of Iquique. The fish-meal factories are centered here.

Fig. 16-4. A fish-meal factory, Tarapaca S.A., in the port of Iquique. In the background are the dry docks.

Fig. 16-5. The port of Iquique, showing the business and fishing sections, with the dry docks. In the center is seen the machinery used for loading bulk nitrate, not in kegs.

Tocopilla. The key ports of this fleet are Arica with 40 boats, and Iquique with 183.

The anchovy fleet is characterized by the fact that more than 80 percent of it was made in Chile, with the rest coming from Peru, the United States, Norway, and other countries. Eleven fishing boats of United States registry are still working in Iquique. Ten boats are of wood construction and 213 of steel; all are relatively new and modern. In December of 1965, 104 were less than two years old, and 123 were between 3 and 5 years old, while only 4 were more than 6 years old. All the boats are propelled by diesel motors and are equipped with radiotelephone, a power block to bring in the net, a metal auxiliary boat, nylon nets, and some accommodations for the crew. More than half the boats are provided with echo sounder and with suction pump for the cargo of anchovies. In general, the length varies between 20 and 25 meters (in 58 percent of the boats), and the weight is from 80 to 140 gross tons. The total capacity of the holds of the ships in 1965 was 30,185 cubic meters (Tilic, 1966a).

The total value of the anchovy fleet was estimated in 1965 at 90,000,000 escudos. The value of a fully equipped ship built in Iquique was approximately $90,000 in 1964, and this cost influenced the great fragmentation of ownership of the fleet. In 1965 the 223 units of this fleet were distributed between 74 shipowning companies that possessed between 1 and 20 units each. Only one possessed as many as 20 ships, some of which worked from Arica and some from Iquique. This multiplicity of ownership of the fleet was greater in Iquique in 1965 where 183 ships belonged to 63 companies. In Arica, on the other hand, 40 ships were owned by only 11 companies. The independent shipowners, who must sell the anchovies at the prices fixed by the factories, owned 44 percent of the ships and 45 percent of the total capacity of the fleet; the rest were owned by the factories. The independent shipowners increase the number of owners of the fleet because they control few ships, while the industrialists tend toward concentrating ownership since they own fleets of several units. In 1965, 41 independent owners worked in Iquique and 4 in Arica (Tilic, 1966a).

The anchovy fishing is carried out according to a system known as the *cerco* or encirclement; it is practiced on the coasts where water is shallow, as is the case on the northern coast. The anchovies lie in schools which are visible as stains on the surface of the water, and this allows the use of planes as a means of detecting them. The ships used are of the *purse seiner* type, where the bridge and the machinery are placed in the prow, leaving enough space in the poop to carry the net and the auxiliary boat, or *panga,* which is used to spread the net around the school. The net is then brought in by means of the power block of the ship; when the net is brought in, it closes from below and is thus transformed into a large sack that brings the anchovies to the side of the ship. From there they are sucked up by a pump that carries them to the hold.

Fishermen and Workers

The quick development of the fishing and industrial activities amply fulfilled the immediate goal of reducing unemployment among the inhabitants of Iquique and the province of Tarapaca. The *pampino,* the miner who worked in the extraction of nitrate, became either a worker in the fish-meal factories or in the dry docks, or a fisherman. The process of adaptation was swift. The first task was the construction of the factories.

The factories did not have problems finding workers, since the work involved is simple and mechanized. But the fishing companies did have problems finding seagoing veterans for the boat crews, since the fishermen of Iquique were few, and there was no maritime tradition. It was necessary to improvise captains for fishing boats. The Gobernación Marítima gave courses, but the improvisation had as a consequence, nonetheless, a loss of equipment because of the inexperience of the crews and the captains in the process of fishing. The type of fishing, near the seashore, and near the surface of the water, in a region of stable climatic conditions, calm seas, and no storms, facilitated the adaptation and work of these inexperienced crews, a task that would have been impossible on the coasts of the central and southern parts of the country.

The intense demand for manpower also affected the women, who were employed to repair nets, thereby earning a higher salary than they would have received working as maids. The workers, fishermen, and clerks came from the north; the technicians, administrators, and business executives came from the central region of Chile. The need for specialized people obliged the companies to pursue a competitive hiring policy, each company trying to offer the best salaries and working conditions in order to win from the other whatever specialized personnel were available.

The salaries and daily wages paid in this industry are high as compared to those paid for other activities in the same region; however, the work is of a seasonal character, and there is a certain degree of instability in it. In normal years there is a period of intense activity and good profit that lasts from December to May; there is also a time of little activity and profit, which brings unemployment for periods varying from two to five months (from June to October). Only the skilled personnel are retained throughout the year by the companies. The personnel who work on the ships earn a percentage fixed on a basis of the fish tonnage obtained; in the fishing season, a captain earned approximately 154 escudos per day in 1964 (U.S. $45.00)); the factory workers earned an average of 7.50 escudos ($2.00) per day, which amounts to 225 escudos ($64.00) monthly, plus payment for extra hours and family allowances. The salaries are fixed yearly, according to the law; they are agreed upon by the companies and the labor syndicates jointly. In parliament, salaries have been criticized as being too low in comparison with the profit and discounts received by the companies. The technical and administrative personnel receive high salaries and participate in the profits made.

The fishing activities and the fish meal factories are an important source of employment. In Iquique in 1964, the Marco dry docks alone employed 350 workers plus 100 clerks and technicians. In 1965 the anchovy fleet employed 2,210 crew members; the 33 fish-meal factories in Arica, Pisagua, and Iquique employed 4,841 workers, clerks, and technicians, 2,665 of which belonged to the factories of Iquique (CORFO, Oficina de Iquique).

Markets and Exports

The first goal of the production of fish meal and oil extracted from fish was the exportation of these products. The national market consumes approximately 15 percent of the annual production of fish meal. An increase in consumption has followed the expansion of the feed industry for poultry and cattle: 9,980 tons in 1957, to 26,217 tons in 1964. The national consumption of oil reached 3,188 tons in 1965, that is, 25 percent of the total production (Tilic and Purcell, 1966).

The increase in exports was sharp, and in 1963 Chile occupied the fifth place among the world's countries exporting fish meal (table 16-4). The demand for fish meal in the world market since 1961 has supported an average price of U.S. $100 F.O.B. per metric ton. This price has had a slight tendency to rise; in 1965 the price per ton reached $140 F.O.B. The oil, obtained as a byproduct during the processing of fish meal, has a good price on the international market: $170 to $180 F.O.B. per metric ton.

These circumstances transformed this industry into the fourth-ranking source of foreign exchange for Chile in 1964; the fish meal and oil earned U.S. $17,787,100 of the total $22,073,000 earned by the export of the marine products. Arica and Iquique monopolize the production, and they enjoy the advantages of their leading role, a good location on the maritime routes to Europe and North America, and a protectionist policy.

The main market for Chilean oil and fish meal is Western Europe, which in 1965 bought 80 percent of the exports: Holland, 37 percent; West Germany, 29 percent; and Belgium, 14 percent. In South America, Argentina bought 8 percent, and the remainder was sold in small quantities to the United States and Japan (Chile, Ministerio de Agricultura, 1965, pp. 4–6).

The Resurgence of Iquique

The improvement of Iquique was rapid and manifested itself in the growth of the population, changes in the urban characteristics, reactivation of trade, establishment of banks, and reactivation of the harbor.

While from 1930 to 1960, as a result of the nitrate crisis, there had been a constant emigration from Iquique, from 1961 on Iquique became transformed into the main center of immigration into the region. In 1964 during the big fishing boom the population reached 70,000. The census in 1930 was 46,458; in 1940, 38,094; in 1952, 39,576; in 1960, 50,824 (Dirección General de Estadística y Censos).

At the beginning, the immigrant population was mainly composed of men seeking employment, which gave Iquique the appearance of a pioneer town for a time. Once they had obtained a job, the men brought their families. The lack of housing gave rise to a belt of shanty towns, locally known as *poblaciones callampas*. Therefore it was necessary to provide the new dwellings with elementary sanitary services, as well as water and electricity.

Commercial activities of the town were renewed with increased vigor. The heavy demand by the newly immigrated workers led to an expansion of stocks of merchandise and to the creation of a system of installments and credit sales. At the same time, the shops were enlarged and modernized so that the growing mass of buyers could be efficiently served. Another sign of the commercial and financial boom in Iquique was the increase in the number of banks, from 3 in 1960 to 11 in 1965.

Important changes were introduced into the urban landscape with the establishment of dry docks, factories, the fishing harbor, and the creation of the fleet. All this lent to the town a new and highly industrialized appearance. The streets became filled with people, places of entertainment were busy in the evening, and shop windows became large and modern. The banks began to invest in the construction of four-story buildings for their own offices and for rent, thus contributing to the new look of the city center. Private and state companies began to build accommodations for workers and clerks. The municipality, with its increased revenues, became concerned with public works for urban renewal, and the state increased or improved health services in the city.

The Fishing Crisis of 1965

Extinction of the Anchovies

The constant and rapid increase in the amount of fishing was suddenly brought to a halt in 1965 by the disappearance of the anchovies, an event that was first noticed in March 1964 and which became more acute

TABLE 16-4
Exports of Fish Meal in Quantity and Monetary Value

Product	1959		1964		1965	
	Tons	Thousands U.S. $	Tons	Thousands U.S. $	Tons	Thousands U.S. $
Fish meal	13,589	1,656	143,330	15,748	66,203	8,014
Oil	110	36	13,713	2,039	7,218	1,242
Total export	13,699	1,692	157,043	17,787	73,421	9,256

SOURCE: Nestler, 1966.

in 1965 toward the month of November. This phenomenon was not observed on the Peruvian coast.

Nobody had foreseen a crisis due to lack of fish. The only antecedent had been the financial crisis of the Peruvian industry, as a result of the decrease of the price on the world market (Tilic, 1968; Dollfus, 1964), a situation that was not feared in Chile because the state protected and supported the industry. CORFO used all possible resources to tackle this situation, which once again threatened the stability of the fragile economy of the Norte Grande in general and of Iquique in particular. Extraordinary loans were granted in order to keep boats and factories functioning at a reduced rate in order to avoid massive unemployment; a moratorium was granted on the payment of the credits that shipowners and manufacturers owed to CORFO; and a postponement and consolidation of credits owed to intermediaries, foreign contractors, commercial banks, and others, was worked out (Nestler, 1966). In order to prevent bankruptcy, purchases of future consignments were made.

Two possibilities were considered in attempting to explain the disappearance of the anchovies: (1) Was the anchovy off the Chilean coast a limited resource that was gradually being exhausted? (2) Was the scarcity of this fish a cyclical phenomenon?

One trend of thought was that the scarcity of anchovies implied the exhaustion of a limited natural resource that had been subjected to intense exploitation. The symptoms of such extinction would be visible too in Peru, where also the anchovies would be less plentiful each year. This would make it imperative to forbid anchovy fishing for a period of time. The natural resources of the Chilean shore are more limited than those of the Peruvian shore, where anchovies can be found along 1,000 miles of the coast, from north of Chimbote to Ilo in the south. Along the Chilean coast, on the other hand, anchovies can be found only for 300 miles, from Arica to Antofagasta. The continental platform in Peru has a width of 60 miles, while in Chile it is only 11 miles across. In Peru the haul of fish reaches six to eight million tons of anchovies, while in Chile the catch is about one million tons at most (Bahamonde, 1966).

The anchovy lives near the surface in the relatively cold water of the Humboldt current, which is rich in oxygen and plankton. However, in occasional years the conditions change, and the temperature of this water rises from 17 to 23 degrees, as happened in 1965. The nature of this phenomenon is not clear. Some believe it is due to the Niño Current when tropical surface waters advance the length of the coast toward the south, between the Humboldt current and the shore. Others believe that there are cycles which last from 10 to 12 years, and which are related to the absence of the south wind necessary to disperse the warm surface waters that accumulate along the shore of the northern coastal region and prevent the up-welling of the deeper and colder waters. These changes oblige the anchovy to seek deeper waters or to leave the shore in search of more favorable conditions. These two cyclical theories are not contradictory, but complementary. The reappearance of the anchovy in 1966 confirms the belief in a cyclical process (Fuenzalida V., 1965).

TABLE 16-5

Evolution of Fishing in the North, According to Metric Tons Caught

Ports	1961	1962	1963	1964	1965	1966
Arica	177,917	219,973	218,551	247,185	142,071	352,491
Iquique	82,176	221,468	321,282	676,947	293,283	685,849
Other ports	13,226	11,648	11,432	15,666	19,982	52,489
Total fishing	273,319	453,089	551,265	939,798	455,336	1,090,829
Total anchovy fishing	259,088	439,652	539,451	934,000	438,495	1,038,385

Quick Growth and Improvisation

The need to solve the problem of unemployment in Iquique resulted in the creation of an excess of facilities as well as improvisation, since what was most urgently needed was the immediate creation of new jobs. Added to this, strong political pressures and those of economic interests prevented the adoption of reasonable limitations in establishing the fishing plan for the north.

The excess privileges granted to Iquique and the lack of control unbalanced the fishing industry of the Chilean coast, concentrating most of the fleet and the fish-meal factories in Iquique. Ships and factories that exploited sardine fishing for fish meal in the central ports (Dichato and Coronel) were transported to the north, and this affected the productive capacity of these ports and reduced the exploitation of rich sources of fish.

It is evident that the development for fishing and manufacturing was not regulated according to the possibilities of the natural resources, the potentials of which were unknown. In 1963 CORFO foresaw the problem of the geographic concentration of the fleet and the factories and tried to stop further expansion by refusing to grant financial aid; despite this, in 1965 there were ten new factories, either in the planning stage or under construction. There also was a noticeable decrease in the average number of hours that each plant operated during a year, adjusted to the available raw material; there was a decline from 3,970 hours in 1962 to 1,255 hours in 1964, considered a normal year. A similar decrease was noted in the average annual haul per ship; this fell from 8,600 tons in 1962 to 5,000 tons in 1964. In 1965 the averages dropped to 395 hours for the factories and 2,000 tons for the ships. These latter averages are below the break-even point that would allow the fishing or manufacturing companies to operate without losses. The break-even points are estimated to be 1,200 yearly hours for factories, and six to seven thousand tons of fish per year, for ships (Cámara de Diputados de Chile).

This change in fishing conditions was not foreseen. The ships were built for surface and coastal fishing only, and they do not have the necessary equipment to catch fish other than anchovies. Their power block can operate only to a depth of 30 fathoms; the ships are not suitable for fishing on the high seas, and they have been criticized for their lack of seaworthiness when the weather is stormy. The factories have only the equipment needed to produce fish meal and oil, and in 1965 only 3 of the 23 factories of Iquique had a diversified production: fish meal, oil, and canned and frozen products. A sort of monoproduction predominates.

The concessions granted by CORFO in Iquique, plus the generous credit terms of 50, 100, and even 300 percent of the capital invested, favored an uncontrolled expansion of the fleet and the industrial plants on the basis of debts. Companies of powerful capitalists were given credits similar to those granted to small owners of limited resources, without considering that the former did not need the help of the state to carry out their investment.

The crisis clearly illustrated the lack of financial ability, on the part of the industrialists and shipowners, to meet yet the economic needs of a long period of stagnation. The inexperience of many of the people involved in this activity did not allow them to confront the situation in a more organized fashion. The operating costs were considerably increased by the economic concessions granted to the technical and administrative personnel in lieu of salaries during the "normal period," and this contributed to the financial weakness of the companies (Nestler, 1966).

Reorganization and Rationalization of Fishing Activities

The crisis originating from the scarcity of anchovies in 1965 underlined the need to adopt measures that would stabilize the fishing industry on the northern coast. The technical and economic aspects should be emphasized. It was necessary to correct the defects of the original improvisation, and CORFO applied several measures to this end.

Financial and structural changes. A first step was to promote the absorption of the fishing companies by the industrial companies, because the shipowners made up the sector most seriously affected, and the one which had the most unreliable income. This measure would lead to the elimination of the independent shipowner, whose role would be transferred to the financially stronger industrialist (Nestler, 1966). The industrialist is better placed economically, since he receives subsidies for the use of home-produced raw material, while the shipowner does not.

A second proposal, the horizontal merging of the industrial companies in order to form more solid economic groups, was also envisaged. This would permit a more effective use of the installations, would reduce the administrative expenses, and thus would reduce the costs. In order to promote the mergers, it has been agreed to exempt them from taxes of any nature; moreover, they are authorized to use the costs of the mergers to fulfilling in part their obligation to capitalize 75 percent of their profits. As a result of these measures, or as a result of bankruptcy, by March 1967 the 41 independent shipowners of Iquique had been reduced to 12, and the 23 industrial enterprises to 18.

Thirdly, it was planned for a marketing corporation to be formed by CORFO, and an association of fishing industrialists. This corporation would sell the fish meal and oil on the world market, obtaining better prices while avoiding competition. It would also establish an organization to reduce the costs of shipment and insurance.

Measures to assure geographic dispersion, and the adaptation of the fleet and the factories to the possibilities of using other raw material. It was decided to prohibit the establishment of new factories and the expansion of the ones that already existed in Arica, Pisagua, and Iquique; also to forbid any increase in the fleet.

Secondly, it was planned to reduce the fishing fleet of Iquique in order to establish a balance between the fishing capacity of the fleet and the quantities of raw material available under normal fishing conditions. This would favor the transfer of ships to other fishing areas south of Iquique and establish a control of ship departures.

Third, it was proposed to encourage the removal of some factories to other ports of the Norte Grande, which offer both possibilities for anchovy fishing and sites suitable for establishing factories. These ports include Tocopilla, Mejillones, Antofagasta, and Taltal. Between 1964 and 1966 seven factories were established at these ports.

Finally, it was decided that those companies formed by the merging of shipowners and industrialists, or by the merging of manufacturers, should establish at least two bases on different parts of the coast so that the functioning of these companies would not depend entirely on the existence of good fishing in a limited area. In 1966 six industrial companies in Iquique merged with four in Arica and one in Tocopilla (I.F.O.P.).

Measures to diversify fishing and the processing of fish. In spite of the fact that this suggestion was not mentioned by CORFO among the measures it considered to be necessary, we believe it is desirable to build a more versatile type of fishing boat, one that could be used both for fishing near the shore and in the open sea, for deep sea or surface fishing, and for catching anchovies or other types of fish used for canning and freezing.

However, CORFO has proposed to diversify the lines of production so that besides producing oil and fish meal, factories would produce also pickled fish for popular consumption, and canned, frozen, and smoked fish. In the late 1960s, only 3 of the 23 factories of Iquique had diversified production.

Technical measures and measures to improve fishing and industry. It has been suggested that investigation be made of industrial systems that would permit the use of proteins, fragments of anchovies floating in the water of the pump, and blood. These products have until now been wasted by many factories.

Research also is being carried out into better ways of managing the catching of fish. Among these are the strict observance of a period in which it is forbidden to fish, the application of better administrative systems and systems to control costs, the incorporation of new technological processes, and the training of specialized personnel (Nestler, 1966). All these measures under study would guarantee the stability of this important industry on the arid north coast of Chile.

Bibliographic References

BAHAMONDE, N.
    1966   El mar y sus recursos. *In* CORFO, Geografía económica de Chile, Apéndice, p. 82, 84, 86. Editorial Universitaria, Santiago. 369 p.

CHILE, CAMARA DE DIPUTADOS
    1965   Acta de sesión sobre la crisis pesquera de Iquique. La Nación, 27 Nov. 1965. Santiago.

CHILE, DIRECCION GENERAL DE ESTADISTICA Y CENSOS
    1956   XII Censo de Población, 1952, I: Resumen del país, p. 45. Edit. Gutemberg, Santiago.
    1964a   XIII Censo de Población, 1960, ser. B: Tarapaca, Antofagasta, y Atacama. Imprenta Dirección General de Estadística, Santiago. 3 vols.
    1964b   Población del País: Características básicas de la población, p. 3. Santiago. 66 p.

CHILE, MINISTERIO DE AGRICULTURA, DEPARTAMENTO DE PESCA
    1965   Informaciones estadísticas sobre pesca. Resumen anual, no. 68, p. 3–27, 29, 95–101. Santiago.

DOLLFUS, O.
    1964   La pêche et l'industrie de la farine de poisson au Pérou. Cahiers d'Outre Mer 17(68):370–385.

ESPINOZA, E.
    1907   Geografía descriptiva de la República de Chile, p. 82–105. Santiago.

FUENZALIDA V., H.
    1965   El mar y sus recursos. *In* CORFO, Geografía económica de Chile, texto refundido, p. 273–286. Editorial Universitaria, Santiago. 885 p.

NESTLER, J.
    1966   Industria pesquera. *In* CORFO, Geografía económica de Chile, Apéndice, p. 185–187, 192. Editorial Universitaria, Santiago. 369 p.

NESTLER, J., AND C. BERNER
    1965   Industria pesquera. *In* CORFO, Geografía económica de Chile, texto refundido, p. 565–570. Editorial Universitaria, Santiago. 885 p.

SANTIAGO DE CHILE, INSTITUTO DE FOMENTO PESQUERO (IFOP)
    1966   Lista de empresas pesqueras en Chile. I.F.O.P. Circular 3:2–3, 6–7.

TILIC, I.
    1966a   La flota pesquera en Chile, 1965. Instituto de Fomento Pesquero, Publicación 17:8–10.
    1966b   La industria de la harina de anchoveta en el norte de Chile. Instituto de Fomento Pesquero, Publicación 18.
    1968   Reseña del estado económico actual de la industria pesquera chilena y sus tendencias generales. Instituto de Fomento Pesquero, Publicación 35, Cuadros B-I y B-V. 52 p.

TILIC, I., AND A. PURCELL
    1966   Consumo y comercialización de la harina de pescado en Chile, Presente y futuro. Instituto de Fomento Pesquero, Publicación 13:3–4.

CHAPTER 17

THE PLIO-QUATERNARY CLIMATIC CHANGES ALONG THE SEMIARID SEABOARD OF CHILE

Roland Paskoff
Department of Geography, University of Tunis, Tunisia

The *Norte Chico,* literally "Little North," refers here to the region of Chile between 30° and 33° south, corresponding to the area of transition from the desert to the mediterranean climate. This is the semiarid part of Chile (fig. 17-1), in contrast to the *Norte Grande,* the extremely arid desert of northernmost Chile.

The *Norte Chico* is a transition zone between aridity and humidity, a zone in which the climatic changes since the Pliocene period have had a fair chance of leaving their mark. There is indeed no lack of traces of climatic fluctuations: fossilized mammals, remains of vegetation, paleosols, and fossil land forms and deposits. The coastal terraces where they have not been distorted by the neotectonic movements provide a good basis for a chronological essay.

The climate of the semiarid Chilean littoral is typical of the climate of the west coasts of continents in subtropical latitudes, with a cold offshore current (the Humboldt current) particularly strong here. The total precipitation varies between 400 millimeters in the south and 100 in the north; 85 percent of it occurs between May and August, as a result of the seasonal northward shift of the polar front system. The great interannual irregularity is explained by the marginal position of the area in relation to the tracks of the polar front perturbations. Relative humidity is always above 75 percent. The frequent coastal fogs are known here as *camanchaca*. Average annual temperatures (14°C) are low for this latitude and do not show great differences along the coast, so the isotherms run parallel to the coast. The annual temperature variation is slight — 6° or 7°. West winds are dominant (Almeyda, 1958).

A number of reasons permit the assumption that this region had a tropical climate during the Pliocene period, with an average annual temperature certainly higher than today, reasonably long rainy seasons, and the total precipitation greater.

A transgression during the Middle to Upper Pliocene period left fossil-bearing deposits along the coast (Horcón, Tongoy, Coquimbo–La Serena). Studies (Herm, 1969) of the fauna of these deposits suggest that the temperatures at the end of the Tertiary period were higher than they are now. The representatives of *Perna, Ostrea,* and *Anomia* — which then lived in the coastal waters of the Norte Chico — have since disappeared, although sometimes they are found further north in more temperate waters. Since that period cooling no

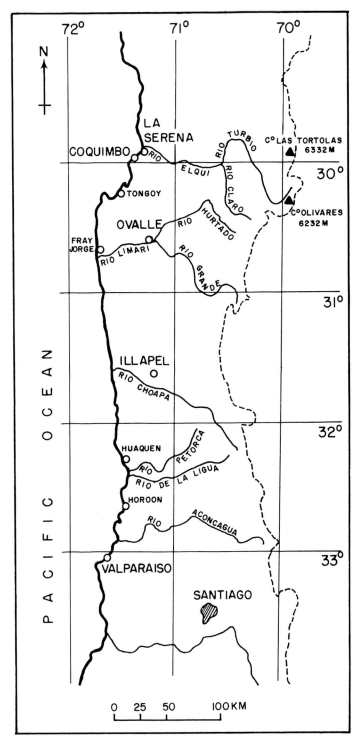

Fig. 17-1. The semiarid part of Chile.

doubt has been in progress, and the Humboldt current, if it existed then, was not as well developed as now. On the whole the climate must have been warmer and sunnier.

The Pliocene was distinguished first by dissection of the great valleys and later by their being filled in to great depth by alluvia deposited by strong floods. Even if the deepening of the valleys is linked to a new system of slopes resulting from the last great Andean uplift, and even if the sedimentation in its final stage is contemporaneous with a transgression, the tremendous effectiveness of exogenic forces is due primarily to particularly forceful and active climatic conditions. They may have no likeness in today's world, but one may imagine for the period of dissection a climate characterized by strong seasonal contrasts in rainfall and by very violent floods, while the deep accumulation could result from a trend of increasing dryness, which could have weakened the strength of the morphoclimatic system (Herm and Paskoff, 1966).

At the end of the Pliocene era the climate must have been semiarid. This would account for the prevalence of lateral erosion actions which, during that period, reworked the former surface of accumulation and even eroded into the base of the relief consisting of weathered granitic rocks. This type of erosion implies the intervention of volumes of water much greater than that available today (Paskoff, 1970).

The Quaternary climate in the Norte Chico is characterized by a double tendency: a decrease in temperature and a reduction in rainfall. But this trend toward dryness was interrupted by several humid relapses, apparently accompanied by a cooling of the atmosphere.

Some studies show that during the Quaternary period a continental faunal association, since vanished, existed in semiarid Chile. From the stratigraphic point of view all the sites belong to the Pleistocene, as far as one can judge, as it is often difficult to locate ancient collections exactly. Four types could thus be identified: *Mastodon, Macrauchenia, Equus,* and *Hippidion* (Hoffstetter and Paskoff, 1966).

Can one establish from these discoveries any paleoclimatic facts? The range of their validity is limited by two considerations. First, the study of the sites shows that these animals could survive the climatic changes of the Quaternary period. The existence of at least two types (*Mastodon* and *Macrauchenia*) until the Holocene period is almost certain. Second, their distribution was very widespread. Bones of the *Mastodon* have been found in the Norte Grande desert and also in the southern lakes area. Does this imply that this mammal was found at one and the same time over a vast area, or that it was forced to migrate following climatic changes? At the moment this question cannot be answered.

Nevertheless, the presence of these animals, known only during the Pleistocene period, indicates that the climate of the Norte Chico was no longer of the humid tropical type. In fact, all the animals appear to have lived outside the warm rainy zone.

The marine fauna provide more reliable indications: the corresponding deposits of the marine Quaternary cycles which cut terraces on the formations left by the Pliocene transgression contain cool-water mollusks. This renewal of the species at the beginning of the Pleistocene period apparently points to a thermal intensification of the cold Humboldt current. One may infer from this that the semiarid coast of Chile was cooled during the Pleistocene period.

About 30°30' South, in the steppe proper, the relict forest of Fray Jorge comprises hygrophylic species (trees, ferns, and lianas) which reappear only about 1,000 kilometers further south in the evergreen forest of Valdivia. Two basic types of plant associations have been recorded here: the one, *Aextoxicum punctatum* (tique)–*Drymis winteri* (canelo) with a herbaceous layer of *Nertera granadensis* and *Peperomia coquimbensis,* and the other *Aextoxicum punctatum–Myrceugenia correaefolia* (olivillo) with a carpet of *Uncinia phleoides* var. *longispica* and *Urtica magallenica* (Muñoz and Pisano, 1947).

The survival of this forest, extending over an area of about 800 hectares in the Altos de Talinay to the north and south of the mouth of the Rio Limari, is easily explained. Being situated on the flattened summit of raised coastal blocks which directly face the ocean, at a height of 600 to 700 meters, the area benefits from the condensation of rising maritime air, which is already very much cooled by contact with the Humboldt current. This "hidden precipitation" increases the total annual rainfall almost tenfold, to about 1,500 millimeters (Kummerow, 1962).

Opinion is divided among botanists on the origin of this forest. Some believe in a large-scale migration of the Validivian forest during a glacial period (Skottsberg, 1950). Others (Schmithüsen, 1956) see it rather as a remnant of the Cenozoic tropical flora, which probably covered a large part of the South American continent prior to the climatic differentiations which occurred at the end of the Tertiary period. The latter might have allowed the individualization of a green austral forest, which despite subantarctic contributions, preserves the floral and physiognomic likenesses to the Peruvian *ceja*. Similarly, the plant associations of Fray Jorge might have been able to survive in a favorable geographical environment.

The discussion based on purely botanical arguments is beyond our frame of competence. But the results of our field observations (a considerable increase of the extent of glaciation in the Cordillera, the thickness of flood deposits, the development of paleosols) provide evidence pointing to important climatic variations during the Pleistocene. Furthermore, the evidence of changes in the vegetative landscape in Western Europe during the Quaternary does not rule out as extravagant a displacement of the austral forest over more than 1,000 kilometers. The migration responsible for the location of the forest of Fray Jorge might well coincide with one of the climatic crises of the Quaternary, marked on the coast of the Norte Chico by a decrease in temperature but principally by an increase in rainfall.

The Fray Jorge forest, morover, is not unique. Other forest remnants, at smaller scales, are found along the semiarid Chilean coast on mountain slopes facing the ocean.

The Quaternary coastal terraces exhibit pedological alterations that cannot be explained under present climatic conditions. These are the calcareous and ferruginous crusts. On the other hand, the existence of the interdunal soils emphasizes the alternation of dry and wet periods.

The calcareous crusts are no longer formed, for the conchiferous sands of the holocene beaches never show even incipient signs of consolidation. However, thick incrustations (*losa*) have consolidated the shell deposits of the marine terraces at La Serena (30° S) and at Tongoy (30° 15′ S) and have caused the formation of impressive coquinas forming veritable crusts preserving the forms of these terraces. They imply, even though the deposit is highly calciferous, climatic conditions favorable for cementation. An ascending mechanism is most likely involved in their formation (Coque 1962), since they suggest (1) a prolonged humid season that is capable either of forming a phreatic nappe or of thoroughly moistening the land, (2) and a dry warm season, during which evaporation causes an ascending movement of solutions charged with calcium bicarbonate, which are deposited close to the surface. One must thus imagine a moister climate than at present, but with a marked dry season in summer and without doubt with a denser vegetation that not only limited the effects of runoff but also increased the role of the carbon dioxide in making the calcium soluble.

The road-cuts of the Pan-American highway, where it is constructed on the littoral platforms, have exposed the frequent occurrence of a reddish crust, under the surface layer of the soil, associated with an ascendant movement of iron and aluminum oxides, deriving from a light colored eluvial horizon. To understand its formation one must assume likewise a climate with much rainfall but markedly dry during the warm season. These ferruginous crusts are apparently the climatic equivalent, on a siliceous parent rock, of the calcareous crusts.

Near Hacienda Huaquen (32° 15′S) a road-cut reveals four dunes piled one upon the other, separated by clayey horizons, indicating a pedologic evolution during alternatingly moist and dry climatic phases (Paskoff 1970).

In the Elqui section of the Cordillera (30° S), at an altitude of about 5,000 meters, the glaciation presents only decaying forms, rock glaciers, or motionless slabs of ice. But two major periods of glacier advance have been recognized here (Paskoff 1970); their traces are particularly clear in the upper part of the Rio Elqui (Rio La Laguna — Rio Turbio). (1) A recent glaciation (Wisconsin — with all due reservations concerning the parallelism of climatic changes between the two hemispheres) left a well-formed terminal complex at an altitude of 3,100 meters. (2) Evidence of an older glaciation (Illinois?) is provided by a deposit of clay blocks close to 2,500 meters altitude.

Outside the high Cordillera these periods of cooling and also of increased precipitation have been marked by torrential floods, of greater violence where the topography was appropriate. In their middle courses the large transverse valleys of the Norte Chico (Rios Elqui, Limari, Choapa, Petorca, and La Ligua) are obstructed by huge gravel fans, formed by tributary *quebradas* (creeks). The fronts of these fans were partly destroyed later on by the main river. A large well-preserved cone is often included in a larger and older one, most of which has been destroyed. Finally, there is occasionally a small accumulation, apparently recent but nonactive, which completes the series of geomorphic features. This is dissected today by a gully in which the rainwater flows.

It seems that one can relate the pluvial phases, which affected the entire coastal strip, to those climatic events. They have left their mark as pediments or as rock fans wherever the topographical and lithological conditions were favorable, for example when fine-grained Pliocene accumulations, whether marine or continental, are found at the foot of the granitic slopes of the coastal range. Thus, southeast of Tongoy, two different pediments have been eroded beneath the marine terrace of the Early Quaternary; over fairly large surfaces of intrusive material also are found indications of lateral leveling, now dissected.

Apart from these signs of cooler and wetter periods, traces of dry periods similar to the present also exist, while at least one of them was even drier. These traces are of dune accumulations or wind-scattered sand, either piled one upon the other and separated by ancient layers of soil or by unconformities, or arranged in parallel lines following the irregular retreat of the ocean during the Quaternary. The formation of one "large dune" (La Serena, Tongoy) should be attributed to a stage of powerful eolic activity, under conditions of marked aridity.

How can one account for the pluvial periods which affected the semiarid parts of Chile during the Pleistocene? One should first remember that the present dryness is due to two factors. The first factor, zonal and most important, is the subtropical latitude (30°–33°S), placing the area on the margin of influence of the southern hemisphere polar front. The second factor, regional and of lesser importance, is the existence of a cold current reinforcing the tendency toward aridity. On the other hand, the analysis of the wind system indicates that the Norte Chico — within the limits drawn for it here — belongs entirely to the zone of temperate circulation, characterized by westerly winds (Schneider, 1969).

The study of exceptional meteorological conditions

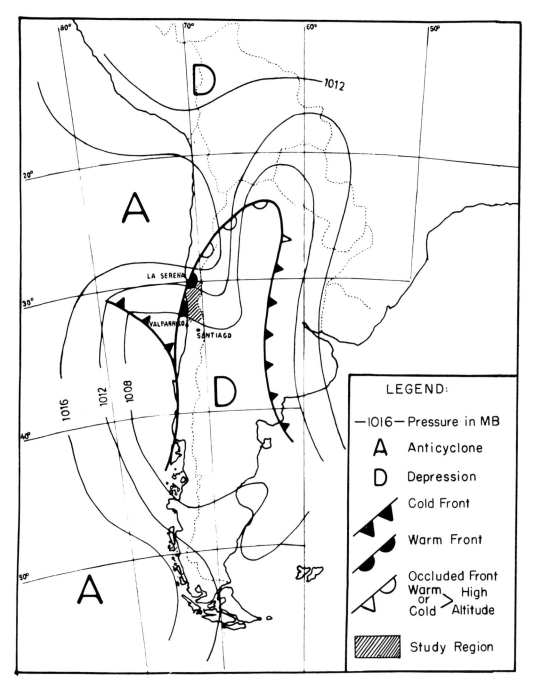

Fig. 17-2. Synoptic map for July 12, 1965, showing rainfall over the coastal area and snow over Cordillera of semi-arid Chile. (M. Collin-Delavaud has kindly drawn the author's attention to the fact that exactly on July 12, 1965, the anticyclone in Peru had largely spread north and lost its intensity.)

presently occurring occasionally — particularly rainy weather situations causing heavy snowfall in the Cordillera, as exemplified in figure 17-2 — provides interesting information for the reconstruction of what might have happened during more humid periods. The moist years — for example, 1963 and 1965 — correspond to a more northerly winter position than is normal for the polar front. At present, in high mountain areas after heavy snowfalls, deep snow accumulates in cirques which were formed by glaciers in the past. Finally, we have seen that the relict forest of Fray Jorge appears to represent a living remnant of an important migration of the Valdivian forest toward the tropics during the Pleistocene period.

All these facts constitute arguments for thinking that the glacial and pluvial periods of the Norte Chico correspond to a northward movement of the southern hemisphere polar front, thereby implying a displacement toward the equator, and perhaps also a weakening of the great anticyclonic cell of the South Pacific. This would explain both the decrease in temperature due to the inflow of cold polar air, able to advance to low latitudes due to the retreat of tropical air, and the increase in rainfall linked not only to the more frequent passage of frontal perturbations, but also to a more active cyclogenesis. So to the simple movement of climatic regions toward the tropics would be added changes that gave to the climate of the Norte Chico its distinctive character: very heavy rains, cold air inroads, and reduced evaporation.

Moreover, the possibility cannot be ruled out that the intensity of the Humboldt current may have varied in accordance with these climatic changes; one might assume (Tricart, 1963) that the periods of glaciation led to its weakening. If this fact were proven, it would have to be taken into account, as it should have assisted in increasing the precipitation; nevertheless this would be a secondary element compared to the still fundamental zonal factor.

This discussion of Plio-Quaternary climatic changes along the semiarid Chilean seaboard leads to two important conclusions: first, the climate at the end of the Tertiary was probably of the warm humid tropical type, with a dry season of varying importance; it changed toward the Quaternary with a decrease in temperature and a reduction in rainfall. Second, the Pleistocene period experienced an alternation between humid phases, accompanied by a cooling of the atmosphere, and dry periods marked by a warming up. These results are in accordance with those obtained elsewhere in the world on the west sides of continents in subtropical latitudes.

Bibliographic References

ALMEYDA, E.
1958 Recopilación de datos climáticos de Chile y mapas sinópticos respectivos, con la colaboración de Fernando Sáez Solar. Chile, Ministerio de Agricultura, Santiago. 195 p.

CAVIEDES, C.
1972 On the paleoclimatology of the Chilean littoral. The Iowa Geographer 29:8–14.

COQUE, R.
1962 La Tunisie présaharienne; étude géomorphologique. A. Colin, Paris. 476 p.

HERM, D.
1969 Marines Pliozän und Pleistozän in Nord- und Mittel-Chile unter besonderer Berücksichtigung der Entwicklung der Mollusken Faunen. Zitteliana 2: 1–159.

HERM, D., AND R. PASKOFF
1966 Note préliminaire sur le Tertiaire supérieur du Chile centre-nord. Société Géologique de France, Bulletin Sér. 7, 8(5):760–765.

1967 Vorschlag zur Gliederung des marinen Quartärs in Nord- und Mittel-Chile. Neues Jahrbuch für Geologie u. Paläontologie, Monatshefte (10): 577–588.

HOFFSTETTER, H., AND R. PASKOFF
1966 Présence des genres *Macrauchenia* et *Hippidion* dans la faune pleistocène du Chile. Bulletin du museum national d'histoire naturelle, Paris (séries 2) 38:476–490.

KUMMEROW, J.
1962 Mediciones cuantitativas de la neblina en el Parque Nacional Fray Jorge. Universidad de Chile, Santiago, Boletín 28:36–37.

MUÑOZ, C., AND E. PISANO
1947 Estudio de la vegetación y flora de los parques nacionales de Fray Jorge y Talinay. Agricultura Técnica (Santiago) 7(2):71–190.

PASKOFF, R.
1970 Recherches géomorphologiques dans le Chile semiaride. Bordeaux, Biscaye édit. 420 p.

SCHMITHÜSEN, J.
1956 Die räumliche Ordnung der chilenischen vegetation. Bonner Geographische Abhandlungen 17: 1–86.

SCHNEIDER, H.
1969 El clima del Norte Chico. Publication of the University of Chile. 132 p.

SKOTTSBERG, C.
1950 Apuntes sobre la flora y vegetación de Fray Jorge (Coquimbo, Chile). Goteborgs Botaniska Tradgard, Meddelanden 18.

TRICART, J.
1963 Oscillations et modifications de caractère de la zone aride en Afrique et en Amérique latine lors des périodes glaciaires des hautes latitudes. Unesco, Paris. Arid Zone Research 20:415–419.

CHAPTER 18

THE SEMIARID COASTAL REGION OF NORTHEASTERN BRAZIL

Manuel Correia de Andrade
Department of Economic Geography, Federal University of Pernambuco, Brazil

The semiarid region of northeastern Brazil covers an area of more than 500,000 square kilometers, which is more than 5.8 percent of the total area of Brazil, and extends into the territories of Ceará, Rio Grande de Norte, Paraiba, Pernambuco, Alagoas, and Bahia. In the north, the region reaches the sea, and in the south it is crossed by the great river São Francisco, 3,100 kilometers long and with a rate of flow varying between 1,000 and 13,000 meters per second. In the east it is located more than 300 kilometers from the sea.

The coast is always low, with sandy beaches stretching several kilometers along it and occasionally interrupted by the discharge of rivers reaching the sea from the *sertão,* for example, the Piranhas-acu, the Apodi-Mossoró, and the Jaguaribe, which is the "longest dry river in the world." Sometimes small mounds of lateritic clay are found on the beach—evidence of relief now vanished through erosion. The climate is warm and dry. The region lies between 6° and 4° South of the Equator, and thus the temperatures are always above 25°C. The annual temperature range is minimal and has no effect on the life of the inhabitants. Rainfall is not very abundant — less than 600 millimeters per year — and is badly distributed, as almost all falls during the summer and autumn, principally in February, March, April, and May. Winter and spring are the driest seasons.

Along the beaches, 2 or 3 meters above sea level, large coconut trees are found, providing copra and oil, which are processed in the towns within the region, especially at Fortaleza, the capital of Ceará, which, with its 500,000 inhabitants, is the main town and port of this semiarid northeastern area. The coconut (*Cocos nucifera* L), a plant originating in the Orient and introduced by the Portuguese during their colonization of Brazil, has since become scattered along all the beaches. The palm tree is grown by small and large landowners alike, since it produces six harvests a year and is thus of great value to the cultivators. Once planted the coconut produces its first crop after five or six years but maintains its yield for many years. The large landowners practice a more intensive type of agriculture, using organic fertilizer (castor-oil cake and animal manure) and mineral salts (calcium and potassium) to obtain an annual crop of 40 to 60 nuts per palm tree. The small landowners who lack the money to buy fertilizer have a smaller yield.

In this region one finds not only coconuts but also native plants such as the cashew nut (*Anacardium occidentalis* L) and clove (*Eugenia michilii* Berg.). The cashew nuts have already been industrialized in the production of a fermented drink — *cajuina* — and in the making of certain cakes.

The agricultural laborers who work in the coconut plantations always live in small towns and coastal villages, coming to the plantation for the harvest. In small hamlets beneath the coconut palms, or in little isolated houses made of palm straw, live the fishermen, who take to the sea every day to fish from their small sailing boats — *jangada*. The fish are put on ice and transported to the towns. This activity is still a craft. The fishermen's wives and daughters spend their time making lace and embroidery. The embroidery from Ceará is famous throughout Brazil.

Along the coast, near the river mouths, large plants for the extraction of marine salt are situated. This region produced more than 592,723 tons of salt in 1964, more than 80 percent of the Brazilian output. The dry climate, almost rainless for seven or eight months of the year, the great variation in tide, and the low coast provide favorable conditions for such production. This industry is but little developed; water is directed into reservoirs where it is evaporated, but the only profitable product is sodium chloride. Shipment is by primitive methods, since the local ports (Aracati, Areia Branca, and Macau) are poorly equipped. The salt is transported in sailing boats from the river mouth to the cargo boats lying 15 to 18 kilometers offshore.

Work in this industry is concentrated during the spring and summer months. The laborers live in the towns and villages and during the rainy season have other employment. They work the small cotton plantations and plots of beans, maize, and millet under a tenancy system for the medium and large landowners. Their work in the salt industry is manual. Because they have neither decent sanitary conditions nor adequate nourishment, there is much occupational sickness among the workers. The Salt Institute is beginning to organize production, and it promises social security for the laborers.

The production and income from this industry are not at all satisfactory, despite the development in Brazil of a chemical industry based on the production of caustic soda. Neither is the communications network satisfactory, as the roads to the large towns, where the industries are situated, are not paved, and the distance by rail is very long. To develop this industry and to supply the national market, adequate transportation to the ports and the mechanization of industrial activity are necessary.

The other important industry is the extraction of carnauba wax. The carnauba (*Copernica cerifera* Mart.) is a native palm tree, over 10 meters (30 feet) tall, which, to protect itself against the dryness in summer, produces a wax in its leaves. The wax is used in industry and in the making of candles, records, and other products. The palm tree is always found in the alluvial plain of lower river courses, near the sea, since it needs water beneath its roots and sun on its leaves. The palm trees form a true forest, 15 to 20 kilometers long and 2 to 3 kilometers wide. The caatinga, the forest of small thorny trees of the Brazilian *sertão,* is rich in cactus such as the *mandacaru* (*Cereus peruvianus* Mill), the *facheiro* (*Cereaus* sp), the *xique-xique* (*Philocereus gounellei* Web.), the Bromeliaceae, the *macambira* (*Bromelia laciniosa* Mart.), and in trees which lose their leaves during the dry season, that is, the *Jurema* (*Pithecalobium tortum* Mart.), the *faveleiro* (*Jatropha phyllacantha* Muell. Arg.), the *juazeiro Zyphus joazeiro* Mart.), and others.

The landholdings always include some palm trees and caatinga, in whose production the landowners are always interested, as they are also in cotton and cattle rearing. There are still many laborers who live in small carnauba straw huts, on the holdings, and retain the right to live there by working for the owners one or two days a week. These laborers also work as tenants growing such crops as cotton, beans, maize, millet, and gourds. Fishermen work in the fish markets near the lakes and pools, and along the river bed in the dry season. Carnauba wax is gathered during the dryest months (October, November, December), and almost all the laborers participate therein. The leaves are cut and taken to the mills where the wax is separated from the leaves. The wax is transferred to the large towns, and the leaves are used to cover the huts, or for making hats, bags, and other items. Much importance is attached to the goods made of straw by the women. Carnauba wax is largely exported to the United States, amounting in 1964 to 7,681 tons, almost 60 percent of the Brazilian production.

Cattle rearing is primitive. The livestock (cows, donkeys, and goats) live in the caatinga and graze on natural pasture. The caatingas do not have fenced fields. During the rainy season pasture is plentiful, but in the dry season the livestock are put in enclosures and graze on natural pasture or seed cake, cotton seeds, millet grains, and maize. As payment, the vaqueiro, or cattle-hand, receives one animal out of every four born during the year.

Nowadays this frugal society, based on the exploitation of natural resources, is being transformed through the industrial development of the country and by road building. The closed society of the area no longer exists, and the region is constantly increasing its participation in Brazilian economic life. Projects for the construction of ports and roads have been drawn up. The importance of the industrialization of coconuts and cashew nuts is increasing, the carnauba wax and marine salt industries are being mechanized, and cattle strains are being improved. Agricultural activity also has developed in the alluvial plains, which are irrigated by windmills and pumps operated by diesel oil or electricity.

This semiarid coastal region of northeastern Brazil is increasingly developing and changing from a patriarchal to a modern economy.

PART THREE

THE OLD WORLD DESERTS

CHAPTER 19

COASTAL DESERTS OF THE OLD WORLD AND THEIR RECLAMATION

Leonid E. Rodin

Botanical Institute, Academy of Sciences of the U.S.S.R., Leningrad

The coastal deserts of the Old World have been studied within the temperate, subtropical, and tropical belts on the shores of the Aral Sea, Caspian Sea, Arabian Sea, Red Sea, and northern shore of the Mediterranean Sea.

The climatic conditions, although appearing rather different in the various areas, nevertheless if examined in detail have much in common: high summer temperatures, low precipitations, a critical water deficit during the hot season.

Within the above-mentioned limits we distinguish the following five desert regions:

1. Northern subboreal (Aral Sea, northern and central parts of the East-Caspian coast)
2. Southern subboreal (southern part of the Caspian coast)
3. Mediterranean (southern Mediterranean coast)
4. Subtropical (northern part of the Red Sea coast)
5. Tropical (southern part of the Red Sea coast and the Arabian Sea coast)

These coastal desert regions are characterized by climatic indices as shown in table 19-1.

Without going into details, we can say that the coastal desert soils are of two main variants: clay salty soils and sandy soils. Both, although situated in different natural zones, show more similarities than differences, and pedologists have not distinguished the zonal types of solonchak or sandy soil. A high level of groundwater characterizes solonchaks, while mobility of the soil material is characteristic of sands.

We are convinced that it is important from the practical point of view to discover the above-mentioned differences that would be valuable in the process of coastal-desert reclamation. The differences between these natural zones may be seen by identifying the leading species representative (dominants and edificators) of vegetation.

1. Northern subboreal deserts. On solonchaks: *Halocnemum strobilaceum, Kalidium gracile*. On sands: *Haloxylon ammodendron, Ephedra omatolepis, Elymus sabulosus, Agropyrum sibiricum*.

2. Southern subboreal deserts. On solonchaks: *Halocnemum strobilaceum, Kalidium caspicum, Halostachys belangeriana, Salicornia herbacea*. On sands: *Haloxylon persicum, Ephedra strobilacea, Nitraria komarovii, Aristida karelinii*.

3. Mediterranean deserts. On solonchaks: *Halocnemum strobilaceum, Suaeda pruinosa, Salsola tetrandra, Limonium tubilforum*. On sands: *Nitraria tridentata, Retama retam, R. duriaei, Gymnocarpos decandrum, Pityranthus tortuosus*.

4. Subtropical deserts. On solonchaks: *Salicornia fruticosa, Haloxylon salicornicum, Suaeda monoica*. On sands: *Hypaene thebaica, Acacia tortilis, Retama retam, Ammophila arenaria*.

5. Tropical deserts. On solonchaks: *Salicornia fruticosa, Suaeda* spp., *Haloxylon salicornicum*. On sands: *Calligonum polygonoides, Prosopis spicigera, Calotropis procera, Cyperus arenarius, Ipomaea biloba*. On littorals (shallow water) the mangroves develop consisting of *Avicennia officinals, Rhizophora mucronata, R. conjugata, Bruguiera caryophylloides*.

The hydrological characteristics of the solonchak soils are the following: the high level of the strongly mineralized groundwater, high degree of capillarity, surface salinity, and easy crystallization of salts.

The hydrological conditions of the coastal sand dunes are characterized by the following peculiarities: relatively low (2 to 5 meters) level of mineralized groundwater, rapid absorption of atmospheric waters and their gradual downward percolation (gravitation process), water-vapor condensation in the upper sandy layer resulting in the under-surface (70 to 150 centimeters in depth) wet horizon.

The sea coasts are distinguished in the increased humidity of the lower atmospheric stratum. Some points (Baku, Alexandria, Marsa-Matruh, and others) are characterized by frequent mists and dew in certain seasons. Experiments carried out in the USSR (Nikitin, 1910; Sochevanov, 1938; Gael, Kolikov, and Maliugin, 1950; Blagoveshchensky, 1940, 1954) and special observations in the United Arab Republic (General Desert Development Organization, Marsa-Matruh) showed that about 30 percent of freshwater accumulating in the above groundwater layer resulted from the atmospheric vapor condensation.

TABLE 19-1

Climatic Factors of Coastal Desert Regions

Desert Region	Temperature (C°)		Precipitation	
	Average Annual	Absolute Minimum	Annual (mm)	Rainless Period (mo)
1. Aralskoie More (USSR)	6.8	—	103	—
2. Baku (USSR)	14.1	−5.0	187	4–10
Krassnovodsk (USSR)	15.7	—	116	4–11
3. Alexandria (U.A.R.)	20.2	2.8	184	3–10
Marsa-Matruh (U.A.R.)	19.1	2.0	158	3–11
4. Suez (U.A.R.)	21.7	1.4	21	2–12
5. Bhuj (India)	26.2	1.1	347	10–6

SOURCE: Walter and Lieth, 1960.

Having been condensed in the upper sand horizons, the water penetrates into the deep sand layers under action of the gravitational effect. The poorly developed capillarity of sandy ground reduces surface evaporation so that it is very low compared to that from clay ground surface. Besides, considerable quantity of freshwater is accumulated through the process of condensation of vapor water that has evaporated in the soil from the groundwater surface.

On the sand soils of coastal deserts (under the conditions of the above-mentioned level of groundwater) the following trees can be grown successfully.

1. Northern subboreal deserts: *Populus diversifolia, Salix* spp., *Elaeagnus orientalis, Robinia pseudoacacia*
2. Southern subboreal deserts: *Morus alba, Ficus carica,* grapes
3. Mediterranean deserts: *Ficus carica,* grapes, *Acacia saligna*
4. Subtropical deserts: *Phoenix dactilifera, Eucalyptus* spp.
5. Tropical deserts: *Cocos nucifera, Prosopis juliflora, Acacia* spp.

The methods of reclamation of the coastal sand deserts are the same for all the zonal regions. Measures are undertaken to make the fresh groundwater horizon accessible for the root systems. In the subboreal deserts, both northern (Aral Sea) and southern (southern part of the Caspian Sea), a method is used that the USSR calls trench farming. A trench is dug 150 to 200 centimeters wide at the bottom and 300 to 400 centimeters wide in the upper part. The depth may reach 150 to 200 centimeters, depending on the level of the capillary border of the fresh groundwater. One hectare of sand area includes as much as 1,600 linear meters of trenches. Then soil rich in organic and mineral fertilizers is placed in the trenches, and annual crops, trees, and shrubs are sown.

In the Aral area, such plants as potato, carrot, beet, cabbage, tomato, cucumber, watermelon, apple, pear, currant and others are planted successfully (Shevtsov, 1910; Vinogradov-Nikitin, 1938; Zonn and Pavlova, 1942; Gael, Kolikov, and Maliugin, 1950; Maliugin, 1955). Small trees already grafted are planted in deep holes.

On the southern Mediterranean coast (United Arab Republic), *Ficus carica,* grapes, and sometimes *Ricinus communis,* the principal crops, are cultivated in the sandy soils (Milad, 1934). Since the roots are spread through the wet horizon, the plants continue to grow successfully and to bear even if afterward covered with sand.

Like the sands, the reclamation of solonchak soils in the coastal deserts is generally the same in all zones. Here the endeavor is to lower the level of mineralized groundwater. This may be achieved by drainage ditches or by pumping out the water. The choice of means depends on the topography of the land and the economic conditions. In any case a stable agriculture is possible only if the freshwater required for leaching salts from the soil as well as for irrigation is available in ample quantity.

The main crops are chosen in accordance with the agroclimatical conditions and the economical requirements. They are: rice in the northern subboreal deserts; cotton in the southern subboreal and Mediterranean deserts; citrus, dates, and coconuts in subtropical and tropical deserts.

In the tropical deserts of the Arabian Sea coast (Kach peninsula), there is a real danger of inundation by seawater (tidal or wind-driven) covering the arable plots. Therefore dams and a system of floodgates in the drainage channels must be constructed.

References

BLAGOVESHCHENSKY, E. N.
1940 Some new data on the intrasoil condensation (translated title). Meteorologia i Hydrologia 3. Leningrad.
1954 Condensed water in the desert soils (translated title). Pustyni SSR i Ikh Osvoenie, Moskva-Leningrad.

GAEL, A. G., M. S. KOLIKOV, AND E. A. MALIUGIN
1950 Trench farming (translated title). Peschanye Pustyni Severnogo Priaralia, Alma-Ata.

MALIUGIN, E. A.
1955 The agricultural utilization of the Aral semi-desert (translated title). Zemledelie 8. Moskva.

MILAD, Y.
1934 Fruit growing on sandy soils. Egypt, Ministry of Agriculture, Technical and Scientific Service Bulletin 132. 20 p.

NIKITIN, S. N.
1910 The contemporary state of the problem of the soil dew condensation in connection with the plant nutrition in the desert-steppe regions and the ground waters formation (translated title). Protokoly zasedanij Vtorogo meteorologicheskogo sjezda, 11–17 January, 1909, St. Petersburgh.

RODIN, L. E.
1961 Dynamics of the desert vegetation (translated title). Akademia Nauk SSSR, Moscow-Leningrad. 227 p.
1963 Desert vegetation of the West Turkmenistan (translated title). Akademia Nauk SSSR, Moscow-Leningrad. 309 p.

SHEVTSOV, N.
1910 On the trees and fruit crops cultivation on sand (translated title). Plodovodstvo 6.

SOCHEVANOV, V. E.
1938 Water exchange and the water vapor condensation in sands (translated title). Trudy Gosudarstvennogo. Hydrologischeskogo Instituta, 7.

VINOGRADOV-NIKITIN, P.
1938 The Apsheron method of tree planting on sands (translated title). Sovestskie Subtropiki 3.

WALTER, H., AND H. LIETH
1960 Klimadiagramm-Weltatlas, Lfg I. VEB G. Fischer, Jena. 82 bl.

ZONN, S. V., AND E. A. PAVLOVA
1942 Melon cultivation on the shell sands in the desert (translated title). Sovetskaia Botanika 4–5.

CHAPTER 20

CLIMATIC-GEOMORPHOLOGICAL ZONES AND LAND UTILIZATION IN THE COASTAL DESERTS OF THE NORTH SAHARA

Wolfgang Meckelein
Department of Geography, University of Stuttgart, Federal Republic of Germany

The Region

The outskirts of the Sahara desert border on the coast of the Mediterranean Sea from southern Tunisia across northern Libya to the Nile delta in Egypt. This is a broad desert-like transition zone of the northern Sahara. It appears in many diverse forms.

In the southern part of Tunisia the geomorphological structure is one where Quaternary deposits form a coastal plain and where lagoons and coastal sebkhas are frequent. A normal steppe vegetation not only spreads over the whole area but also over the border region, inland of the adjoining cuesta of Dahar, consisting of Triassic to upper and middle Cretaceous rocks. The highlands, however, are soon immersed by the extensive sand masses of the Great Eastern Erg. North of it the depression of the shotts is marked by great salt plains and halophytic vegetation.

At the border of Tunisia and Libya the coastal plain widens considerably and changes into the Tripolitanian Jefara. In this area are loess-like soils in part, also more extensive dunes. The rest is covered by bush and salt steppe. *Artemisia* and *Retama* grow on sandy and clay soils; characteristic are the nebka of the *Ziziphus* lotus. In contrast the border of cuesta of the Jebel Nefuza (over 900 meters) ascending in the hinterland has still mediterranean sclerophyllous vegetation in part, including maquis, because of a higher rainfall. The transition southward to the Cretaceous highland of the Hamada el-Hamra rapidly becomes distinct. On the infertile rock desert only a scanty desert-steppe vegetation of low bushes thrives, except for the wadis, where tamarisks, acacias, and other trees or bushes grow now and then.

This desert steppe closely approaches the Mediterranean Sea in the region of the Gulf of Sirte, which penetrates deeply to the south. Only directly on the coast does there remain a narrow strip of steppe, which is interspersed with bigger and smaller coastal sebkhas and long coastal dunes. In contrast, in the northern Cyrenaica, the Highland of Barka, with its Karst topography, is covered by steppe vegetation to a greater extent. However, the highest parts of this highland, the Jebel el Akdar (870 meters) and its gorges have a distinctive mediterranean vegetation with cypress, stone oaks, juniper, Aleppo pines, and so on. Naturally the macchia is not missing either. Terra rossa soils are to be found on the entire highland, which consists mainly of tertiary limestone.

In general, tertiary sediments dominate the Saharan coastal border from the Gulf of Sirte to the Nile delta, except for usually small Quaternary deposits near the coast. Toward the east the rainfall is constantly diminishing, and the desert steppe advances toward the sea everywhere in the United Arab Republic (U.A.R.). The coastal deserts here are rock and stony deserts, with bigger or smaller depressions set into them, consisting, in part, of extensive salt plains (especially the Quattara depression). The latter lies in the wholly arid region, because the desert steppe belt becomes very narrow, until it ends near Alexandria in the cultivated landscape of the Nile delta.

So the northern Saharan coastal deserts are preponderantly dry steppe regions, that is, semideserts. On average, the transition into the desert proper does not set in until 120 to 250 kilometers south of the Mediterranean coast.*

While in the desert proper only an isolated oasis cultivation is possible, the rather broad strip of land north of it up to the coast offers a good chance for widespread cultivation. Success, however, and truly optimal economic utilization, depends on the consideration of the whole complex of natural geographic factors. Among these factors the climate plays the most important guiding role. Unfortunately, the finer climatic differences are difficult to comprehend.

Climatological Division and Land Utilization

The differences in average annual rainfall may appear to be the most important factor in land utilization. Nevertheless the rainfall is only a limited indicator. For example, the course of the 200-millimeter isohyet, which can be regarded as the border between the dry variant of the Mediterranean climate and the sub-Saharan desert climate, oscillated in the mean maximum more than 150 kilometers in a north-south direction in Tunisia in 1931–1934 and 1944–1947 (Tixeront, 1956, p. 91). The fact is that in regions with deficient annual rainfall, the

---

*P. Meigs (1953) uses another aridity scale for his world map of arid climates: semiarid, arid, extremely arid. The latter approximately corresponds to the term of desert proper. But there are reasons here to use a somewhat different classification in order to be able to exclude the extreme parts of the desert proper (Meckelein, 1959).

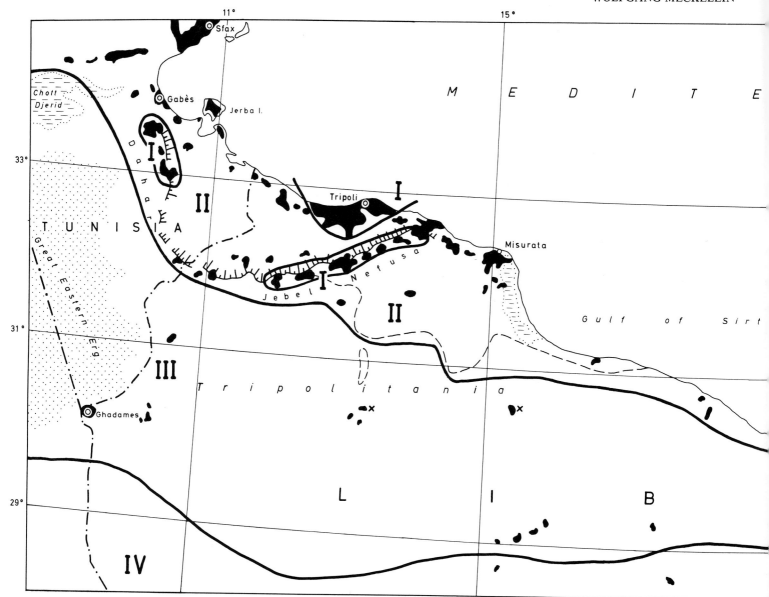

Fig. 20-1. Zones I, II, and III are climatic-geomorphological areas, as described in the text. Zone IV is the desert proper, that is, the full and extreme desert. In Libya, the interrupted line indicates the southern border of the old Roman farms, and crosses designate important Roman forts.

total per year is very uncertain. The customary mean over longer periods in the climatological records obscures the true conditions.

For the division into climatic zones and their utilization possibilities, it is well known that the temperature conditions are not altogether useful. This may be shown by an example: In the village of Azizia, in the Jefara, the highest temperature of the Sahara was read at 58°C. Here, it is not so important that the reading may not have been quite correct, and that in the meantime a hot spot has been found in the Ahaggar mountains which has been read with greater accuracy (Dubief, 1959, p. 235). The crucial point is rather that in the hot region of Azizia, which undoubtedly exists, local cultivation is practiced, partly with the dry-farming system, which is not possible in many considerably "cooler" border zones of the Sahara. The situation is that temperature is of most importance indirectly, as one of the factors on which evaporation depends. Indeed for well-known reasons, representative and comparable readings are extremely difficult to get. However, one experiment by Dubief (1950) for the Sahara should be mentioned. He used the "evaporation physique" of De Martonne (which does not take into consideration the evaporation by vegetation) and found that the theoretical possibility of evaporation is between 2 and 4 meters per annum in the regions in question.

The practical significance of these evaluations becomes evident on investigation of the soil formation. An annual evaporation within the range of 2 to 3 meters results in crust formation. Naturally there must exist suitable minerals in the soil. In these areas average rainfall varies from 70 to more than 300 millimeters per annum. Above the evaporation limit of 3 meters evap-

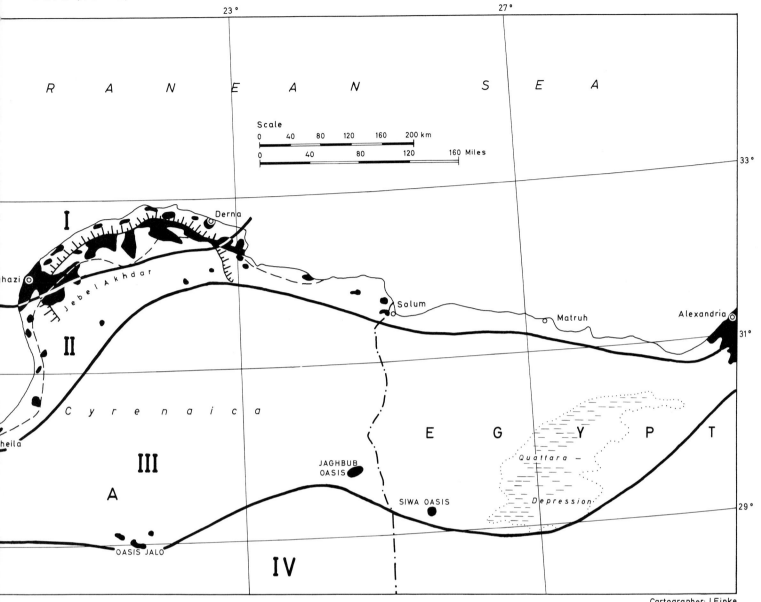

Cartographer: I. Finke

oration is too strong to make possible a sufficiently deep penetration to give enough time for dissolving the minerals. Also, the rainfall will probably be insufficient. Thus the crusts that partially exist there are distinctly fossil (Meckelein, 1961). Incidentally, these are always climatically formed crusts which are independent of the groundwater.

In this connection, a division according to hydrological conditions might seem logical. But the presence of groundwater naturally varies greatly, depending on the geological conditions. When there is no fossil water, groundwater depends very much on the runoff conditions. All three possible types of runoff for arid areas are present in the investigated regions. In extensive parts of northwestern Libya, intermittent streams have an outlet into the sea. But there are also areas with interior drainage, as in southernmost Tunisia and Cyrenaica. In the remainder of the area, however, streams are largely absent (Dubief, 1953). The consequences for the geomorphology are clear, but will be touched on later.

Climatic-Geomorphological Zones

The discussion of the individual factors of climate and the water situation demonstrates how difficult it is to undertake a division into uniform zones. The next step proceeds from the idea that the actual effects of climate and the water situation on the surface of the land must be observed. These are reflected, in fact, in geomorphological development and soil formation as well as in the distribution of certain plant families and hydrological phenomena. Investigation by this geomorphological method produced three zones whose boundaries can be cartographically fixed quite accurately.

Zone I. The first zone is of a typically mediterranean climate. It is restricted to parts of the Tunisian and Tripolitanian Jebel, the near surroundings of the town of Tripoli, and the Highland of Barka. Everywhere the rainfall is over 300 millimeters per annum. Characteristic are the great differences in runoff each year if there are no intermittent springs which feed the streams, lineal erosion, terra rossa soils, and the distribution of mediterranean sclerophyll plants, including maquis.

Zone II. The second zone is much more widely spread. It should be regarded as a transition zone from the mediterranean to the subtropical dry climate. Its vegetation is dry steppe and partly halophytic. As a marginal strip of the desert it possesses fine-grained sands and loess-like soils. Linear erosion is still to be found; the wadis, however, carry water periodically only, and for short spells of time. In areas where the rainfall is 200 millimeters (8 inches) and less, the landscape is frequently covered by calcareous crusts, the result of relatively high temperatures and corresponding evaporation. These crusts are 30 to 50 centimeters thick, and sometimes a series of them lie one on top of the other. Due to these crusts, the relatively high winds blowing during the dry months are restricted in their activity on the land surface. This zone is widely spread through southern Tunisia, Tripolitania, and northern Cyrenaica; on the south coast of the Gulf of Sirte, however, and in Egypt, it is found very little or is almost entirely missing.

Zone III. Whereas the third zone in the western part of the investigated region is spread over the hinterland farther away from the coast, in the eastern part it constitutes the greater part of the area where utilization is possible. Here we have a true semidesert. The vegetation of low bushes and bunch grass spreading over the area covers less than 50 percent of the soil. Therefore the formation of the surface is to a considerable extent determined by the wind over the whole year, especially since crust formations are to be found only as fossil remains. Thus the sands that have been worked upon by the wind are blown together into quickly moving dunes and small barkhans. The low rainfall with an average of 50 millimeters per annum permits only an occasional intermittent runoff. The wadis lose themselves in depressed areas without outlet. Because of the high evaporation they often have the form of great flat salt pans.

An analysis of these climatic-geomorphological zones is important for land utilization purposes.

Present Land Utilization

Because of the diversity of the natural landscape there is no uniform or exclusive land utilization in the northern Sahara littoral. One part of the population practices farming and horticulture, another animal husbandry. However, the boundaries between the two groups are not fixed. The farmer sometimes possesses herds; the stock farmer cultivates fields temporarily and often owns fruit trees. We find all transitions from the sedentary farmer to the genuine nomad, and thus the most diversified ways of husbandry. This does not eliminate the distinct contrast between farmers and nomads as social economic groups. It intensifies, through mutual dissension — often in connection with barren years — the instability of land use which already exists through natural causes. This fact has played an important economical role.

The areas permanently used for agriculture can be divided into coastal and highland oases according to their topographical situation. They are situated mainly in the Mediterranean region. Coastal oases can be found at different places in southern Tunisia, including the island Jerba. In Tripolitania the coastal oases constitute a more or less compact, if often only narrow, zone from the western border of Libya to the district of Misurata-Tauorga. In the adjacent desert of Sirte there are only small isolated oases. The same applies to the Egyptian littoral, where, with the exception of the region of the Maryut Project southwest of Alexandria, there is little horticulture and only occasional farming in the few small settlements along the 450-kilometer coast. In contrast there are several bigger oases in Cyrenaica, which partly graduate into the highland oases of the Jebel Akhdar. The chief region of the highland oases is the Jebel Nefuza in Tripolitania, and its continuation into southern Tunisia to the Matmata Highland. Here are larger and smaller oases in places which are favorable by relief and climatic exposure.

Farming without irrigation is practiced only by settled farmers in the coastal oases of Tripolitania and in the highlands. Even here, usually additional moisture from irrigation is necessary to guarantee good yields. In the highland oases of the Tunisian and Tripolitan Jebel, grain (barley and a little wheat) is grown in the valleys, where dams collecting the winter rainfall have been constructed and where the groundwater also is used to some extent. For horticulture (vegetables, legumes, also tobacco) we find well or cistern irrigation. In Cyrenaica, horticulture and farming depend on the soil deposited on the hollows and basins in the limestone tablelands. In some places there are wells for additional irrigation, but usually only cisterns. The former Italian colonization landscape, which disappeared from the intensive land use after World War II, was supplied by aqueducts from distant intermittent springs providing plenty of water.

An important part of the cultivation in all oases is arboriculture. There are mainly mediterranean-type fruit trees (especially olive, but also fig and more rarely almond trees, citrus fruits, apricots, and others) and date palms. Above all, in the oases of the coastal zones of southern Tunisia and Tripolitania we find extensive orchards of dates and olives which, at a distance, have the appearance of woods. The fruit trees are fed by groundwater, except for special situations. In some regions, however, unirrigated plantations have been set up in areas with an annual rainfall of less than 200 millimeters, provided that the distance between the trees is 20 to 40 meters. In this way enough groundwater is guaranteed for each tree. Arboriculture, horticulture, and agriculture are normally a joint activity both in the coastal oases and the highland. Everything is irrigated, though this is not always really economical (limited water supply, high work intensity). Usually irrigation is carried out from wells.

On the Tripolitan coast there is an upper, abundant though completely utilized, water table at a depth of 2 to 15 meters. In the desert of Sirte this is to be found only locally, and the subterranean flow of groundwater of several wadis, which reach the sea here, also is used. On the Cyrenaica coast groundwater is to be found only at a depth of 10 to 20 meters, if at all. Often one has to make do with water from cisterns. The same applies to the Egyptian littoral, where the few wells near the sea are 1.5 to 4.5 meters deep; those farther inland, if there are any at all, are up to 65 meters deep. Since the upper water near the coast has salt and brackishness beneath it, the useful water supply is limited.

Because of the scarcity of water, the great amount of hand work, splitting up of property, and ancient cultivation methods, the grain field generally is very small. A start toward a modern agricultural landscape was made in Libya by the Italian administration before World War II. In addition to the already mentioned colonization area in the highland of the Cyrenaica, a new territory for cultivation was developed in Tripolitania. Some colonists went — sometimes without adequate preparation — into the dry steppe. There a new freshwater table was drilled at 300 to 400 meters depth (artesian water), water was distributed by aqueducts, and shelterbelts were planted. Also the dry-farming system was tried out in part successfully, at several places in the steppe of Jefara.

Unfortunately these achievements have been lost to a great extent through postwar changes. However, the Libyan government has endeavored to expand the agricultural area by generous encouragement. To this purpose the National Agricultural Settlement Authority (NASA) was founded in 1963 and has started numerous projects. The financial means have been covered for the greatest part by the proceeds of Libyan oil production.

Since the areas of permanent agriculture are only the more or less limited regions of the oases, farming and nomadic animal husbandry intermix in the same area. The latter has considerably increased in Libya since World War II. In general seminomadism and part-nomadism play the greatest role. In Tunisia there is also a sort of migratory peasantry, or transhumance. Cattle and camels are relatively rare. Sheep and goats, which are the most important domestic animals, graze in the steppes near the coast and in the highlands of southern Tunisia and northern Libya. In the rather dry Egyptian littoral the possibility of pasturage depends solely on sufficient winter rain and is therefore rather uncertain and restricted to the months from November to April.

The seminomads are sometimes also owners of fruit trees. They always have at least periodic crops. Usually it is barley which is sown in moist winters in the basins and flat hollows fed by wadi water. This "shifting cultivation" is practiced on very small fields; the preparatory ploughing is often done with camels as draught animals. The migratory peasants in southern Tunisia sometimes irrigate their fields, cultivated on the coastal plains in winter, by assembling and distributing the runoff of rainwater. An additional employment of the seminomads is, in some regions, the "harvest" of the wild esparto grass.

The real desert steppes in the hinterland are little used. Only a few camel nomads come here seasonally, partly from the far south. The genuine nomads do not cultivate anything. Suffice it to say that although nomadic utilization dominates large areas and is always extensive, it is economically an insignificant factor at present (UNESCO, 1963; Monteil, 1959; Abou-Zeid, 1959; and Herzog, 1963).

Past and Future Land Utilization

The types and growth rates of land utilization have always fluctuated in the northern Sahara coastal regions. A natural border region reacts with particular sensitivity to human interference. An example occurred in Libya, with the retrogression of permanently cultivated land after 1945. Today Libya is busy with projects that aim mainly at cultivating the formerly colonized land. Egypt, on the other hand, has developed bigger land-reclamation projects, also in the Mediterranean littoral, since World War II. The Egyptian General Desert Development Organization has been developing the Maryut-Project. Southwest of Alexandria 12,000 hectares of land have been reclaimed for farming by deep wells. The Northern Coast Project, the opening up of small strips of land, along the Mediterranean Sea to the Libyan border, is at a preparatory stage.

The maximum expansion, however, of the cultivated landscape, with the most intensive utilization of the whole area, existed in old Roman times. This level has not been reached again until in modern times. A change of climate can hardly be responsible. This conclusion derives from the great care the Romans took to insure sufficient water supplies. The remains of dams, aqueducts, wells (partly used today) and cisterns, long since silted up, are to be found from Egypt to Tunisia, even in regions completely unsettled in modern times. Furthermore certain boundaries have been observed, graded according to different forms of utilization. It can be demonstrated that the greater towns and the more densely populated settlements were restricted to the zone of the typical mediterranean climate. The same applies to the important country houses, or villas. The type of the simple, if fortified, Roman home extended far into the dry steppe. Though there are no trees today, it is interesting to note that almost all the Roman farms possess the remains of antique olive-presses. We find no settlements to the south; only small forts existed in the region of the desert steppe. These had to protect the cultivated land against the nomads. (Compare the sheets Lepcis Magna and Cyrene [1:1 Mill.] of the "Tabula Imperii Romani" published by the Society of Antiquaries of London, Oxford, 1954.)

The recognition of the borders of these zones may have resulted from traditional experience during a long time of colonization in antiquity; or possibly instinct may have been responsible; or perhaps just the fact that only certain forms of land use could be employed in certain areas for any length of time. In any case, the importance of various natural factors becomes clear, the combination of which is most easily comprehended by the division of the area into the climatic-geomorphological zones, as previously outlined.

The differentiation of the three zones, specifying the most important natural geographical factors to be observed, with certain inferences drawn from each zone, are presented in table 20-1. In addition, there are indications of the water resources which cannot be defined according to climatic-geomorphological zones but only according to local geology.

Zone I belongs clearly to the Mediterranean region because of climate and geomorphology. Therefore, the same problems exist here as in the other areas with the dry variant of the Etesian climate. Cultivation of crops everywhere requires irrigation. The water resources may be increased by better use of the winter runoff, that is, by erecting small dams and reservoirs (if possible cemented) in the valleys. Drawing more water from the upper water table is hardly possible in many cases, and the danger of penetration of saline water increases near the coast; therefore only deep wells are of use. Animal husbandry should be practiced only in connection with farming. The goat herds are a special danger to the growth of trees, which are indispensable for soil conservation, and for the necessary afforestation. The sedentarization of the nomads in this zone is essential.

Zone II takes up the greatest part of the northern Sahara coastal region and is the proper area for the expansion of farming. Because of climatic-geomorphological conditions, this zone is particularly prone to soil erosion. Only in a few parts will dry-farming be possible, and sensible field cultivation or orchards with irrigation will bring the best results. The drilling of deep wells and the establishment of a system of pipes is important. Irrigation by shallow wells alone is too uncertain because of the dependence on irregular and insufficient rainfall. Additionally, the winter runoff should be gathered in wadis here; the wadis' subterranean flow of groundwater may be used all the year. Because of the relatively high evaporation, special irrigation systems are to be recommended (among others, irrigation at night) as well as shelterbelts around the fields. This way the heat effect of the desert winds and drifting sands can be diminished. Plants selected for cultivation in the shelterbelt should not require too much groundwater.

Extensive nomadic grazing is especially disastrous in zone II, and the sedentarization of the nomadic herdsmen is indispensable. Nomadic grazing increases the danger of soil erosion. Many completely useless areas of drifted sands are witnesses to this. Grazing has to have some connection with cultivation, and in this way an intensification of animal husbandry can be developed. Success requires a wise choice of domestic animals (in Libya satisfactory experiments have been made with fattail sheep), as well as improvement of grazing land by the sowing of fodder plants and establishment of new watering places. The grazing areas have to be recognized and to be distinctly separated from the agricultural land.

TABLE 20-1

Climatic-Geomorphological Zones and Land Utilization of the North Sahara Coastal Deserts

Factors	Zone I	Zone II	Zone III
Weathering and soil development	Lack of topsoil weathering; terra rossa formation (in limestone areas)	Chemical weathering; sandy and fine-grained soils, to some extent loess-like; climatic crust development; salt-contaminated soils, and crusts if groundwater near the surface. Corrective measures: mechanical breaking up of crusts; deep drainage to clear salt from contaminated soils.	Physical weathering prevails to a small extent; Serir and Hammada soils; fine-grained soils in flat depressions; salt-contaminated soils in depressions without outlet, and crusts if groundwater near the surface
Erosion (by water)	Heavy seasonal rainwash. Corrective measures: soil conservation by afforestation; cultivation in mountainous areas: terraces on the slopes	Linear erosion; in regolith soil sometimes quick development of deep water gullys. Corrective measures: barricading and marginal shelterbelts.	Because of scarce rainfall, largely sheet erosion; shallow rills only
Deflation (by wind)	Relatively small to moderate. Corrective measures: shelterbelts good protection against heat and duststorms.	Except for rainfall periods, considerable deflation, especially on fallow cultivated land, therefore frequent drifted sands. Corrective measures: shelterbelts, fixing of dunes by plants; preservation of natural vegetation on waste areas	Deflation during the whole year because of the scant growth; quickly moving dunes typical; oases often threatened by shifting dunes
Runoff conditions	Exoreic, fluctuating according to the season (runoff almost entirely during the winter months). Corrective measures: storage of surface water; better use of intermittent springs with abundant water (also by aqueducts)	Preponderantly exoreic, mostly periodic during winter months; runoff of wadis often only for a short time and in few sections only. Corrective measures: storage of surface water; use of groundwater flow present all year in bigger wadis	Endoreic; very little runoff, intermittent, and for short distance; cultivation in oases may be possible only with irrigation from springs and wells.
Evaporation according to Dubief, 1950	Moderate per year (1.5–2 meters potential evaporation), heavy during the summer. Corrective measures: water storage, subterranean, if possible (cisterns, etc.)	Potential evaporation 2.5–3 meters per annum, with maximum in summer. Corrective measures: special irrigation systems with good drainage, subterranean water storage, covered aqueducts, and irrigation channels	Potential evaporation 3–4 meters per annum; in cultivation in oases, danger of oversalting of the soil if scarcity of water and insufficient drainage.

In the climatic-geomorphological zone III, farming is normally not wise and may be impossible. Special attention has to be paid to this dictum because this zone approaches and occasionally reaches the coast at the Gulf of Sirte and in Egypt.

In the Northern Sahara coastal semidesert, the oases are generally small; the bigger ones are found only near the desert proper, outside the area included here. Farming is possible in this zone only at these real oases with plentiful water resources, which could be created by pumping only in exceptional cases. Moreover, every economically useful horticultural and agricultural possibility is subject to special conditions, which are briefly indicated in table 20-1. The semidesert is, on the other hand, a natural sphere of existence for the nomads. The danger of overgrazing is insignificant because of the great areas at their disposal. A disadvantage is the lack of camels, since small animal husbandry is possible only in the best but limited parts of the semidesert (better pastures, for example, because of frequent heavy precipitation in the form of dew). The establishment of watering places, and perhaps local introduction of new plants with better fodder quality, could be of some help. The optimal utilization of this zone is reached by extensive nomadic grazing. It would be a mistake to try to settle the nomads here. On the other hand nomadism should be restricted to this zone. A territorial separation of the different forms of living (a social disengagement) should be attempted which would advance the safety and the economical use of the Saharan coastal deserts.

Conclusion

The discussion herein has been limited to the potentials for food-producing land utilization, as well as the conditions pertaining thereto. Otherwise, mineral resources (oil in Libya), possibilities of industrialization, and transport problems as well as the importance of fishing and spongefishing at the coast, might have been discussed. The aim, here, was to show the practical importance of using geographical methods in order to recognize the complicated combination of natural factors in arid regions. It is not necessary to achieve, with technical and financial means, a maximum utilization, that is, the greatest expansion of farming and animal husbandry. Not only would this be uneconomical, it also would result in discouraging regressions. The knowledge and consideration of the closely related climatic geomorphological zones offer the possibility of an optimal utilization under the present conditions. This could help to increase the economic prosperity and stability of this border region of the greatest desert of the world.

Bibliographic References

ABOU-ZEID, A. M.
   1959 The sedentarization of nomads in the Western Desert of Egypt. International Social Science Journal 11(4):551–558.

DUBIEF, J.
   1950 Evaporation et coefficients climatiques au Sahara. Institut de Recherches Sahariennes, Travaux 6:13–44.
   1953 Essai sur l'hydrologie superficielle au Sahara. Algerie, Direction du Service de la Colonisation et de l'Hydraulique, Service des Etudes Scientifiques, Clairbois-Birmandreis. 457 p.
   1959–1963 Le climat du Sahara. Université d'Alger, Institut de Recherches Sahariennes, Mémoire (hors série). 2 vols.

HERZOG, R.
   1963 Sesshaftwerden von Nomaden; Geschichte, gegenwärtiger Stand eines wirtschaftlichen wie soziologischen Progresses u. Möglichkeiten der sinnvollen technischen Unterstützung. Westdeutscher Verlag, Köln. 207 p.

MECKELEIN, W.
   1959 Forschungen in der zentralen Sahara. I: Klimageomorphologie. G. Westermann, Braunschweig. 181 p.
   1961 About the problem of the climatic geomorphological structure of the desert. Zum Problem der klimageomorphologischen Gliederung der Wüste. Paper read at the 19th International Geographical Congress, Stockholm 1960 (Commission on the Arid Zone). Reprinted: Geographisches Institut, Technische Hochschule, Stuttgart. 15 p.

MEIGS, P.
   1953 World distribution of arid and semi-arid homoclimates. In Reviews of Research on Arid Zone Hydrology. Unesco, Paris. Arid Zone Programme 1:202–210. [Maps dated 1952 were revised in 1960]
   1966 Geography of coastal deserts. Unesco, Paris. Arid Zone Research 28. 140 p.

MENSCHING, H.
   1963 Die sudtunesische Schichtstufenlandschaft als Lebensraum. Fränkische Geographische Gesellschaft Erlangen, Mitteilungen 10:82–93.

MONTEIL, V.
   1959 The evolution and settlings of the nomads of the Sahara. International Social Science Journal 11(4):572–585.

TIXERONT, J.
   1956 Water resources in arid regions. In G. F. White, ed, The Future of Arid Lands, p. 85–113. American Association for the Advancement of Science, Washington, D.C., Publication 43. 453 p.

UNESCO
   1963 Nomades et Nomadisme au Sahara. Unesco, Paris. Arid Zone Research 19. 195 p.

UNESCO-FAO
   1963 Bioclimatic map of the Mediterranean zone. Unesco, Paris. Arid Zone Research 21. 58 p.

CHAPTER 21

THE CRISIS OF THE SAHARAN OASES

Jean Despois
Department of Geography, University of Paris, France

The Sahara, the largest and one of the most arid hot deserts, has long remained unknown, or little known. But during the 1960s it has held an important place in the world because of its underground products: oil in particular, but also gas and iron ore.

Since 1956 oil has been extracted from the Algerian Sahara near Ouargla (Hassi Massaoud) and along the Libyan border (Edjele, In-Amenas); gas was discovered the same year at Hassi-Rmel (south of Laghouat); oil has been extracted since 1962 in the Libyan desert, south of the Gulf of Sirte, and iron ore since 1963 in western Mauritania. All this has revolutionized the Sahara and the life of its inhabitants. Many pipelines and feeders have been laid, and a broad-gauge mining railway in Mauritania. Also, a complete network of modern roads and tracks in regular use crosses part of the desert, while airports have multiplied.

But the contact between the modern techniques and economy introduced from the outside and the traditional ways of life, culture, and agriculture has been brutal. The competition between the meager incomes from a subsistence economy based on oasis farming and nomadic pasturing on the one hand, and the higher wages provided by the oil and mining companies and the public works contractors on the other hand, soon proved disastrous.

This competition and this contact surprised a society and an economy already in a state of crisis for the preceding fifty years. Robert Capot-Rey has clearly shown in his classic book *The French Sahara* (*Le Sahara Français*), published in 1953, how the traditional Saharan economy "rests on the combined services of the slave and the camel." In the oases, the black slaves, brought in the past from the Sudan, along with the half-breeds, worked for owners who were more or less white: they drew water from the wells, maintained the *foggaras*, (underground water channels), irrigated the gardens, fertilized the date palms, and did all the cultivating and harvesting work. Transport was provided by dromedary camels (one-humped) led by nomads; while these same camels carried both the tents and the other equipment of the pasturalist during his moves from one encampment to another.

Slavery, though theoretically abolished in the middle of the nineteenth century, was eliminated from the European-settled areas of the Sahara and its margins only in the first half of the twentieth century. Most of the former slaves became land-tenants. Others, taking advantage of the peace and of the progress in transport, migrated north, or toward the Sudanese borders, in even greater numbers. Since 1925, with the opening of the first automobile roads, this movement has increased. Recruitment of tenant farmers has become increasingly difficult.

On the other hand, aid has been given in the gradual replacement of the slow small-scale caravan transport by automobile transport of ever greater capacity and speed. It was this rather than the suppression of the *razzia* (pillage raids) and the tribute imposed on the oasis cultivators and on the *vassals,* which was the main cause of the impoverishment and loss of prestige by the nomads.

Nevertheless it is in the oases, despite the abolition of the tribute paid in kind to the nomads, and the establishment of administrative centers, that the crisis appears to be greatest. It is not simply that the tenants have fled, but that the agriculture, without exception, has hardly progressed. The traditional subsistence economy, slightly opened up by barter (the exchange of dates for crops and goods brought by the caravans), was upset by the introduction of a money economy and the provision of salaries and better treatment. Merchants, officials, and members of the military were the only ones who knew how to profit from the situation, and the social inequalities were thus emphasized.

At the same time the population grew dangerously, faster than the resources of the oases, despite the increase in the number of artesian wells — at least in the lower Sahara (south of Tunisia and eastern Algeria). Therefore, the social inequalities grew as well. At Laghouat, for example, an oasis of 320 hectares, 350 kilometers south of Algiers, the land was divided into tiny lots, and the gardens into 874 parcels. In 1963 half the palm trees belonged to 74 landowners, while 57 percent held less than one-quarter hectare. There were 150 to 250 palm trees per hectare. In the Oued Righ (Touggourt area), a modest holding, varying from 20 to 200 palm trees, accounted for nearly two-thirds of the total; very small holdings (1–20 palm trees) accounted for about 20 percent and the large holdings (200 palm trees and more) for 13 percent. In the Djerid (Tunisian Sahara), 9 percent of the landowners held 44 percent of the palm trees, while 62 percent of the landowners had not even a quarter of that. In the largest oasis — Tozeur — one-seventh of the water is monopolized by a noble, *marabout* family (presumed holy) which does not even live there, and most of the gardens belong to officials, merchants and to pensioners, even more than they did formerly.

In the old oasis of Ghadames, nearly two-thirds of the family groups own fewer than 50 palm trees, while two groups own more than 500. More and more the oases, notably the major ones, are being populated by a non-landowning miserable group of people, the true sub-proletariat of small tenants, daily workers, and partially unemployed.

On top of this potential crisis in the economy of the Saharan oases, already half a century old, came the violent impact of the most modern techniques and of revolutionary types of economy, and this has caused deep confusion. Prospecting; the extraction of oil, gas, iron and various secondary minerals; the laying of roads and tracks; the construction of airports and canals; the formation of new settlement agglomerations; all occurred at the margin of oasis life. But several thousand Saharan dwellers, most of them sedentary, were attracted by the cash wages, which were much higher than the average received from oasis agriculture or from cattle rearing. Thus, while a small landowner made a hard living from the yield of a few palm trees, and while a tenant farmer was happy to receive a monthly salary equal to 100 francs (the French franc of 1958–69), a simple workhand employed by the oil or transport companies earned 250 to 300 francs a month, and nearly always received board and lodging.

The result was a drop in the emigration from the Sahara, and instead substantial internal migrations (toward several port agglomerations as well), and the quick abandonment of cultivation and cattle raising. The new wage-earners, perhaps for the first time working continuously, soon tended to form a sort of superior proletariat in comparison with the sub-proletariat of the small landowners, the tenant farmers of the small estates, and the partly unemployed. In fact, several approached the status of the long-standing aristocracy of merchants, officials, and pensioners.

But the savings of the new wage-earners were of no help to agriculture or to cattle raising; there was simply far too little "return to the land." The money was mainly used for buying bicycles and motorbikes, transistors, cigarettes, new clothes, and more varied food; it was also used for building houses, for the establishment of small businesses, or for buying trucks.

Unfortunately, most of the camps are temporary; most of the immigrants were unable to work as expected and were sent back; others could not withstand the continuous effort required and abandoned their new work. These unemployed withdrew to their oases but did not wish to take up their former occupations. The bravest looked for new camps, sometimes a few hundred kilometers away, for the affluence from the new wages led to a serious rise in the cost of living.

The problems, therefore, are (1) how to save the traditional economy of the oases; (2) how to integrate it into the modern economy; and (3) how to aid the farmers (and also the cattle breeders) so they may benefit from the new economy and techniques, instead of becoming their victims (as happens in most cases).

The new mining centers and the various camps need fresh produce which the oases can provide. Despite their water requirements (20,000 to 30,000 cubic meters per hectare per year), the crops could quite often be extended, especially into the lower Sahara, where the number of artesian wells has already been increased, and where a rich aquifer of deep water has been discovered in the sandstone of the "Continental Intercalary" chiefly belonging to the lower Cretaceous period. But it will be difficult to appreciably increase, and above all regularize, the flow of water from the wadis of the Atlas Mountains. Since the abolition of slavery, foggaras are no longer dug; the users find it difficult to maintain them, and their construction by modern methods is not at all worthwhile. Finally, the oases which can be irrigated only by well water seem destined to disappear, for the tiring job of lifting the water, whether by a weight mechanism or by other means using an animal, does not allow irrigation of more than a few acres or a fraction of a hectare. Instead, the water must be provided by simple gravity, because a mechanical pump is a machine usually much too expensive and tricky to handle and to repair for the frustrated isolated peasants. Furthermore, there is the risk of exhausting the less plentifully supplied aquifers.

In addition to the natural and technical obstacles, there are more serious problems within the social order, which confront the attempts to extend, improve, and diversify agriculture. Moreover, in most of the attractive oases, the property is crumbling into tiny lots, and a large part of the water and many of the palm trees belong not to those who irrigate and cultivate the land but, as has been seen, to the merchants and officials — be they active or retired — and only occasionally to the nomads, for whom the gardens and palm trees provide a modest but safe investment.

The progress of Saharan agriculture requires technical and financial help from the state. But would the latter help the absentee landowners and those who refuse to work and irrigate the land themselves, scorning as "slave laborers" those who actually do it themselves? Does not the second part of the alternative thus demand a complete revolution in the social structure, and therefore comprise a political choice?

However, perhaps the most formidable obstacle facing the rebirth of the oases is the lack of interest in agriculture on the part of the youth, and their emigration from the oases. Also there is the absence of a large number of active men because these have been attracted by the wages of the modern enterprises. This lack of care for the oasis sometimes attains disastrous dimensions, notably in Libya. In the Fezzan, in Brak, the water from the artesian sources is only half used and forms dangerous ponds. The palm trees of Ouad el-Adjal are less in danger. Already in 1962 L. Eldblom estimated that in Ghadames gardening was practicable for no more than one-third of the active male population. In Cyrenaica the oases that extend along both sides of the 29th parallel — Marada, Audjila, Djalo — are becoming deserted. In Algeria, Chr. Verlaque notes that "one no longer finds anyone to maintain the oases, such as Souf and Tidikelt, which entail considerable amounts of work."

The economy of the oases is thus in a severe state of crisis. Meanwhile, the causes are known and methods of expanding the agriculture are being developed. The obstacles are known and are all the more serious as they are primarily human and psychological, and they demand the revival of the oases. This therefore necessitates not only a large amount of capital and the presence of many agricultural technicians to assist the cultivators of the oases in raising their standard of living, despite the increase in population, but also a competent and comprehensive administrative framework.

It must be admitted that the recently recognized independence of the various states comprising the great desert has only aggravated the situation. The colonial countries, in particular France, had a sufficient number of specialized civil and military administrators in the Sahara, and a number of experienced people were included. The elite and the cadres of the newly independent states are chosen citizens. For the North Africans, as for the Negroes south of Mauritania and north of Mali and Niger, the Sahara seems a foreign land where posts are feared because of their distance and isolation. The new officials often feel exiled and lack the experience and the means, and therefore lack authority. In Mali and Niger, where the Saharans represent the black states, the officials are despised for their color by the Berber Arabs, and above all by the white nomads of the desert. Serious incidents have occurred with the Tuareg.

The financial resources, on the other hand, need not be lacking in a Sahara with so great a production of oil, gas, and iron. A small percentage of the state income from these products is set aside for the economic development of the oases. But the countries on which the Sahara depends are poor and in a state of crisis, for their budgets are ill-balanced and they generally run deficits. The only important production in Mauritania is iron ore; oil and gas are the main resources in Algeria, where the agricultural production is weak; and oil is almost the sole export from Libya. Libya however is the exception; always a poor country, she has risen in several years to fifth position among the oil-producing countries.

The major benefactors are, therefore, the countries whose underground resources are mainly responsible for the maintenance of the budget. The beneficiaries especially include the inhabitants of the towns and ports wherein the central services and administration are sited. These people are predominantly urban merchants, but also include those of the Sahara, who reap most of the benefit from the laborers, and create a new business bourgeoisie. But even if the tertiary sector of the economy has sometimes received a great push, such as in Libya, neither the secondary sector (industry) nor the primary (agriculture) have really benefited.

As regards the Saharans of the oases, they appear to have been forgotten. The youth and the active males abandon agriculture, and the underground wealth of their countries helps only a few families, 15,000 at the most, of about 500,000 who populate the Sahara from the Atlantic Ocean to Egypt.

Bibliography

ATTIA, H.
   1965 Modernisation agricole et structures sociales. Exemple des oasis du Djerid. Revue Tunisienne de Sciences Sociales 2:59–79.

CALCAT, A.
   1959 Etat actuel et possibilités de l'agriculture saharienne. Institut de Recherches Sahariennes, Travaux 18:133–159.

CAPOT-REY, R.
   1953 L'Afrique blanche française. II: Le Sahara Français. Presses Universitaires de France, Paris. 564 p.

CLARKE, J. I.
   1963 Oil in Libya: Some implications. Economic Geography 39(1):40–59.

DESPOIS, J.
   1965 Problèmes techniques, économiques et sociaux des oasis sahariennes. Revue Tunisienne des Sciences Sociales 2:51–57.

DESPOIS, J., AND R. RAYNAL
   1967 Géographie de l'Afrique du Nord-Ouest. Payot, Paris. 571 p.

ESTORGES, P.
   1964 L'irrigation dans l'oasis de Laghouat. Institut de Recherches Sahariennes, Travaux 23:111–137.

FOSSET, R.
   1962 Pétrole et gaz naturel au Sahara. Annales de Géographie 71(385):279–308.

MARBEAU, V.
   1965 Les mines de fer de Mauritanie, Société Anonyme des Mines de Fer de Mauritanie (M.I.F.E.R.M.A.) Annales de Géographie 74:175–194.

MARTHELOT, P.
   1965 La révolution du pétrole dans un pays insuffisamment développé: La Libye. Cahiers d'Outre Mer 18(69):5–31.

MERLET, M.
   1957 Enquete sur le revenu de la population musulmane de la commune mixte de Laghouat. Institut de Recherches Sahariennes, Travaux 15:113–143.

MIEGE, E.
   1958 L'agriculture au Sahara et ses possibilités. Bulletin Economique et Social du Maroc (Rabat) 22:25–75.

NESSON, C.
   1965 Structure agraire et évolution sociale dans les oasis de l'Oued Righ. Institut de Recherches Sahariennes, Travaux 24:85–127.

VERLAQUE, C.
   1964 Le Sahara petrolier. Comité des Travaux Historiques et Scientifiques, Paris, Section Géographique, Mémoire 1:9–452.

CHAPTER 22

EILAT: SEASIDE TOWN IN THE DESERT OF ISRAEL

David H. K. Amiran
Department of Geography, The Hebrew University, Jerusalem, Israel

Eilat, the southernmost town of Israel, is the nation's only maritime outlet to the Red Sea, and through it to the Indian Ocean. It shares common characteristics with many towns in arid environments. It has to balance precariously the limited advantages of its location against the limitations imposed by the strict frame of its arid environment. Furthermore, compared with nonarid areas, it has but a rather limited choice of alternatives. With many towns of the Near East the area of Eilat shares a long history of occupance. However, the town of Eilat proper is one of the "development towns" established only after Israel achieved independence. Many interesting features of a seaside town in the desert can, therefore, be studied at Eilat. It is the purpose of this chapter to discuss problems of infrastructure and of choices of alternatives at a desert location. It does, therefore, not deal with port activities and the mining industry, both of which play a major role in the economy of Eilat.

The Regional Setting

Mainly for reasons of the availability of groundwater, the major settlement of the area was located nearly throughout history at the northeast corner of the gulf, from Biblical Ezion Geber to present-day Aqaba, the sister town of Eilat in the Kingdom of Jordan (fig. 22-1).

The road distance from Eilat to Beersheba, the nearest town of consequence in Israel, is 235 kilometers; the distance to Tel Aviv is 342 kilometers. Road distance from Aqaba to Ma'an, the nearest town in Jordan, is 115 kilometers; to Amman, the capital of Jordan, it is 360 kilometers. The paved road to Eilat was completed only late in 1957. It passes over 200 kilometers of arid desert, including some stretches with very steep and rugged topography. From the point of view of the economic geographer, it is desert land too, hardly generating any traffic en route.

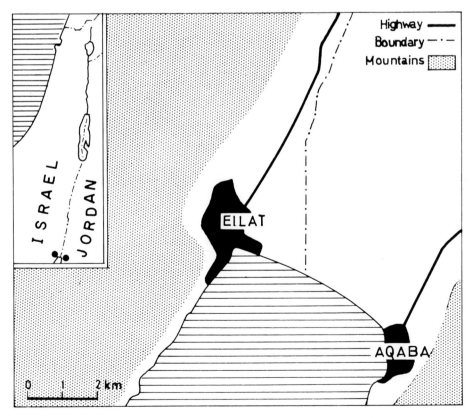

Fig. 22-1. The port cities of Eilat (Israel) and Aqaba (Jordan). In the inset the cities are indicated by dots at the south end.

After a military garrison had been established at Eilat in 1949, followed initially by a lively influx of population amounting to some 500 people, Eilat had a slow beginning. By 1956 it had but some six hundred inhabitants. It was the opening up of regular shipping through the Straits of Tiran and down the Red Sea, as a result of the Sinai campaign of 1956, which boosted the development of Eilat. The population reached 2,000 by the end of 1957 and grew steadily to about 13,000 by 1969.

The Physical Setting

Topography. The location of Eilat within the rift valley and at the shore of the gulf implies ipso facto that its connections are either due north through the Aravah, the southern section of the rift valley, or by sea down the gulf.

The town was sited on an intermediate, flat, and slightly inclined surface above the western shore. Its altitude is 25 to 90 meters above sea level. This site provides satisfactory building ground and lessens disruption of life in town when one of the rare cloudbursts occurs. It furthermore evades the high groundwater table in the Aravah, sets apart the flat land on the north shore for the resort industry and the airport, and permits the remaining flat land to the north to be cultivated by the suburban kibbuts-village of Eilat, which some 3 kilometers to the north of the town grows winter vegetables, date palms, flowers for the northern and export market in winter, some fruit for local consumption, and alfalfa as fodder for its dairy herd.

Climate. Eilat has a typical arid climate. Maximum temperatures are above 30°C from April through October and may exceed 40°C in every one of these months. The absolute maximum temperature recorded so far was 47.4°C on June 8, 1961. Even in midsummer, late-night temperatures cool down to about 25°C. Whereas noon in winter is still quite warm with temperatures in the twenties, night temperatures are distinctly chilly and minima down to +1°C have been recorded.

As is typical for a desert station, it is rather difficult to give a meaningful numerical expression for rainfall at Eilat. Its occurrence is restricted to the winter months from October to April or May. The rainfall record for the year 1965/66 may serve as an adequate illustration. On January 1, 60 percent of the total annual rainfall was recorded, and four of the ten days on which measurable rainfall occurred accounted for 88.5 percent of the annual total. Annual totals vary considerably, but concentrations of moister and drier years stand out in the record. Clearly, "averages" are of no meaning here. Significantly, since observations were started here in 1940–41, no year produced less than 2.5 millimeters (or one-tenth inch) of rainfall. By contrast, intensive rainfalls are by no means exceptional. If we list all daily rainfalls in excess of 5 millimeters we find that these occur in 17 of a total of 23 years for which data are available (Aqaba excluded), that is, in 74 percent of the years. As usual in a desert environment, the intensive rainfalls are a curse rather than a blessing. The record rain of March 31 to April 1, 1953, was a deluge of 50.5 millimeters (1.99 inches) within 21 hours, or 65 percent of the total rainfall for the year, making this the year with the highest rainfall recorded at Eilat. Sheetfloods of mud and water came down from the surrounding mountains, spreading out over the fans, and coating the streets with a slimy film of mud, as a result of which no vehicle could move in town during three days.

The Development of Eilat

The military occupation of Eilat in March 1949 brought about the establishment of a permanent garrison here. In 1950 civilian population established itself, and by 1951 Eilat was accorded municipal status. In those early years and until 1956, Eilat was but a small town which tried a variety of activities to create an economic foundation. Fishing, small industry, the port, and resort industry were all tried in turn, and mining was added later. All of these activities were boosted greatly by the Sinai campaign late in 1956, which opened up the gulf and the Red Sea to Israeli shipping. All of these activities bear the clear imprint of arid environmental conditions, and it is from this point of view that they shall be considered here.

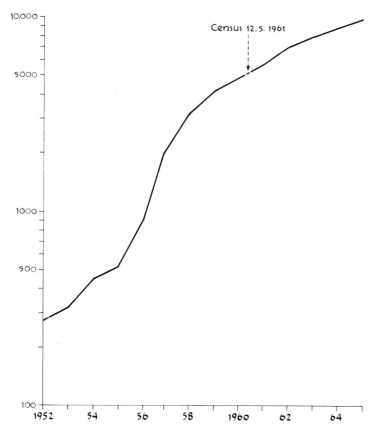

Fig. 22-2. Population growth of Eilat from 1952 to 1965.

A variety of efforts were made to develop industry in Eilat as a reliable base for employment, all of these of rather small scale. Basically they can be classified into two groups, the one trying to develop special resources of the Eilat area, the other service industries to serve Eilat itself. These included construction firms, a plant producing prefabricated parts for houses, stone quarries, carpenter shops and shops of a variety of other craftsmen, garages, a dairy product and ice-cream factory, a bottling plant for soft drinks, and a large laundry. Although all these service industries must be rated as nonbasic or secondary sources of employment, they have been one of the main bases of employment for the population of Eilat.

Much less prominent is the role of industry based on resources available locally but not producing for local consumption. The first such plant was established in the very early days of Eilat, late in 1950, some four kilometers south of the town at the western shore of the gulf. It was to utilize one of the obvious resources of the area — fish — to provide Israel's agriculture with fish meal, a commodity mainly imported. Its rated capacity was one ton per hour. Neither the volume nor the steadiness of fish landed throughout the year was sufficient to justify the operation of the plant, and it was eventually dismantled without ever having been put into operation. Only slightly better fared another plant based on Eilat's fishery, a fish-processing and canning factory. It has employed up to 30 workers, with considerable seasonal fluctuations.

A different type of raw material was to be used by a sizable processing plant for decorative stone located at the site of the dismantled fish meal plant. The raw materials were granite and certain other plutonic and metamorphic rocks quarried nearby. Its products were gravestones and ornamental stone for building purposes. This plant operated for a number of years, nearly all the time at financial loss, due in part to the high cost of quarrying, to the uneven quality of rock, and in part to the high transportation charges, as products exported had to pay for the overland haul to Haifa, about 450 kilometers. The plant finally ceased operations in 1962. However, decorative stone found in the Eilat area, especially malachite, forms the base of a small and stable jewelry industry.

The Water Supply

To provide a fair water supply for a town in the desert of 15,000 inhabitants and possibly more in the future is a problem both in quantity and quality. As no water of acceptable quality is available at Eilat, the town was supplied in its early years by a supply piped from springs and wells as far distant as Yotvatah, 38 kilometers north of Eilat. Not only were the amounts available from these sources limited, but their mineral content proved to be considerable and troublesome, reaching 500 milligrams per liter at Yotvatah and 740 elsewhere. Furthermore, some of these sources contain unpleasant amounts of sulfides, especially magnesium sulphate.

With the inauguration of a dual-purpose power and distillation plant in April 1965, the situation improved vastly. By mid-1966 about 60 percent of the water consumed at Eilat originated from the distillation plant, while 40 percent was groundwater. Water was supplied to consumers at a salinity of 200 to 220 milligrams per liter. Eilat, therefore, has nearly the best water supplied to consumers anywhere in Israel.

A considerable part of the water consumed in Eilat is used in air-conditioning appliances. In order to economize in water use, pumps for recirculation were installed in every flat. Their cost (about 150 pounds, or $50) was borne by the consumer and financed by government credits. There resulted a considerable economy in water use, as shown by the average daily consumption in the peak months of 1965 and 1966 respectively, with 10,500 cubic meters in August 1965 and 7,125 cubic meters in July 1966. Average overall consumption per day in 1966 was about 530 liters per person.

An important experiment in seawater conversion is part of the Eilat situation. A four-unit desalination plant operates with the Zarchin process, which separates saline brine and pure water by a freezing and unfreezing process. According to figures which must definitely be considered preliminary, the Zarchin plant has quoted a cost price of about IL (Israeli pounds) 0.8 (24 U.S. cents)* per cubic meter if 400 cubic meters per day are treated. This would decrease to IL 0.6 (18 cents) if production increases twofold. By contrast, one cubic meter of distilled water from the dual-purpose power plant costs IL 0.3 (9 cents), not accounting for the power generated by the same plant.

Both these prices are excessive for sea-level locations, which do not include lifting to higher altitudes. Even the lowest of the prices quoted is five to six times higher than the cost of water in northeastern Israel. The Zarchin-plant price is similar to the prices charged to consumers at the highest locations in Israel, involving lifting to 800 meters altitude and more. Here is a clear example of the cost of amenities due to the arid environment.

Alternate Choices of Economic Activity

Eilat was founded in 1949 or 1950 without any direct antecedent settlement, save for a lone small desert police post. After twenty years it had a population in excess of 13,000. A survey and assessment of its development and future prospects are of particular interest, as they pose possibilities, limitations, and conflicts rather typical for many arid areas.

---

*Prices quoted are those of the mid-1960s.

Possibilities. Eilat, like many a coastal arid-zone site, is endowed with three natural assets: mineral resources, fishery, and the desert climate.

Although the basement complex endows the Eilat area with a considerable variety of minerals, at present only copper ore is exploited, at the Timna mines. Recent prospecting has shown the ore body to be of considerably larger volume than known hitherto. Nevertheless, Timna is a minor copper producer and therefore exposed to the economic fluctuations of copper on the world market. With a working force of about 900, the Timna copper works employ approximately 20 percent of Eilat's working population. It is the largest single employer. The majority of these are residents of Eilat. The families of a small minority reside in the north, mainly Tel Aviv; the man working at Timna flies home for his weekends. This group of people employed at Timna and not resident at Eilat comprises mainly higher-grade technical or managerial staff.

The fishery, which at first was considered an important resource and base of employment, has become unimportant, the main reason being the rather high temperature of the waters of this subtropical gulf. Whether intensive research will provide the basis for significant improvement remains to be seen.

At first reasoning the gulf would not be regarded as a particularly rich fishing ground, considering its warm subtropical water. This might be illustrated by a comparison of the average surface temperatures of the offshore waters at Eilat (29°30′N) with those off the coast of Peru (14°30′S) — the average temperatures of the former being 26°–27°C in summer and 20°–21°C in winter, and the latter being 16°–19°C in summer and 14°–15°C in winter.

The same factor that makes fishing poor in the Gulf of Eilat is at the root of its resort industry. The dry and healthy climate of this seaside resort prevails at different levels of temperature throughout the year. Even in midwinter swimming, bathing, and all types of water-sport are pleasant at Eilat. Its appeal is enhanced by the colorful scenery of its desert mountains. Eilat, as a resort and tourist center, appeals to two types of foreign tourists. The majority come for a short visit while touring Israel; they enjoy a swim, the view of the coral banks from a glass-bottom boat, and possibly a short drive over one of the mountain trails. The others go to Eilat for an

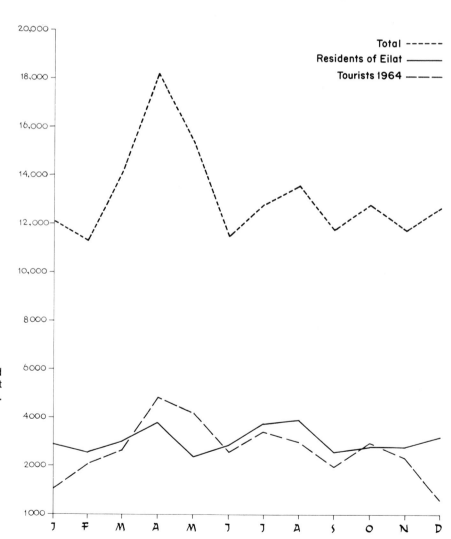

Fig. 22-3. Number of passengers carried per month in 1965 on Tel Aviv–Eilat flights.

extended vacation for anything from two weeks to two months or even more. Many of these foreign tourists, especially those from northern and central European countries, time their vacation for the winter, to get away from the most unpleasant season at home.

Israel makes the lesser contribution to the tourist trade of Eilat. However, there is a certain percentage of Israeli vacationers who do go to Eilat, either for a winter vacation, or in large numbers during the holidays, especially Passover and the Feast of Tabernacles. Eilat's special attraction at these times is, once more, climatic. Whereas elsewhere in Israel weather is still bound to be rather cool at Passover, it is warm and pleasant for bathing at Eilat. But apart from this, the desert scenery attracts tourists to Eilat throughout the year.

Ten hotels were catering to tourists at Eilat in 1969; four of these were rated class A and six class B. Their combined capacity was some 500 rooms or about 1,000 beds; of this number 220 rooms were in class A hotels. Planned additions to existing hotels, as well as a new hotel, will increase the number of rooms to about 700 and the number of beds to about 1,400. To this number should be added the youth hostel, which can accommodate another 165 people.

Its location as the sole port of Israel with access to the Indian Ocean provides Eilat with an important economic function. However, its volume depends on economic and political considerations, and the port activity, therefore, fluctuates as influenced by them.

Limitations. The distance of Eilat from the centers of economic activity imposes considerable limitations on its economic activity. The local market is too restricted to sustain many industries catering to the needs of Eilat only. On the other hand, transportation charges to Tel Aviv and the rest of the national market make Eilat's industry noncompetitive. We find, therefore, two types of rather small industrial plants at Eilat: those catering to the need for food and to the need of construction.

Unfortunately, efforts to establish diamond-polishing at Eilat have been unsuccessful. The special advantage of the diamond industry would be its footloose character because of the irrelevance of transportation costs. Isolation and the unavailability of competent managerial manpower were the reason for the failure of this industry at Eilat.

Due to the limitations of employment potential in an isolated town such as Eilat, the municipal authorities have made it their carefully considered policy to use temporary labor from the north for construction work, rather than inducing an appropriate number of people to settle down permanently at Eilat. As construction work is an ephemeral phase in the economic life of the town, especially considering the lack of a hinterland requiring or absorbing part of its labor force, the use of itinerant labor prevents the formation of considerable unemployment or underemployment as construction work slackens or ceases. Eilat, as a result of this policy, has one of the lowest unemployment ratios in Israel, about 1.2 percent in the summer of 1966 at the height of an economic recession, and practically none since 1968. As a principle, such a policy seems wise in its application to isolated towns in general and those in arid areas in particular.

Conflicts of Interest and Alternate Choices. The limitations in economic opportunity in an isolated desert environment create the desire to broaden its economic base. The creation of a sizable industry holds, therefore, particular attraction for Eilat. However, such industry should not interfere with, and should have no nuisance value detrimental to, the established economic base of Eilat, the resort industry.

Eilat experienced in this area a rather typical conflict of interests. A group in which oil interests and the municipality cooperated planned the construction of a refinery at the oil terminal of Eilat with a view to basing a petrochemical industry on the refinery. Actually, work started on the site of the refinery. A rather heated discussion arose between those in favor of this project and a vociferous opposition who claimed that a petrochemical industry would be detrimental to tourism and that the effluents of the refinery would destroy the coral banks and fish-life in the gulf. The protagonists of petrochemical industry maintained there would be no air pollution, a somewhat doubtful claim as inversions are a meteorological feature of arid areas. The antagonists pointed out that, unproven facts aside, one could not expect tourists, foreign and local alike, to spend their vacation in view of — and in smell of — a sizable petrochemical combine. Eventually, the government ruled against the continuation of work on the refinery site at Eilat and insisted that the petrochemical industry in the Eilat area should be located well north of the town, out of view, at a site where it could not affect the tourist industry.

This is a clear case of difficult decision-making. One choice was the acceptance of the obvious location of a refinery and an accessory petrochemical industry at a maritime oil terminal, providing steady production and employment. This choice, however, was incompatible with the tourist industry, well established and with a potential for expansion. This resort industry is based not only on the scenery of Eilat's desert mountains placed against the background of the sea, but on its clean and dry air and on clean water for swimming.

This conflict required a clear-cut decision and did not permit of compromise. Rather often man, in an arid environment, has to choose between such mutually exclusive alternatives. The harsh nature of the desert does but rarely permit of compromises.

CHAPTER 23

THE UTILIZATION OF THE NAMIB DESERT, SOUTH WEST AFRICA

Richard F. Logan
Department of Geography, University of California, Los Angeles

The Namib, a cool coastal desert of southwestern Africa, a nearly rainless area virtually devoid of vegetation, was occupied in prehistoric times by primitive hunting and gathering peoples who sometimes herded small flocks of goats and cattle. Modern developments focus chiefly on two ports that serve the better-settled interior and are the bases for a fishing fleet. Of lesser importance are scattered mining operations, both subsistence and commercial grazing, and a rapidly expanding tourism.

Physical Setting

The Namib Desert parallels Africa's Atlantic Coast from Mossamedes in Angola across the full length of South West Africa to the mouth of the Olifants River in the Çape Province of South Africa — a distance of nearly 1,200 miles (fig. 23-1). Its width is much less, seldom exceeding 100 miles, and probably averaging between 60 and 80 miles. Its eastern border in the north and center is the foot of the Great Western Escarpment, atop which more humid lands prevail; but in the south,

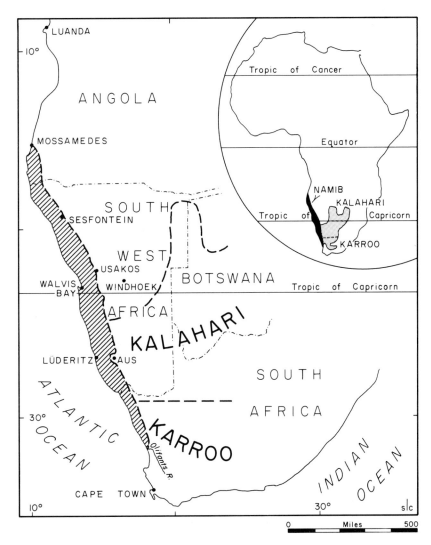

Fig. 23-1. The Namib Desert.

Fig. 23-2. Landforms of the Namib.

a gradual transition merges the Namib with the Kalahari and Karoo Deserts of the interior.

Climate. Climatically, the Namib is a contradictory area: it is almost rainless, yet its air is normally at or near the saturation point, and fog is very common. Temperatures in the coastal area are mild at all seasons, ranging between a maximum of 67°F and a minimum of 58°F in the warm season (December) and between 57° and 46°F in the cool (July). Inland, more continental conditions prevail, with summer temperatures reaching the eighties each day and dropping into the low forties on the average winter night. At the coast, humidity is at 100 percent for 19 hours per day in summer and for 11 hours in winter. At the inner edge of the desert, the air is much drier, fog is virtually unknown, and the humidity seldom exceeds 50 percent. This is nevertheless fairly high for a desert region.

Violent weather is practically unknown. Rare thunderstorms bring the precipitation, which averages less than two inches annually inland, and about one-half inch at the coast. Strong winds blow a few days a year from the interior, bringing heat and discomfort to the coast. But storms such as characterize midlatitude areas are totally unknown.

Landforms. The Namib is characterized by several distinct types of landforms: a vast sand sea, monotonously flat plains of gravel and bedrock, mountains of bare rock, and areas with surfaces fretted into strange sandblasted forms.

Most of the desert consists of a broad platform, eroded into bedrock, of monotonous flatness, and rising in a very gradual slope from the coastline to an elevation of some 3,000 feet at the foot of the escarpment. South of the tropic, much of this platform is veneered with sand, in the form of sheets, waves, and dunes, the latter sometimes exceeding 800 feet in height. Scattered sand areas also occur in the northern third. Elsewhere, the rock platform is sometimes mantled with a thick layer of water-laid gravels, often cemented with gypsum. Over large areas, the bedrock itself (largely mica-schist, granite, and gneiss) comprises the surface.

In the southern and central portions, mountains rise above the smooth platform only as isolated individuals and short chains. In the north, the platform is narrower, and in some areas a broad belt of mountainous terrain intervenes between the platform and the escarpment. Most of the mountains are rugged, steep-sloped, and almost soilless.

Cut below the platform in the northern half of the Namib are canyons of intermittent streams, which extend westward from the escarpment toward the sea (fig. 23-3). Steep-sided and rugged in the extreme, these canyons nevertheless have broad flat floors which, although subject to attack from sudden floods, have some potentiality for irrigated agriculture.

Vegetation. In general, the Namib is a barren waste, with very sparse vegetation. Large areas, notably the gravel flats, the bedrock platform, and the dunes, are almost totally barren. In the zone of heaviest fog near the coast, especially in the south, low succulent bushes grow scatteredly. Along the eastern border, a thin to moderate cover of annual grasses appears in most years, supporting for a time herds of antelopes, zebra, ostrich, and their attendant predators. Riparian vegetation, largely in the canyons of the major streams, consists of acacia trees and various shrubs, usually growing in a broken, scattered pattern.

Soil. Over the mountains, bedrock platforms, canyon sides, and dunes, soil is essentially absent. On the gravel flats and in coastal areas, high alkalinity, salinity, or presence of gypsum, renders the soil toxic to all but the hardiest halophytes. Only on the interior plains and some river bottoms are arable soils to be found.

Water Supply. Potable water is found only as subflow beneath streambeds — chiefly of the larger streams that rise in the rainier plateau east of the escarpment. In some cases, dissolved salts render the water unpleasant. In other cases, such as the Kuiseb and the Koichab, which supply Walvis Bay and Lüderitz, the quality is good to excellent. Untapped waters probably occur in association with other streams, especially those that terminate in the great sand dunes south of Walvis Bay and those that flow to the sea across the northern Namib.

Before the introduction of Koichab water to Lüderitz in 1969, that port was supplied entirely by thermal distillation of seawater, using coal brought by rail from the Transvaal.

Experiments made in the time of German colonial administration at both Swakopmund and Lüderitz indicated a great water-producing potential in the moisture content of the air. Thin copper sheets, exposed to the air, had such rapid radiational heat loss after sunset that their temperature dropped quickly below the condensation level, and considerable quantities of atmospheric moisture were condensed upon them. This resource has never been utilized.

Settlement History

Before its occupation by Europeans late in the last century, the Namib was sparsely occupied by primitive native peoples. Small wandering bands of Saan or Strandloper (Beachrunner) Bushmen subsisted by strand-gathering and hunting, and by the gathering of tubers. Because of the scarcity of both food and water, the population density was low. Small bands of Topnaars, a Nama (Hottentot) subtribe, lived along the lower Kuiseb River, getting water from shallow wells in the river bed, grazing their flocks of goats and herds of cattle on the

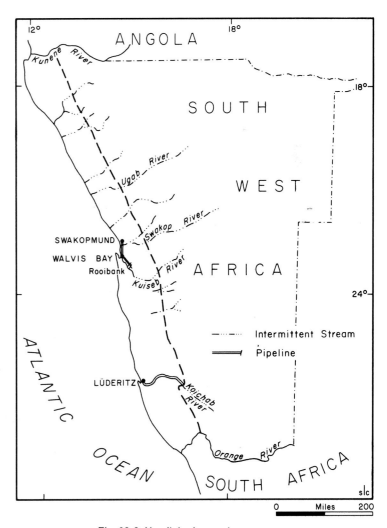

Fig. 23-3. Namib hydrography.

limited riparian vegetation, gathering nutritious *narra* seeds from melons growing amid the dunes, and fishing in the shallow coastal lagoons.

Between 1484, when a Portuguese expedition first explored the Namib coast, and 1876, when the British laid claim to Walvis Bay and established a customs post there, little of lasting importance occurred in the Namib. At Walvis Bay, whalers, many of them from New England, early maintained temporary bases. By the 1870s, traders of several nationalities were bartering for ivory and hides with the tribes of the plateau, and hauling their goods by ox wagon to and from Walvis Bay.

In 1884, Germany formally proclaimed a protectorate over the whole of South West Africa (except the British enclave at Walvis Bay). Because the only deep-water harbor on the entire coast was in British hands, the Germans of necessity built an artificial harbor nearby at Swakopmund. From it, a railroad was constructed across the desert to the interior, where a commercial grazing and mining economy was soon established.

Swakopmund grew rapidly during the first couple of decades of German control. It had an important port and transportation function, with the attendant administrative functions. In addition, the cool conditions prevailing at the coast during the summer proved attractive to Germans sweltering in the heat of the interior. Early in the German period, there was initiated the custom of moving the seat of government from Windhoek to Swakopmund for the period December through February, which in turn established the port as the social capital of the territory during those months. Curiously, too, Swakopmund established a brewery, which gained a very high reputation for its product, in spite of the fact that the quality of the local water is so bad that it is barely drinkable at times. Water for the brewery was, at first, brought from Germany by ship; later, for some forty years after World War I, it was brought by railroad tank-car from Walvis Bay, which in turn obtained it by pipeline 23 miles from the underflow of the Kuiseb River.

Lüderitz, established by a Bremen merchant in 1884 continued as a minor port until the early years of the twentieth century. At that time, in the construction of the rail line connecting it with the interior, diamonds were discovered, bringing on a mining rush in the desert and a short-lived boom at the port.

Post World War II Situation

Since World War II, South West Africa has enjoyed a period of growth, development, and prosperity that has advantageously affected all segments of its population: the manifestations of this are clearly evident in the coastal desert area (fig. 23-4).

Fisheries. The cool waters of the Benguela Current, flowing northward along the Namib coast, are particularly rich in plankton, making this one of the richest fishing grounds in the world. Its potentiality was rather recently realized, however, with the result that the fisheries have undergone a striking development in the last two decades.

The most important fish taken is the pilchard or sardine. Formerly these were almost entirely canned, but in recent years world demand for fish oil and fish meal has resulted in an increasing proportion being allocated to those products. In 1966, 72,000 tons were canned, but 175,000 tons of fish meal and 34,000 tons of fish oil were produced. Since the early 1960s, there has been increasing exploitation of the whitefish population, especially hake. While some of the fisheries are shorebased, operating chiefly from Walvis Bay, a great proportion of the catch is made by vessels of many foreign nationalities, some operating in conjunction with factory ships which completely process the fish at sea, producing frozen fillets, canned fish, fish meal, and fish oil. In many cases the operations are highly efficient, with the schools of fish being located by scouting planes or helicopters and tracked by sonar, and with the catching done by fleets of smaller vessels which deliver their catch to the factory ship.

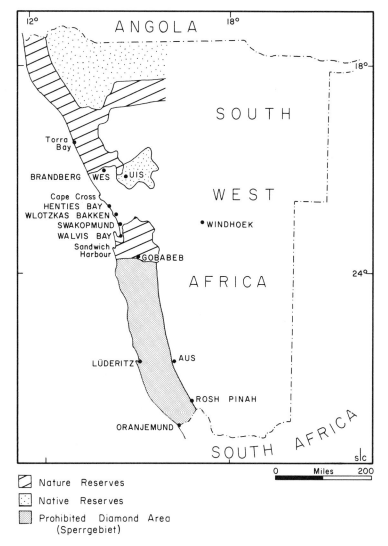

Fig. 23-4. Namib land allocations.

The rock lobster or crayfish industry, based chiefly at Lüderitz, exports large quantities of live lobsters, canned lobster meat, and frozen lobster tails.

Some of these products are marketed within South West Africa and South Africa, but the bulk of them are marketed overseas, in almost every part of the world.

Port and Urban Developments

Walvis Bay. Walvis Bay has the only harbor on the South West Africa coast accessible to deep water vessels. The harbor is naturally sheltered from all but northwest winds and swells by a long sandspit (Pelican Point) built from the southward. Beyond that, however, the place originally had little to recommend it. The margins of the bay are extremely shallow, with mud flats extending far seaward; fog is prevalent; and sulphurous submarine emissions cause severe damage to paint, hardware, and seawater distillation apparatus.

Today, however, a dredged channel makes it possible for large freighters and tankers to come alongside the modern, efficient wharves. From the center of the city northeastward, fishing factories line the shore of the bay, each with its own wharf for the unloading of the catch through a suction line that conveys the fish directly into the factory. Within, the fish are cleaned, cooked and canned, or cleaned and frozen, or converted directly into fish meal or fish oil, depending upon the kind of fish and the state of the market. Thanks to the rainless climate, sacked fish meal is stored out-of-doors alongside the factories, sometimes for months, until it is transported by truck to the deepwater wharves for loading onto a freighter. Fish oil is stored in large tanks, which resemble petroleum storage tanks, until transferred by pipeline to seagoing tankers.

The port also functions in various relationships to the foreign fishing vessels working offshore. Ships of many nationalities call there for watering and provisioning, for repairs, for hospitalization of injured or ailing crew members, and for transshipment of their catch to other vessels for transport to markets. Industries accessory to the fisheries are located at Walvis Bay, including rope and net supply firms, marine engine companies, a marine railway, and instrument installation and maintenance firms.

As the only deep-water railhead of the territory, Walvis Bay serves as the principal entryport for the interior. In addition to fish products, exports are chiefly minerals, wool, hides, and pelts, destined for all parts of the world. Imports consist chiefly of manufactured goods, petroleum, and foodstuffs, chiefly from northwestern Europe and the United States. The bulk of the ocean traffic of the port is overseas, although some coastwise traffic is carried on, chiefly with Lüderitz and Cape Town.

As a city, Walvis Bay has a most unprepossessing physical situation. It sprawls on a mud flat at the mouth of the Kuiseb River, which has repeatedly flooded it during years of extraordinary rainfall on the interior plateau a hundred miles or more to the eastward. It is subjected to much wind, drifting sand, a high frequency of fog, high humidity at all times, which results in much mildewing, and salty soil, which makes gardening impossible. Many of these handicaps have been overcome, or even used to advantage, however: the roads are surfaced with salt dug from the nearby salt pans and mixed with clay, and the deliquescent impurities in the salt extract moisture from the air and keep the road surface firm and slightly moist; dikes hold out the floods; the municipality has erected a stout fence along the windward side of town, which catches the drifting sand in its eddy, thereby creating a "municipal sand dune" more than thirty feet high and several miles long; and soil is brought from the plateau atop the escarpment more than one hundred miles away to be used for lawns and gardens.

The original buildings of the towns were raised on stilts to evade the floods and the drifting sand. Today, well-built modern homes, some with attractive yards and gardens, spread far out from the center of town, evidencing both the rapid recent growth of the community and its prosperity: a reflection of the expansion of the fisheries and the growing trade through the port.

The European (white) population today numbers about 5,500, and the total population (all races) about 18,000. In accordance with the prevailing custom throughout southern Africa of separating the residence areas of the various ethnic groups, the Coloreds (mixed bloods) have their own section, Narraville. Most of these people are Cape Coloreds (Malay, Hottentot, White mixtures from the vicinity of Cape Town) engaged in the fishing industry, both on the boats and as foremen in the fishing factories, and the high wages received by them are reflected in the quality of their homes.

The labor force employed in the fishing factories consists largely of Ovambo males, Bantu (Negro) tribesmen who came from their homeland astride the South West Africa–Angola border to work for twelve or eighteen months under a voluntary cash-wage contract. They are housed in compounds (barrack-style housing with group feeding arrangements) operated by the cannery or by the municipality. Other Ovambos constitute the labor force of the wharves. The Bantu population of the city totals 10,835, of which 8,579 are Ovambo.

The water supply for Walvis Bay is derived from wells tapping the underflow of the Kuiseb River at Rooibank, 23 miles to the southeast, where the river is still confined to a well-defined channel. It is delivered to the city (and on to Swakopmund) by pipeline. The quality is good, and the amount sufficient to supply the city to several times its present consumption of over one million gallons per day.

Petroleum products reach Walvis Bay by sea, but coal comes by rail from the Transvaal. Packaged and canned foods come from overseas by ship and from South Africa by rail. Some fresh vegetables are available from small farms on the Kuiseb Delta and in the lower reaches of the Swakop River valley, and some fruit comes from the northern interior of the Territory, but most of the fresh food supply comes in by rail from South Africa.

In short, except for its water, all of the supplies of Walvis Bay are obtained from great distances; and most of the goods handled at the port have originated at great distances and have yet a great distance to travel to their ultimate destinations. Even the fisheries utilize the port for their operations merely because it happens to be nearby. The town merely occupies space in the desert — a desert in which it has no real interest of any sort. However, the desert poses certain restrictions and conditions upon the community, to which the city has to adjust or adapt. And thus the desert has imparted a certain atmosphere and character to the city which sets it apart from other port and fishery communities in more "ordinary" environments.

Lüderitz. In many aspects, Lüderitz is a small-scale replica of Walvis Bay; but it has many strongly distinctive characteristics. Its harbor is well sheltered from all seas, but its rocky islets and submerged bedrock reefs prevent its use by larger vessels. A twenty-foot channel has been dredged to the wharf, but larger vessels must be worked from lighters in the bay.

The chief role of Lüderitz is as the base for the rock lobster or crayfish industry. Boats based here operate for several hundred miles along the rocky coast, as well as on some offshore sea-mounts. The catch is handled in canneries and freezing plants, and the product is shipped by coaster to other ports (Walvis Bay and Cape Town) for transshipment, or loaded from lighters onto larger vessels in the bay. As a cargo port, the role of Lüderitz is minor; its hinterland is poor, and such trade as generates there is more likely to be handled by Walvis Bay or Cape Town. Its inland connections are via a rail line and a single graded road.

The town has a most irregular pattern, rambling in an elongated fashion along the much indented coastline and up over rocky hills that border the bay. Roads are narrow, crooked, and sometimes steep. The town is plagued by high winds, dust, sand, and all-pervading dampness. It has long been inadequately supplied with water produced by thermal distillation of seawater, but since May 1969 an abundant supply of freshwater has become available from the underflow of the Koichab River through a 67-mile pipeline across the Namib.

Lüderitz has a population of about 7,000: 2,400 Europeans, 1,600 Coloreds, and 3,000 Bantu (largely contract laborers at the factories).

Swakopmund. Although it was the chief port of the Territory during the German time, Swakopmund ceased to exist as a port shortly after the take-over of the country by South Africa in 1915. A rail line constructed from the German railhead at Swakopmund across the dunes to Walvis Bay resulted in the immediate transfer of all port functions to that place, with its better harbor and facilities. Today the harbor facilities at Swakopmund are entirely dismantled and/or silted up. The old Customs House is a museum, and the office of the main shipping line is a school hostel. The lighthouse (emblem of the town, appearing on the town seal) still functions — but as one of the navigation aids for Walvis Bay. Only two industries exist at Swakopmund: a tannery and the brewery. The resort function will be discussed later.

Möwe Bay. At Möwe Bay, a minor indentation in the rocky coastline 250 miles north of Swakopmund, a future harbor development is being considered. Its purpose would be to serve certain mining areas of the adjacent desert, and as a fishing harbor with fishing factories; it is also possible that it might be connected with a larger hinterland by a transdesert rail route. Water is available in considerable quantity from the underflow of two rivers in the vicinity.

Mining

Diamonds, produced in South West Africa only in the Namib, constitute by far the most important mineral product in the territory. Except for sporadic "discoveries" at other coastal points, all are produced in the south, and almost entirely in the extreme south. The diamonds are alluvial, being found in beds of gravel, usually under a considerable overburden of sterile materials. The large-scale operation thereby necessitated in diamond recovery, together with the desire for close control over production and sales, has resulted in the concentration of the entire operation under one corporation: the Consolidated Diamond Mines (CDM). Because of the high value and light weight of the product and the resultant ease of theft (combined with the difficulty of rescue operations for retrieving "poachers" stranded in this extremely arid region), the entire diamond-bearing area and much of its surroundings has been proclaimed a Prohibited Area *(Sperrgebiet),* totally closed to public entry at all times. Diamond sales are completely controlled by the government, and illicit diamond buying (IDB) is one of the most heinous crimes that can be committed in South West Africa.

The CDM company town of Oranjemund, at the mouth of the Orange River, is the center of diamond mining operations. It is a highly developed, progressive, modern community, with a good water supply from wells near the river, and with gardens supplying fresh produce. Offshore dredging operations for the recovery of diamonds from the sea bed have been carried on, partly under American capital and direction, since the mid-1960s.

The history of mining of metals in the Namib is similar to that the world over. Some mines have been operated recurrently since about 1890, working when market prices are high and wages low, and closing at other times. Many are very remote, posing great problems of transportation — both of supplies to the mine and of ore from it. Scarcely a year passes without the discovery of another "spectacular" deposit, many of which prove to be quite worthless.

Important operations include the Rosh Pinah mine (zinc) in the remote southern Namib near the Orange

River, eighty miles from the coast; and the Uis and Brandberg Wes mines (tin and wolfram) near the inner edge of the Namib 120 miles north of Swakopmund.

Semiprecious stones have been mined sporadically at various places during the last couple of decades. Most operations are seriously handicapped by the problems of supply, transportation being poorly developed in many areas.

Large-scale production of salt is carried on in the coastal pans and lagoons north of Swakopmund, using modern mechanical equipment. Approximately 2,500 tons of guano are produced annually by scraping that bird excrement from the rocky islets off the coast between Walvis Bay and Lüderitz; and another 1,000 tons is removed annually from an artificial "Bird Island" (a wooden platform) between Walvis Bay and Swakopmund.

Oil prospecting has been carried on in various parts of the Namib and the adjacent sea bed for a few years without success. Obviously, the discovery of oil reserves would be of great value to the country, which now has to import all of its petroleum products from overseas.

Grazing

Vast areas of grasslands appear in most years along the eastern margin of the Namib. On them, since time immemorial, great herds of game have grazed. Today these have been replaced, in part at least, by domesticated animals — chiefly cattle and karakul sheep.

The cattle, raised chiefly for beef, are usually driven overland (trekked) to the nearest rail line and shipped thence to markets in South Africa alive. Karakul sheep are raised for the wavy pelts of the newly-born lambs, which must be killed and skinned in the first 24 hours of life if the pelt is to be of marketable quality (soon thereafter the curl tightens, becoming kinky and hence valueless). This removal of the lamb before it begins nursing eliminates a major drain on the ewe's vitality and allows her to exist and reproduce on grazing too meager for ordinary sheep raising. Quality of the pelt is determined genetically by the ram, and the breeding is carefully controlled by the European ranch owner. The flock grazes under the watchful eye of a native shepherd, who carefully collects lambs as soon as they are born. Skinning and curing is done at the ranchstead. Pelts are sold either to itinerant buyers or through marketing cooperatives, ultimately making their way to the London auctions. The flock is also sheared annually, the coarse wool (used chiefly for carpeting) being shipped overseas from Walvis Bay and Lüderitz.

Before the advent of the Europeans, these desert margins were almost unused, owing to the lack of water. Beginning in the German time, ranches (farms, in local parlance) of 10,000 to 25,000 hectares were developed by European settlers, who, after developing water resources by drilling wells, have been able to carry on commercial grazing. Unfortunately, particularly during the wetter cycles, this practice has been allowed to extend too far westward into the arid zone, and many ranchers have suffered acutely from the inevitable droughts. All of the outermost row of ranches should be bought back by the administration and added to adjacent game reserves, present or future.

Ranching here is similar to modern ranching in the western United States and in Australia, except for the presence of the karakul and the use of native herders. Travel is by truck, pick-up or land rover; drilled wells are pumped by windmill and/or diesel or gasoline engines; ranges are usually fenced; animals are dipped, innoculated, sprayed, and otherwise protected against parasites and disease; and a predominantly cash economy prevails. Ranch houses are modern, with running water, paraffin (kerosene) refrigerators, and are usually served by telephone. Children of school age live most of the year in school hostels in some larger town (such as Swakopmund or Windhoek) and return home only for the longer holiday periods. Native workers and their families occupy dwellings of their own construction, either in the immediate vicinity of the European rancher's home or at the detached outposts where the particular herd for which they are responsible is grazed. They receive cash wages plus a weekly ration of food, and keep their own flocks and herds as well.

In the northern half of this marginal area, seminomadic native herders seasonally graze their flocks of goats and herds of cattle. Over large areas, watering points are very widely scattered, and much potential grazing goes untouched. Some new water sources have been developed by the administration, but much of their extensive water-seeking efforts in the area have proven fruitless. The population density (both of herds and of humans) accordingly remains low. Some of the better-favored areas were occupied by European ranchers earlier in the twentieth century, but have been bought back by the administration within the 1960s and added to the existing native reserves.

These native herders live under tribal organization and on a purely subsistence non-cash economy. Their cattle are a small wiry native stock, and while some herders have crossed their animals with improved strains provided by the administration, many have refused to do so, preferring to keep the blood lines "pure." The people live chiefly on the milk of the cattle, which is prepared in several ways, but not on the meat: the numbers of cattle are expressive of the owner's social status, and cattle are slaughtered only ceremonially. On the other hand, goat meat is commonly eaten. Several tribes are represented here: Ovahimbas and Ovatjimbas (Herero splinter groups), Damaras and Namas (Hottentots). Most of these groups have been in the general area for at least a century, and consider it as their traditional grazing ground.

Agriculture

Agricultural development is restricted, occurring only on the floodplains of two of the rivers: in small patches on the Kuiseb delta near Walvis Bay and on the lower Swakop River near Swakopmund. Most of these patches are small, being scarcely more than large gardens. Much could be done, as about the settlements of the Topnaars on the Kuiseb, where water and fairly good land are both available. It is, however, not in the tradition of these people to carry on irrigation agriculture. Were this the Arab world, water would be lifted from the shallow wells by *shadouf* or *saqia*, and lush gardens would flourish on the floodplain. But here, goats graze at will, seeking out edible shoots on the already overgrazed riparian vegetation.

Dry-farming is nowhere attempted, and rightly so, since the combination of deficient rainfall and thin soils would make such an attempt disastrous.

Transportation

Two rail lines traverse the Namib (fig. 23-5), both being attenuations of the rail net of South Africa, operated by the South African government on the "standard" (3.5-foot) gauge of that system. One line connects Lüderitz on the coast with the main line at Seehiem, near Keetmanshoop, in the interior. The other crosses the Namib in a northeasterly direction from Walvis Bay and Swakopmund to Usakos and eventually to Windhoek. Daily passenger trains operate on both lines, as well as freights.

A tarred road crosses the Namib from Usakos to Swakopmund and Walvis Bay, roughly paralleling the rail line. A graded road, usually fairly well maintained, parallels the Lüderitz rail line, and a similar road extends south near the coast from Lüderitz to Oranjemund. Two graded and maintained roads cross the desert from Windhoek to Walvis Bay and to Swakopmund. Another skirts the inner edge of the desert from the Windhoek-Walvis Bay road southward to the Lüderitz road and continues southward to the mining camp of Rosh Pinah. Three graded roads cross the north-central portion of the desert: one from Usakos directly to the coast at Henties Bay; one from Outjo directly to the coast; and one from the vicinity of the Brandberg Wes and Uis mines diagonally across the desert southward to the coast. A salt-surfaced road runs north up the coast from Swakopmund for about 140 miles. There are no roads of any sort in the area between the Kuiseb River and the Lüderitz rail line, seaward of the north-south road mentioned above, and only a few unmaintained truck trails in the northern desert.

Regular air services connect Walvis Bay with both Windhoek and Lüderitz, and Lüderitz with Alexander Bay, in South Africa at the mouth of the Orange River. There are a number of scattered strips, usually unmanned, for administrative and emergency purposes.

Recreation and Tourism

The cooler summer climate remains the chief reason for the existence of Swakopmund. Since the early years of the German time, the administration of the territory has moved its offices from Windhoek to the coastal town for the midsummer period, and with it comes much of the social life of the country. The town's population soars from its permanent figure of about 2,400 Europeans to more than 7,000 during Christmas week, when every hotel room, municipal bungalow, and pension is full, and people are even housed in railway sleeping cars parked at the railway station.

The town retains a quite Germanic atmosphere in many ways: German foods are served in the restaurants and cafes; German music sounds from loudspeakers at the beach; German books and magazines predominate in the shops; many of the buildings, with their half-timbering, their towers and turrets, balconies and ornamentation, seem to belong more on the shores of the Baltic than the edge of the Namib; and in the language heard and the clothing seen on the streets, the high quality of the Scientific Society, the energy and drive of the people, the performances in the theaters, a Germanic culture and attitude prevails.

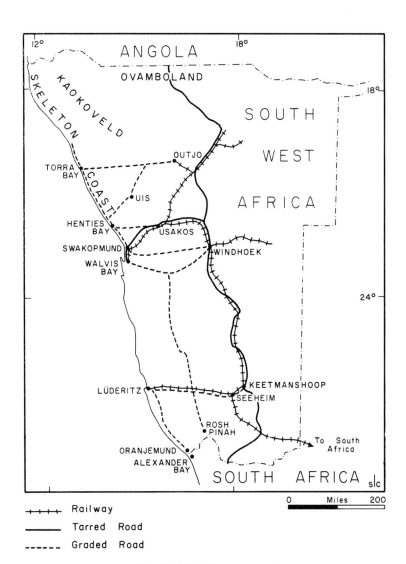

Fig. 23-5. Namib transportation routes.

Poor quality of water long plagued the town (Swakopmunders, long used to their salty coffee and tea, were widely reputed to salt these beverages when served them in other towns!), but the recent connection by pipeline with Walvis Bay's high quality water supply has quite altered the situation. Reconstruction and remodeling of some of the town's hostels, development of the municipal bungalow "colony," and the opening of new cafes, combined with the general prosperity of the territory, increased leisure time and a shorter work week, the paving of the main road across the desert, and the filming of several motion pictures in the adjacent desert have resulted in greatly increased popularity of Swakopmund.

The Skeleton Coast stretches north from Swakopmund to the Angola border — foggy, bleak, and barren. For over a century, it has been notorious for its shipwrecks and the sufferings and deaths of the voyagers cast up on its shores. Virtually inaccessible by either land or sea, it was seldom visited by anyone. During the 1960s, however, the construction of the coastal road and the road from Usakos directly to the coast have made the area accessible, and its cool moist summer climate and excellent sports fishing have made it attractive to many inlanders. Many camp out along the coast, while others have built structures ranging from crude, simple shacks to very comfortable houses. Two communities, Henties Bay and Wlotzkas Bakken, have developed rather amorphous agglomerations of scattered bungalows and embryonic commercial developments.

The Department of Nature Conservation and Tourism of the South West Africa Administration is beginning to exert controls over certain portions of the Namib coast, and to constructively plan for its development in an enlightened manner. Camping areas are being designated and regulated. The northernmost of these is at Torra Bay, about 180 miles north of Swakopmund, near the point where the road from Outjo reaches the coast. Sandwich Harbour, 25 miles south of Walvis Bay, a popular fishing spot, much frequented by residents of the port city, is also under the department's supervision.

Lüderitz has considerable attraction for the residents of the southern interior — an area of great summer heat and dryness. The remoteness of the town, its poor access road, and the inadequate water supply (later remedied) have been serious handicaps to its development.

Nature Reserves. A large part of the Central Namib, easily accessible to both Walvis Bay and Windhoek, has long been a proclaimed game reserve (fig. 23-4). It is, especially along its eastern border, the natural grazing ground for large flocks of springbok, gemsbok (oryx), zebra, and ostrich for at least part of the time — they are exceedingly mobile and migrate over long distances. These forms of wildlife have suffered great decimations from both drought and illegal hunting between the late 1930s and the mid-1960s. Little control was maintained over the area due to lack of both personnel and financing, as well as the fact that game was generally considered to be unlimited and inexhaustible. The area is now well controlled, and rest camps, nature education, and research programs are being developed. In addition, the scientific research station of Gobabeb is situated within the area.

The northernmost part of the Namib within South West Africa (the Skeleton Coast and the Kaokoveld) is quite inaccessible and almost completely undeveloped. A narrow belt along the coast is designated as a nature reserve, and the remainder is Native Territory.

The Future

Since Walvis Bay and Lüderitz exist solely because of their dual roles as bases for the fishing fleet and ports for the interior, they are particularly sensitive to any forces that affect these functions. Thus any overseas situation that increases or decreases the foreign trade of South West Africa affects the prosperity of these ports; and their growth, even their very existence, depends upon the continued prosperity of the rest of the territory and the continued existence of reservoirs of merchantable fish in the neighboring ocean. The prosperity of the territory fluctuates with the price of karakul (a luxury item on the world market, and subject to the whims and fads of fashion), local droughts, and the presence of certain cattle and sheep diseases (an epidemic of foot-and-mouth disease in 1960–62 caused a total cessation of animal exports). It seems almost inevitable that the fish population will decline markedly in the near future. Overfishing will undoubtedly deplete the reserves. South West Africa has imposed stringent quotas and closed seasons upon its fishing industry, but the operations of the large foreign fleets offshore go on without control. No international organization controls fishing on the open seas; thus far, meetings to consider the fisheries situation in this region have attracted only a few of the participants, with the large operators being conspicuously absent.

Mining will probably continue to be carried on in its present sporadic, intermittent fashion, with little change, barring the possible discovery of oil.

Agriculture has little potentiality in the Namib, due to a lack of both water and soil suitable for cultivation. Even with large-scale, long-distance diversions of water, agriculture does not seem likely to increase greatly in importance in the future. There is always the possibility of gardening developments, either hydroponic, or using the system developed by the University of Arizona at Puerto Peñasco in Mexico and Abu Dhabi in the Arabian Trucial States (Hodges, Chapt. 7 herein); but even these would be greatly reduced in efficiency by the incidence of coastal fog in the very areas where the need is greatest.

Grazing should be rigidly curtailed throughout the Namib and its border areas. In good years, large areas can be safely grazed; but to encourage ranchers to settle in the area is exceedingly dangerous, since it means that they and their herds will also be there in drought years, with consequent excessive damage to both the land itself and its vegetative cover, as well as the impoverishment of the people involved. It is far better to rigidly resist any attempts at "taming" the marginal areas, and to let them revert to their natural state, thus preventing the waste of both the natural and the human resources.

On the other hand, tourism and recreation can be developed far beyond their present state, allowing them to replace grazing as the major use of the land and field of employment. On an intra-territorial basis, the annual "urge to the sea" produces a great potential for Swakopmund, which can use more and better hotels and additional entertainment facilities. The first 150 miles of Skeleton Coast north of Swakopmund have great potentials for further development as a recreational area. However, careful planning and control are needed to prevent the creation of a recreational slum, dotted with shacks and littered with debris. Semi-residential communities should be designated, planned, and controlled by administration authorities, with minimum standards of housing set and maintained. Camping areas should be designated, provided with water and sanitary facilities, and properly maintained; indiscriminate camping should be forbidden; and hotels, motels, and rest camps with good facilities should be provided. Carefully selected coastal reserves should be established and stringently regulated, as is done at the Cape Cross seal colony, to protect the fish, seals, birds, and flora of this unique coast.

Tourism from the rest of southern Africa and from overseas can be attracted in a large and profitable way if the proper attractions and accommodations are provided. The various aspects of local color should be preserved and even accentuated, ranging from the Teutonic image of Swakopmund to the colorful garb of the local native women. The great game herds of the inner border of the desert should be preserved and enlarged by natural reproduction and should be made available to tourists through road improvement, air strip construction, and the provision of strategically placed rest camps. More areas should be designated as Nature Reserves: all of the southern Namib, from the ocean eastward to the escarpment, and including some of the present farms (but excluding, of course, the Sperrgebiet); and all of the northern Namib, beyond the Ugab River, from the coast to the inner edge of the desert in the central Kaokoveld. Such vast empty unspoiled areas have a great attraction in our modern world, and as such can become a real economic resource. Closed to indiscriminate public entry and left entirely underdeveloped, these reserves could be opened to carefully regulated safari-type expeditions operated basically for tourists. All shooting of game, collecting of minerals or artifacts, or interference with local native groups should be prohibited. At the same time, the Kaokoveld natives should be encouraged to continue their use of the area for grazing — to do otherwise would disrupt their economy and their seasonal migrations and at the same time would deprive the tourist of the opportunity of visiting their camps.

The Namib, despite its extreme physical handicaps, has a modest potential for future development, but only if handled carefully to prevent the destruction of those features which are attractive yet exceedingly fragile.

* * *

NOTE: The field studies upon which this report is based were made possible by grants from or contracts with the Foreign Field Research Program of the National Academy of Sciences-National Research Council (1956–1957), The Social Science Research Council (1961–62), the National Science Foundation (1965) and the University of California. In addition, the writer owes much to many officials of the South West Africa Administration, without whose friendliness and cooperation this work would have been impossible.

Bibliography

HINTRAGER, O.
   1956   Südwestafrika in der deutschen Zeit. Kommissionsverlag, München. 255 p.

LOGAN, R. F.
   1960   The Central Namib Desert, South West Africa. National Academy of Sciences/National Research Council, Washington, D.C., Publication 758. 162 p. (ONR Field Research Program, Report 9)
   1961   Land utilization in the arid regions of southern Africa. II: South West Africa. In L. D. Stamp (ed), A History of Land Use in Arid Regions. Unesco, Paris. Arid Zone Research 17:331–336.
   1969   Geography of the Central Namib Desert. In W. G. McGinnies and B. J. Goldman (eds), Arid Lands in Perspective, p. 127–143. American Association for the Advancement of Science, Washington, D.C.; University of Arizona Press, Tucson. 421 p.

McGINNES, W. G., B. J. GOLDMAN, AND P. PAYLORE, EDS.
   1968   Deserts of the world. University of Arizona Press, Tucson, 788 p.

MEIGS, P.
   1966   Geography of coastal deserts. Unesco, Paris. Arid Zone Research 28. 140 p.

SOUTH AFRICA, COMMISSION OF ENQUIRY INTO SOUTH WEST AFRICAN AFFAIRS
   1964   Report, 1962–1963. Government Printer, Cape Town. 557 p.

SOUTH AFRICA, DEPARTMENT OF FOREIGN AFFAIRS
   1967   South West Africa Survey, 1967. Government Printer, Pretoria. 190 p.

VEDDER, H.
   1938   South West Africa in early times. Oxford University Press, London. 525 p.

PART FOUR

AUSTRALIAN DESERTS

CHAPTER 24

ECONOMIC DEVELOPMENT OF THE AUSTRALIAN COASTAL DESERTS

Joseph Gentilli

University of Western Australia, Nedlands, Western Australia

The Northwestern Littoral

The Eighty Mile Beach leads to the northern limit of the northwestern coastal desert region of Australia. The name *Ninety Mile Beach* was commonly but not exclusively given to the smooth coastline reached by the Great Sandy Desert until the year 1946. It was then officially changed to *Eighty Mile Beach* in order to avoid confusion with a similar toponym in Victoria.

The prevalent anticyclonic circulation which occurs at the source of the trade winds must have, in a slightly drier climatic phase of an earlier era, driven large amounts of desert sand from the interior to the sea. This accounts for the smooth flat coastline and for the local widening of the continental shelf. Under the present conditions of rainfall, the dunes of the interior are mostly stabilized by vegetation. It is estimated that a decrease of 125 millimeters in the annual rainfall would bring about the extension of the desert vegetation for some 60 kilometers further north, past the northern end of the Eighty Mile Beach. An increase of only 25 millimeters would create a continuous semiarid corridor along the shore, linking the northwestern highlands (Hamersley, etc.) with the wetter Broome and Kimberley region. This corridor would be narrow and frequently affected by drought. An increase of 50 millimeters annual rainfall would almost eliminate this coastal desert, reducing it to a narrow coastal strip from Onslow to Roebourne (Gentilli, 1951).

This is one of the few desert areas affected by tropical

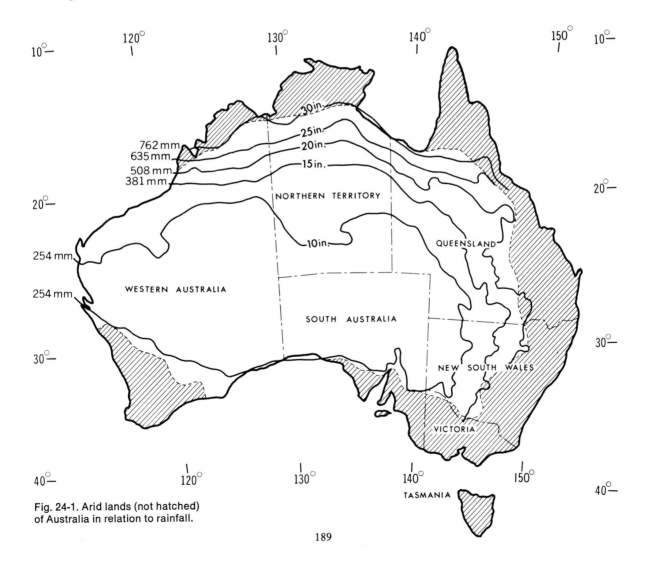

Fig. 24-1. Arid lands (not hatched) of Australia in relation to rainfall.

cyclones, which, on the average, occur about twice a year, but only about once a year cross the coast and penetrate inland. Throughout this region the month of March on the average brings more rain than either February (which is wetter further north) or April (which is wetter further south). East of about 116° East longitude, the coastal desert is so affected by tropical cyclones that March receives more rain than February and three to four times more rain than April. Even a cyclone which dies out before reaching latitude 22° S, with its center not traveling beyond 150 to 200 kilometers inland, can bring substantial falls of rain, possibly more than 200 millimeters in 24 or 36 hours over a limited area, perhaps 100 millimeters or more over some 30,000 to 50,000 square kilometers. Within a few days a cyclone may bring 200 or even 300 millimeters to 150 or 200 kilometers of coastline, where the average *annual* rainfall, including rain from tropical cyclones, is only 250 millimeters. Without these cyclones, the annual rainfall varies between 100 and 150 millimeters; the enormous value of these tropical cyclones to this otherwise extremely arid region, in terms of rain, is quite obvious. This is enhanced by the fact that the vegetation is conditioned to respond almost immediately to any heavy rain, irrespective of season.

These heavy cyclonic convergence rains usually fall on the north side of the Hamersley highlands; a peculiar pattern of drought emerges immediately to the south and west. The area of heavy downpours may be taken as a criterion for the inland delimitation of at least one section of the coastal desert, from Port Hedland westward to about longitude 116°. The combination of tropical cyclonic and winter frontal rains is useful as a criterion for further subdivision.

The regional subdivisions of the whole coastal area are outlined in table 24-1. Since most of the region lies beyond the normal reach of the frontal winter rains from the southwest and the monsoonal thunderstorms from the northwest, its main source of water is provided by the erratic tropical cyclones. No locality with a long and reliable meteorological record qualifies for the definition of extreme aridity currently fashionable, namely that it has at least one stretch of 12 rainless months, but many localities have been absolutely without rain for 9 or 10 months. If insignificantly light falls of rain are disregarded, Mardie shows a dry spell of 13 consecutive months, against its absolutely dry 10 consecutive months. Whim Creek's dry period increases from 10 to 17 dry months. At most other localities, however, prolonged droughts have been broken by heavy falls of rain, and the distinction made above does not apply.

It must be stressed that at least from Exmouth Gulf to the Eighty Mile Beach, the rainfall conditions differ radically from year to year, according to whether there has been a tropical cyclone or not; it is therefore necessary to recompute the rather meaningless often quoted averages. If the years without cyclones are considered, this coastal desert is arid indeed, failing by very little to qualify for the conditions of extreme aridity. Carnarvon, for instance, received only 74 millimeters of rain during 1966. In contrast tropical cyclones may deliver 250 or 300 millimeters of rain in one month. The effect of the influence of such rainfall patterns on soils and vegetation has not been studied, but it must be considerable. The soils of the western subdivisions show a siliceous hardpan of enormous thickness which is very probably due to the alternation of prolonged drought and occasional deep flooding. There is an apparent discrepancy between the normal dryness of river beds and their thorough development; this is due to the fact that after some of the heavier cyclonic downpours the rivers carry enormous amounts of water, albeit for a few days or even just a few hours. The western and central subdivisions are therefore normally areic, but occasionally exoreic, although they have a fully developed drainage network (Gentilli, 1952). The eastern subdivision is wholly areic, covered as it is by desert sands; fortunately, it lies on a large sedimentary basin, from which artesian water can be obtained. (An *exoreic* area has normal drainage to the sea, whereas an *areic* area has no regular drainage at all.)

The population of the coastal desert region and its hinterland was about 7,800 persons of predominantly European descent in 1961, and over 15,600 in 1966. This extraordinary increase was due almost entirely to the exploitation of enormous deposits of iron ore found in the interior. A limited increase in the Shark Bay area was due to the development of prawn fisheries, partly offset by the end of whaling. The aboriginal population in 1966 was about 3,800, less than half being of pure aboriginal descent. Because of the good pay available at construction and mining sites, many aborigines have now left their pastoral employers, and live in the new towns.

For many years the main link of this coastal region with Perth, the capital of Western Australia and the main port at Fremantle, was by sea. The small local

TABLE 24-1
Subregions of the Coastal Desert Area

Subregion	Geographic Location	Rainfall Patterns
I. Shark Bay & Carnarvon	To about 23½°S	Frontal winter rains absolutely predominant
II. Point Cloates to North West Cape	23½° to 21°S	Frontal winter rains predominant; autumn tropical cyclonic rains conspicuous
III. Exmouth Gulf to beyond Onslow	To about 115°40′E	Autumn tropical cyclonic rains dominant; early winter frontal rains conspicuous
IV. Roebourne and Port Hedlund	To about 119°45′E	Autumn tropical cyclonic rains dominant; late summer rains (some due to earlier tropical cyclones) conspicuous
V. Eighty Mile Beach to end of region	To about 121°15′E	Summer rains dominant; autumn cyclonic rains almost as important

Australian Coastal Deserts

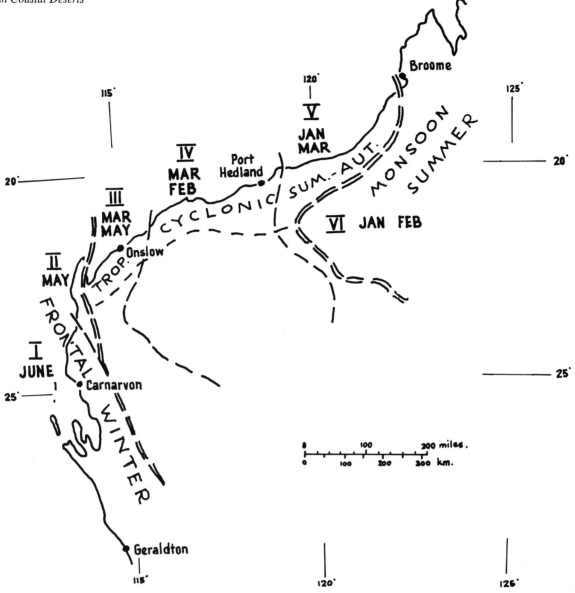

Fig. 24-2. The rainfall subdivisions of northwestern Australia. Frontal winter rains predominate in the extreme western fringe, part of which, subregion I (Vlaming Head, Carnarvon, Shark Bay), receives just enough rain to be classified as semiarid rather than arid. On the Exmouth Peninsula (subregion II), the wettest month is May; further south (Subregion I) the wettest month is June. Tropical cyclones of late summer and early autumn bring erratic but torrential rains to most of this area and make March the wettest month, followed by the frontal rains of May in subregion III, and by the monsoonal or cyclonic rains of February in subregion IV. Further east monsoonal rains become prevalent, but March rains are still conspicuous in subregion V, while they are much less relevant in subregion VI, which is quite distinct. Further inland (outside the coastal belt) summer rains are more important.

ports were primitive, their jetties had to be unusually long because of the very shallow beaches, but the volume of traffic (building materials, later on petroleum products, and few other goods landed, and wool shipped) did not justify any great expenditure on harbor improvement. Damage from cyclones was a permanent threat. Even now, with the exception of Port Hedland, these ports have practically no interstate trade. The goods discharged are mostly gasoline and other oil products, building materials, refrigerated foodstuffs, machinery, vehicles, and live animals to restock the sheep stations. The goods shipped are minerals and wool. A regular shipping line to northwestern and northern Australia is provided by the State Shipping Service, operating since 1912, with a fleet of seven ships in the late 1960s, of which the four "K" carry cargo and passengers, and the three "D" cargo only.

Road transport is fast increasing in importance, with improved roads available, and the increase in the number and capacity of vehicles. The North-West Coastal Highway links all the main coastal centers; the Great Northern Highway links Port Hedland southward with the interior, and eventually Perth. Most of these highways are surfaced with gravel only and need regular grading. Hard-surfacing is not considered economical, except for the coastal tract from Carnarvon to Port Hedland, which should be completely paved by 1975, at a cost of 17 million dollars. Heavy trucks and trailers do much damage to the road surface, especially at the prevailing high summer temperatures. There are several tributary roads

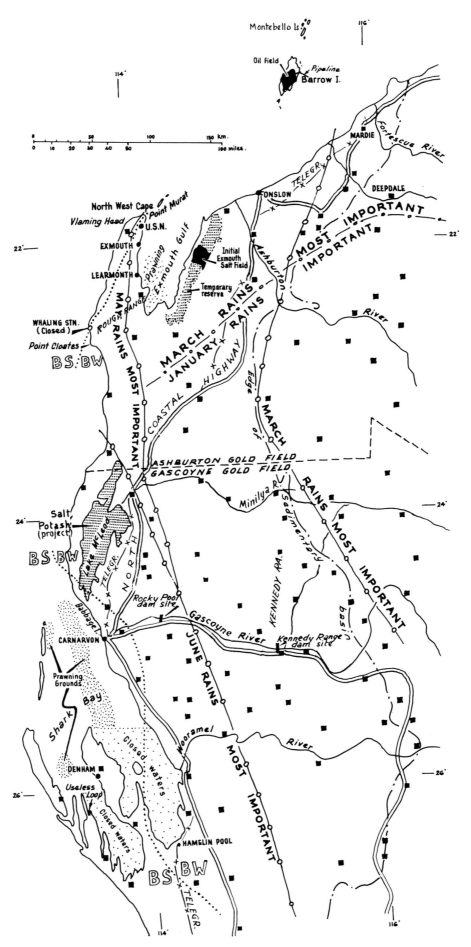

Fig. 24-3. Subregions I, II, and III (see fig. 24-2) are separated by dash-and-circle lines, according to rainfall. Sheep stations are shown by small black squares. The dotted line separates the small semiarid (BS) areas to the west from the arid (BW) areas. Notice the prawning grounds within Shark Bay and Exmouth Gulf, and the oil field on Barrow Island. The eastern boundary of subregion III coincides fairly closely with the edge of the Carnarvon Basin and its sedimentary rocks. The borders of the gold fields shown have only historical interest. The telegraph line is shown where it runs at some distance from the North Coastal Highway.

and many local roads, unsurfaced but graded occasionally. All unpaved roads, including the highways, may be closed to traffic for some days after heavy rains. By the end of June 1965 there were 362 kilometers of paved roads, 1,062 kilometers of gravel roads, and 8,521 kilometers of unsurfaced roads. In addition, considerable lengths of "unformed roads," which might better be classified as tracks and for which statistics are incomplete, are very dusty in dry weather and impassable in wet weather. The motor vehicles are predominantly trucks, four-wheel-drive vehicles, and pick-ups; long-wheel-based heavy cars are preferred. A recent inflow of workers seeking employment in the new mining, building, and mechanical activities brought a great increase in the number of private cars of all types, ages, and state of repair.

This region was one of the first in the world to have a commercial airline; commercial aviation began in 1921, when the Commonwealth Government subsidized the airmail service from Geraldton to the northwestern ports as far as Broome. The company was West Australian Airways Limited, founded by Major Brearley (Colebatch, 1929). A new company, MacRobertson Miller Aviation, began a service from Perth to Daly Waters, in the Northern Territory, in 1934; this provided a new link with the chief northwestern centers. The two companies merged in 1955 to form MacRobertson Miller Airlines Limited, which was later acquired by the Ansett-ANA interests. These services first ensured a speedy postal link between this remote region and the rest of Australia, then enabled passengers to reach any northern town in a very short time and in comfort, and more recently even provided means of transporting heavy and urgently needed machinery to building or mining sites. All airports are under the supervision of the Australian Department of Civil Aviation and are regularly inspected and frequently improved. There are also numerous landing fields at the smaller centers and at nearly every pastoral station. Regular air-coach tours, available from May to September, are gradually becoming popular, but the great distances and the high cost of food and accommodations keep tourist numbers down.

As elsewhere in Australia, all first-class mail travels by air. Ground connections, however, remain slow because of the great distances involved; second-class mail may go by road or by sea and may take a fortnight to reach its destination. Telegraph lines along the two highways ensure a very efficient service, occasionally disrupted by tropical cyclones. Telephone communications are less efficient, especially toward the northern half of the region, where lack of power is the main handicap. A broad-band telecommunication system was being installed between Perth, Carnarvon, and Port Hedland, between 1967 and 1970, at a cost of 12 million dollars. Because of the enormous output of iron ore, there may be a need for some 500 megawatts of electric power by 1977, probably to be met by natural gas.

Shark Bay and Carnarvon (Subregion I)

Subregion I is characterized by the absolute predominance of winter frontal rains, which actually begin at the end of autumn, mostly in May. Morphologically, its outstanding feature is the large and complex "Bay," which in fact is a series of composite gulfs bounded by littoral structures, with no freshwater stream except the usually dry Wooramel River, and therefore of high salinity. North of 26°S latitude fiscal concessions are allowed to residents, because of the hardships imposed by climate and isolation.

The waters of Shark Bay are inhabited by the small Shark Bay Pearl Shell (*Pinctada carchariarum*), which at the beginning of the century supported a small pearling industry. The crustacean population consists mainly of the Tiger Prawn (*Penaeus esculentus*) and the Western King Prawn (*P. latisulcatus*); in 1961 a thriving fishery was established, and during the 1966 fishing season 750 tons of prawns were caught. The processing works have been at Denham on the Peron Peninsula and at Carnarvon; at the end of 1967 part of the Denham activities ceased, but the factory continued its supply of electricity and ice to the small town. The main Shark Bay grounds are immediately north of the Peron Peninsula, tapering off, with interruptions, northward through Geographe Channel. The average depth of the grounds is 15 meters; the trawling nets are between 12

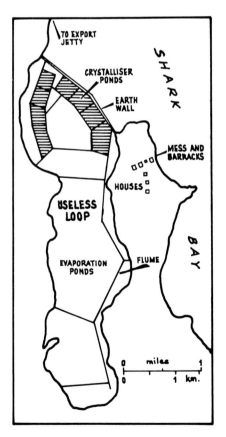

Fig. 24-4. Shark Bay salt ponds at Useless Loop.

and 30 meters wide. In order to avoid too great a depletion of resources, only 30 prawn fishing licenses had been granted toward the end of the 1960s.

From North-West Cape southward to the Murchison River extend the grounds where Snapper (*Chrysophrys auratus*) a most valuable edible fish, is caught by handline. This fishery yields 500 to 1,000 tons per year. The Westralian Jewfish (*Glaucosoma hebraicum*) occurs further south. Some of the boats are based at Fremantle, some at Geraldton, some at Carnarvon or Denham. Much fish is landed at Denham or Carnarvon, filleted at the works, and sent south by refrigerated road trucks.

Just at the southern edge of the arid region, or possibly beyond it, was established an important solar salt industry, which aimed at exporting 1.5 million tons of salt to Japan within seven years, beginning with July 1967. The producing company, Shark Bay Salt owned 60 percent by the Adelaide Steamship Company and 40 percent by Garrick Agnew of Perth, had spent about 4 million dollars in the late 1960s. The construction of the necessary structures at the inlet called Useless Loop took about four years, from 1963 to 1967. An earth wall was built to seal off the long inlet from the sea, and the various ponds were separated. The crystallizer ponds have an area of some 130 hectares; because of their permeable bottom they had to be lined with some 15,000 tons of plastic sheeting, welded on the spot, and covered by a thin layer of compacted sand. The final brine is pumped through a two-kilometer pipeline to a small island where the loading takes place. Salt can be loaded at a rate of 70 tons per hour. The same company closed off part of Useless Inlet, west of Useless Loop, in order to increase the yearly output to 500,000 tons by 1969, and possibly to one million tons a few years later. The mechanization of all phases of the industry has allowed the employment of only 25 men. The Shark Bay area contains large reserves of gypsum which also is shipped from Useless Loop. At the end of 1969 Shark Bay Gypsum (owned 82.5 percent by Garrick Agnew) bought all the assets of Shark Bay Salt, which had been operating at a loss.

Apart from fish and salt production, wool has been the main source of income, with about 100,000 sheep and 30,000 lambs shorn every year in the Shark Bay Shire, and wool production just under 500 tons per year. The area used for grazing is 1.6 million hectares, about 80 percent of the total land surface, and is divided among 13 sheep stations. In all, in 1967 there were 71 shearing machine stands and only 9 tractors in the whole shire. The average weight of wool obtained from each sheep is one of the lowest in Western Australia.

The main town, Denham, has about 400 people; it is an interesting center because of its location on the Peron Peninsula. It has limited port facilities, and a moderate water supply from two artesian wells. Near the extreme southeast of Shark Bay, on Hamelin Pool, is the Hamelin Pool telegraph station, one of the earliest, and well known for its long series of meteorological observations.

Much larger is Carnarvon Shire (formerly Gascoyne-Minilya). The suitability of the area for pastoral occupation was made known by F. T. Gregory after his explorations in 1857–58; the sheep population has grown, with various setbacks due to drought, to over half a million in the 1960s. The average weight of wool per sheep reaches 5 kilograms per year, more than the Western Australian average; this is one of the best areas for pure merino wool, with the shearing of between 400,000 and 500,000 sheep and marketing 2,000 to 2,500 tons of wool each year. The stations in the shire have 290 shearing-machine stands, a higher ratio than that of the Shark Bay Shire.

The population of Carnarvon grew from 1,809 in 1961 to 2,957 in 1966, and that of the shire as a whole from 3,212 to 4,671. The main source of income in the Carnarvon area is irrigated agriculture.

Commercial agricultural production in this region began in 1929. It is limited to some 800 hectares along a narrow strip of irrigated land along the banks of the Gascoyne River, for some 20 kilometers above the river mouth and the town site (Teakle and Southern, 1935). About half the water comes from pumped and shallow wells tapping groundwater on the farms. The remaining half is drawn from concrete-lined wells in the Gascoyne's bed. Most wells tap the so-called "top water," stored in the upper 3 to 7 meters of sands in the superficially dry river bed, sands which are underlain by a thin layer of clay. Several wells reach the "second water" stored in sands underneath this clay. Efforts at conserving some of this water and preventing penetration of seawater upstream by building an underground clay barrier near the river mouth have not been very successful. Some water is pumped from the river sands further upstream. The wells have provision for rapid sealing by bolting down strong lids in order to avoid damage from sudden floods, occasionally caused by tropical cyclones. Cyclonic wind damage to banana plantations can be disastrous, and in 1961 a "Banana Industry Compensation Trust Fund" was created with a levy of 20 cents per case of bananas and a contribution of half that amount by the state government. The fund compensates banana growers for losses caused by cyclones, floods, pests, or diseases which may threaten the existence of the industry. The plants also suffer from the water loss caused by the persistent southerly winds of summer; windbreaks consisting of rows of trees have been planted by the growers, with varying success (Lawson, 1961). On the other hand, the remoteness of this area keeps it relatively free of crop pests and diseases.

Scientific and experimental work is done at the Carnarvon Tropical Research Station (Nunn, 1960); farmers are advised to water the fields once a week if on loamy soil, twice a week if on sandy soil. In winter the intervals

TABLE 24-2
Agricultural Activities in the Carnarvon Area

Crop	Planting Dates		Harvesting Dates	
Vegetables	March–September		May–November	
Bananas	September		September–May	

	Area		Production	
Crop	Hectares	Western Australia (%)	Tons	Western Australia (%)
Beans	259	60	2934	63
Bananas				
Bearing	147	100	3680	100
Not yet bearing	40	100	–	–
Pumpkins	50	13	1144	25
Tomatoes	30	11	1120	15

are doubled. Experiments have been made with several live windbreaks; Yellow Tecoma (*Tecoma smithii*) is recommended, in preference to Tamarisk and River Gum, which produce an invasive and competitive root system.

The calendar of activities in the plantation area is given in table 24-2. The yield of the banana crops is high by world standards, one farm having attained 80 tons per hectare; the district yield of 25 tons per hectare is well above that of the Canary Islands (Lawson, 1961).

The economics of this type of agriculture are unusual: all the bananas are sent to Perth by road, taking 22 hours in the large diesel trucks of the Gascoyne Trading Company. Early beans and tomatoes also are trucked to Perth, then most are shipped to Adelaide and Melbourne by air. Carnarvon bananas (worth $950,000 in 1964–65) supply between one-half and three-quarters of Western Australia's needs, the remaining needs usually being shipped from New South Wales. Bananas bring about 65 percent of the farmers' income. About 500 persons live on nearly 150 farms, all situated on alluvial soils. The area of the farms ranges from 1.5 to 40 hectares. Most farms have access to wells on the river bank.

Carnarvon has fish-processing works which use in part the plant set up in 1951 at Babbage Island, near the town, when the Australian Whaling Commission began shore-based whaling there. The catch consisted mostly of the small Humpback Whale (*Megaptera nodosa*). In 1956 the Commission sold the works to the North-West Whaling Company, which was then operating from Point Cloates. All operations were transferred to Babbage Island, and by using the two companies' combined quotas, hundreds of whales were caught. The total quota allowed was 750 until 1960. That year only 440 whales were caught, producing less than 4,000 tons of whale oil worth nearly $700,000. For the years 1961 and 1962 the annual quota was reduced to 450, but in 1963 only a few whales were caught and the International Whaling Commission declared the species protected. Whaling from Babbage Island ceased altogether, and the North-West Whaling Company directed its activities to prawn fishing, which then became flourishing. The main export markets have been Japan and the United States, followed by South Africa and France. In the year 1964–65 exports (including some from Exmouth Gulf, outside this subregion) reached 331 tons worth over $500,000.

As a port, Carnarvon is affected by tropical cyclones, which caused grave damage in 1889, 1893, and worst of all in 1960. The jetty is especially vulnerable because of its unusual length, made necessary by the shallow coastal waters. The tide is no problem, having a spring range of less than 2 meters. The water supply is good, and normally amounts to 12,600 hectoliters per day. Carnarvon had the only broadcasting station in this region, until the new Port Hedland station began transmissions in 1968.

In 1964–65 a total of 81 ships with a displacement of 99,000 tons left Carnarvon; 50 were bound for western Australian ports, 24 for other Australian ports, and 7 for overseas. The 21 ships, displacing 17,000 tons, which entered the port in 1967–68, unloaded 17,000 tons of goods, and loaded 75,000 tons of cargo including salt loaded at Useless Loop, which made up most of the cargo. This decrease in the number of ships is proof of the increasing competition from road and air transport. The trucks of the Gascoyne Trading Company ensure a regular road link between Carnarvon and Perth; each truck, with two drivers, carries 20 tons; in 1964 the fleet carried 109,000 cases of bananas, 160,000 bags of beans, 75,000 cases of tomatoes, 20,000 bags of pumpkins, and some other produce, traveling nearly 2.5 million kilometers. Air transport carried 10,040 passengers and 177 tons of goods in 1966, against 4,569 passengers and 73 tons of goods in 1955–56.

The northern end of this subregion may yet acquire a new industry. Texada Mines, of Canada, agreed early in 1967 to carry out research into the possible production of potash from Lake McLeod, the 2,000-square-kilometer coastal lake north of Carnarvon. It is hoped to extract 3 million tons of salt per year, from which 200,000 tons of potash may be obtained, enough to satisfy Australian needs. The total capital investment may have to reach 13 million dollars.

Point Cloates and North West Cape (Subregion II)

Subregion II comprises the extreme north of the Carnarvon Shire and the western part of the newly formed Exmouth Shire, which occupies most of the Exmouth Peninsula. It is at the western margin of the tropical cyclonic belt and is affected by cyclones only occasionally, if at all, toward the end of the cyclone season, about March. Light frontal rains occur regularly between May and July, that is, from late autumn to midwinter; by August the rains are so light as to be ineffective. Tectonically, this subregion is at the northern end of the Carnarvon Basin (formerly called the North-

West Basin), along the western axis where anticlinal folding has been conspicuous. It is occupied by a few large sheep stations (separate statistics are not available), and because of its tectonic structure is being thoroughly searched for oil, along both the Rough Range and the Cape Range further north. Since the isolated oil flow at Rough Range No. 1 well, in 1953, about 12 million dollars was spent drilling another 27 wells nearby, which all proved dry, or nearly so.

Whaling was carried out during World War I by a Norwegian company from Point Cloates. This activity was resumed in 1949 by the newly founded North West Whaling Company, until the gradual depletion of whale numbers forced the company to move to Carnarvon.

Near the northern end of this subregion a radio station for communication with ships was built between 1962 and 1967, its tallest steel mast being 423 meters high. It is a U.S. naval installation costing 70 million dollars and supporting a town named Learmonth, of some 3,000 people. The power station and VLF radio transmitter are at Point Murat, a few miles further north. The town is a well-integrated Australian-American community. Learmonth derived its name from a wartime base of the Royal Australian Air Force.

The area has fish-processing works, where the catches of prawns from Exmouth Gulf are treated. The species caught are mainly the Banana Prawn (*Penaeus merguiensis*) and the Tiger Prawn (*P. esculentus*); the total catch in 1966 reached 566 tons. The normal rate of catch is 40 to 75 kilograms per hour; however, during some weeks, after a tropical cyclone has brought copious rain and some localized coastal runoff, the rate has risen to 75 to 200 kilograms per hour. To the west of Exmouth Peninsula are the fishing grounds for Snapper (*Chrysophrys auratus*), some of which are taken to Learmonth for filleting. Rock Cods of various species and the giant North West Groper (*Epinephelus* spp.) are also caught. A groper, which usually dwelt under the Exmouth pier, was estimated to weigh over 400 kilograms. There also is a good potential for tuna fisheries offshore.

The airport at Learmonth has become increasingly busy; it handled 5,364 passengers in 1955–56, and 12,695 in 1966, more than any other airport in the region with the exception of Port Hedland. Its role in goods traffic is less important; it handled 89 tons in 1955–56, and 169 in 1966.

The availability of a permanent beacon at the U.S. station has rendered the well-known Vlaming Point lighthouse superfluous.

Exmouth Gulf to Onslow (Subregion III)

In subregion III, the month of March, most critical for tropical cyclones, brings more rain than any winter month. There are records of storm damage and at times loss of life within this subregion from the years 1875, 1880, 1893, 1897, 1905, 1909, 1911, 1918, 1923, 1945, 1953, 1958, 1961. In the last four years mentioned the damage was severe indeed. The cyclone of 1923 destroyed the modest timber jetty at Onslow. A 990-meter jetty was built at Beadon Point, and after its completion in 1925 the whole township was moved there. In 1961 a cyclone demolished the last 300 meters of this jetty and caused such damage to installations and buildings that it was suggested that the town be moved further inland. The jetty has not been rebuilt, and lighters are used. During the year 1964–65 Onslow handled 79 ships and a total of 3,000 tons; 43 of these ships left for eastern Australia, 34 for destinations within the state. The 68 ships calling in 1967–68 discharged some 2,000 tons of goods, while only 1,000 tons were loaded; the goods landed were mostly supplies, including fuel; and wool comprised most of the outgoing cargo. In 1966 only 2,960 passengers used the airport, and 81 tons of goods were received or sent by air.

Most of the activities have been pastoral. F. T. Gregory reported in 1861 that the country around the Ashburton River was suitable for sheep, and a few years later the first stations were established. In the late 1960s there were 28 stations in the Ashburton Shire, using some four million hectares of land for grazing some 300,000 sheep. The weight of wool per sheep is generally slightly below the Western Australian average. The stations have 175 shearing machine stands and 28 tractors.

There are a few new activities. Exmouth Salt, a Western Australian company, has spent $80,000 investigating salt production on the east coast of Exmouth Gulf. The possible output may be 1.5 million tons per year. A license issued to establish a pearl-culture farm near Giralia Landing, at the southern end of the Gulf, led to a first modest harvest of pearls in 1965.

The most important mineral find in this subregion has been offshore, at Barrow Island, 100 kilometers north-northeast of Onslow, where a good flow of oil was obtained by Wapet in 1964. The oil was held in Jurassic sands at a depth of 2,200 meters. After the discovery of more oil at a depth of only 700 meters, the island was declared a commercial oil field in May 1966. The first shipment of crude oil was made in April 1967. The oil reserves at Barrow Island are estimated at 114 million barrels. Production in the late 1960s was nearing 20,000 barrels per day. It is believed that about 240 wells will tap the whole field; water injections facilitate the rise of the oil in the wells. The Wapet Company (Western Australian Petroleum) is two-sevenths owned by California Asiatic (Caltex), two-sevenths by Texaco Overseas Petroleum, known as Topco (Caltex), two-sevenths by Shell Development (Australia) which is jointly owned by British and Dutch interests, and one-seventh by Ampol Exploration which is Australian owned. The Australian Tariff Board has ruled that any crude oil found in Australia must be bought by oil-marketing companies in proportion to their imports into Australia. Some companies have arranged for the refining of their Barrow Island quota at Kwinana, W.A.; others use their own tankers to take the crude oil to refineries in the Eastern States.

The problem of landing equipment and loading oil at

Barrow Island has been solved. The west coast of the island has deep water but is heavily pounded by the sea. The tidal range is nearly 4 meters. The most suitable site was chosen at what became known as Wapet Landing, on the northeastern shore, at a spot partly protected by a line of coral reefs. For the shipment of crude oil a mooring terminal has been built further south, connected to the island by 10 kilometers of 50-centimeter pipeline laid on the sandy sea floor. Oil is stored in two 200,000-barrel tanks on the shore and pumped through the pipeline to the tankers. During the year 1964–65 the island was reached by 23 ships displacing 24,000 tons; no details of shipments are available. In 1967–68, 96 ships unloaded 18,000 tons of cargo and loaded 1,089,000 tons of crude oil. Supplies of water must also be taken to the island. There is a double landing strip, which in 1966 was used by planes transporting 2,223 passengers and 73 tons of goods.

Montebello Islands, which continue the partly submerged tectonic structure of Barrow Island, also have an arid climate. They carry a further impoverished flora and fauna, and were used in October 1952 and May 1956 for the testing of a British atom bomb. They were surveyed by a team of scientists before the explosions, and have been studied since, but remain closed to the public as a precaution in case of residual radioactivity that may still be at dangerous levels (Serventy and Marshall, 1964).

Roebourne and Port Hedland (Subregion IV)

Subregion IV is by far the most extensive subregion in this belt of coastal desert, extending over some 450 kilometers. It corresponds closely to the shires of Roebourne and Port Hedland respectively, but its inland boundary is somewhat irregular because of the proximity of the various residual highland systems. The Fortescue River, flowing between the Hamersley and Chichester ranges, has deeply dissected the landscape, but all the rivers of this region, as far as the DeGrey, last one before the wholly areic tract, have well-defined beds, even if there may be no water in them for months or years at a time.

The population of the Roebourne Shire was 568 in 1961 and 1,699 in 1966. That of the Port Hedland Shire was 1,120 in 1961 and 2,966 in 1966. Such remarkable increases are due to the development of mining on a scale totally unknown in Australia until very recently. This has resulted in the establishment of much closer links between the littoral and the rugged interior, without the raw wealth of which no change would have taken place.

The westernmost part of this subregion, including the entire course of the Robe River and as far as the left bank of the Fortescue River, is in the Ashburton Land District. The boundary of the Ashburton Gold Field, on the contrary, is a straight meridional line which excludes the upper Robe River but includes the lower right bank of the Fortescue and, on the coast, Cape Preston. These differences are significant because the mining rights over large deposits of limonite near Mount Enid, on the upper Robe River, 130 to 150 kilometers from the coast, are held by Cliffs Western Australian Mining, a partnership owned 51 percent by Cleveland Cliffs Iron of the U.S., 35 percent by Mitsui of Japan, and the rest by Western Australian interests. A contract provides for the export to Japan of 88 million tons of iron pellets over 21 years and 52 million tons of fine ore over 15 years. A deep sea port probably will be established at Cape Lambert, near Port Samson, so that the railway to be built will cross the existing Hamersley line. Similar deposits near Deepdale, further downstream, are leased to the Broken Hill Proprietary Company, an Australian company, which has an almost complete monopoly of iron and steel production in Australia.

So far, this area had only been used for sheep grazing; its carrying capacity is less than that of the subregions further south, and the amount of wool per sheep is also less.

The Dampier Archipelago is the most conspicuous feature of the coast in this area. It consists of many islands, of which perhaps Rosemary Island is best known because the North-West Big Game Club operates from there, generally as a holiday camp. Game fish are plentiful, the most attractive species being the Spanish Mackerel (*Scomberomorus* spp.)

The most outstanding development here is the establishment of an entirely new deep seaport at Dampier, on the mainland opposite the southernmost island of the archipelago. The exact site was named King Bay, but the name has not been adopted for the township because of the likely confusion with the many similar place names in Australia. The port is the outlet for the Hamersley ore. Its site was chosen after extensive hydrographic surveys, a study of wave conditions, and sample boring of the ocean floor. The islands of the Dampier Archipelago shelter it from the west. The port has a 230-meter jetty leading to a 220-meter loading wharf. With a dredged depth of over 13 meters, the wharf takes ships of 65,000 tons, or even more at high tide (the spring tide range is about 5 meters). Later dredging will enable 100,000-ton ships to berth. The port is well equipped, with a loading rate of 6,000 tons per hour. It possibly holds a world record, having loaded 96,000 tons of ore into one carrier at the end of 1967. In 1967–68 some 206 ships unloaded 154 tons of cargo and loaded 6,926,000 tons, mostly iron ore. So far, it has been reserved for ore carriers and other vessels authorized by the company. Other ships, including fishing boats, are not admitted. At Dampier is a 40 million dollar ore pelletizing plant, which has its own power station. Waste heat from this station is used to evaporate seawater in the largest desalination plant in Australia, which cost about $750,000 and produces over 18,000 hectoliters of potable water per day, much of it used by the pelletizing plant. Another 3,000 hectoliters per day is pumped through a pipeline from Miaree Pool, 32 kilometers away. There is enough water to allow for the planting of small gardens, a luxury in this arid climate. More water comes by pipeline over 120 kilometers away, from Millstream.

The ore loaded at Dampier comes from Mount Tom Price, in the Hamersley Range, 283 kilometers away. It contains over 65 percent iron; the amount shipped during 1966 was 206,000 tons, worth 1.9 million dollars. The first shipment, to Rotterdam, took place in August 1966. Sized lump ore is stockpiled at the port and loaded on the ships. Fine ore will be stockpiled in two 150,000 ton heaps, waiting to be ground to a fine dust and then pelletized in the plant, which is reputed to be the world's largest. Shipments of pellets to Japan were to begin in April 1968 and continue at a rate of two million tons per year. The contract is for 18 million tons over 10 years. The deposits at Tom Price are estimated to contain 610 million tons of high-grade hematite. Shipment of ore totaled 13 million tons in 1969. The development of other deposits at Paraburdoo was begun, then halted, during the slight recession of the early 1970s.

The railway is made of rails weighing 60 kilograms per meter, welded into 333-meter lengths. It crosses the Fortescue River on a low steel-girder bridge 400 meters long. The total construction outlay by the company is about 150 million dollars. Each train carries about 12,000 tons in some 120 hopper cars and is constantly linked to both terminals by radiotelephone. The company itself, Hamersley Iron, is owned 54 percent by Conzinc Riotinto of Australia, 36 percent by Kaiser Steel, of the United States, and 10 percent by 26,000 private shareholders. The net 1969 profit was over 25 million dollars after provision of 21 million dollars for taxes and 16 million dollars for depreciation.

At the 1966 census the population of Dampier was 1,024. The airfield handled 2,937 passengers and 51 tons of goods during 1966, with traffic expected to increase. Comalco Industries, an aluminum producer owned by Conzinc Riotinto and Kaiser Steel, has considered a possible 67 percent interest in a total 8-million-dollar investment in a solar salt industry to be located at Dampier. The evaporating ponds would be built on 10,000 hectares of tidal marshes between Dampier and the main coastal area. Annual production could be nearly 500,000 tons. The limited site of Dampier allows a maximum population of 3,000; a new township has been built at Karratha, 22 kilometers away on the mainland shore of Nickol Bay.

Only a short distance further east is Nickol Bay, on which a small fishing settlement was started in 1855. Pearls and trepang were sought; Aborigines were mostly employed to do skin diving. By 1860 there were some 10 boats, and the settlement was officially named Cossack. Within a few years the more accessible fishing grounds had been exploited, and larger boats were needed to go farther along the coast, looking for the Golden Lip Shell (*Pinctada maxima*) which yielded the best pearls. Diving gear had to be used to reach greater depths, and Malay and Kupang divers were engaged.

In 1874 there were some 80 boats, which made a haul of shell worth £62,000 and pearls worth £12,000. Cossack then handled all the supplies for the northwest, and shipped the wool which the area was beginning to produce. In the 1890s Cossack was still the home port of a sizeable pearling fleet and served the new settlement at Roebourne, and the West Pilbara Gold Field. In that period it was endowed with some fine buildings, such as the Customs House and the Court House, which have since become national monuments. The repeated fury of cyclones however caused such loss of life and so much damage that by the end of the century Cossack was already a ghost town.

Roebourne's port was then established at Port Samson, actually a little further away than Cossack. The jetty built there was destroyed by a cyclone in 1925, and Cossack became once more Roebourne's port, but without many facilities, as it had to be served by lighters. A new timber jetty, 756 meters long, was completed at Port Samson in 1938 and has continued in operation since. The port is inadequate and has no bulk fuel storage. A serious handicap is the lack of water; all groundwater in the vicinity is saline, and so it is necessary to bring potable water by ship. Since 1960 storage tanks were connected to the 22 houses by pipes and a maximum of 1,100 hectoliters per house per year has been allowed. Roebourne is slightly better off, having a modest supply of water pumped from wells near the Harding River nearby. Thus, while the 80 inhabitants of Port Samson are restricted to 72 liters of water per head per day, the 550 inhabitants of Roebourne can afford 440 liters per head per day.

The only nonmineral export of Port Samson is merino wool, of which the Roebourne Shire produces over 500 tons per year, from some 170,000 sheep divided among 12 stations which use about 1.7 million hectares of land for grazing, and operate 112 shearing machine stands.

Between 1945 and 1967 Port Samson was the port of export for the crocidolite (blue asbestos) mined at Wittenoom Gorge, in the Hamersley Ranges. The fiber was sent by truck over a distance of 370 kilometers; the trucks carried an average of 27 tons of bagged fiber, and up to a maximum of 33 tons. Total production was 9,374 tons in 1965 and 11,580 tons in 1966. In 1967, however, Australian Blue Asbestos Limited, which was owned by Colonial Sugar Refining Limited, closed the mine at Wittenoom.

Since 1966 prawn fisheries have been established in Nickol Bay (nearly 600 tons in 1967). Offshore there is a good potential for tuna fisheries. In 1970 the first houses were built at Wickham, 8 kilometers north of Roebourne; this is the residential town for Cliffs W. A. Mining's industrial area at Cape Lambert, another 8 kilometers further north. It is planned as a satellite of Karratha, over 38 kilometers to the west.

With the attraction of well-paid jobs at Dampier, and with the limitations imposed by the poor water supply, it is likely that the Nickol Bay and Port Samson area

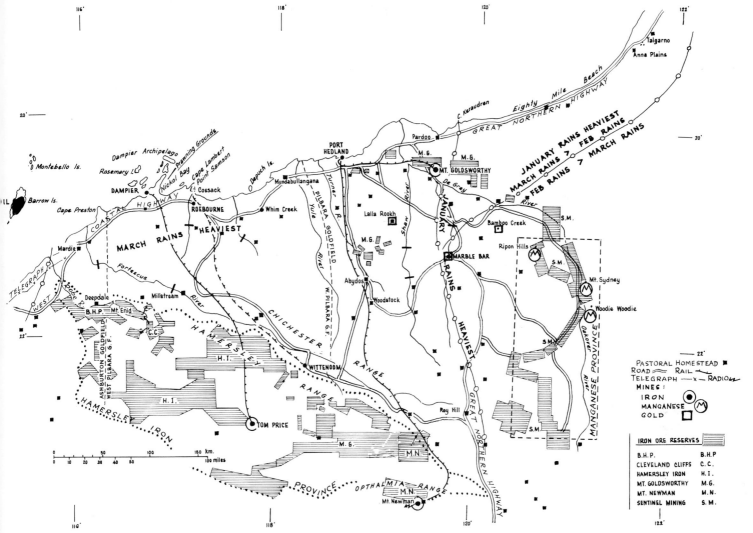

Fig. 24-5. Subregions IV (Mardie to Cape Keraudren) and V (Cape Keraudren to Anna Plains). Of the Iron deposits, only those near Deepdale on the Robe River to the west and Mount Goldsworthy to the east may come within the coastal desert belt, the other deposits being too far inland. The manganese and gold mines are also inland, their mineral wealth is taken to the coast for shipment. Most of the potential dam sites, shown by dashes across the (mostly dry) river courses, also are relatively far from the coast, but the economic utilization of this immense region may succeed only as a result of integrated planning.

will not develop any further, until some new factor intervenes on a large scale. This is possible, for instance, with the development of the Robe River iron deposits, which consist of limonite, an ore of low grade, which must be enriched for economic shipment overseas. If a pelletizing plant is used it becomes economic to desalinate seawater, as is being done at Dampier. It follows then, that should the Robe River deposits be exploited, an entirely new port town would be established with its own supply of desalinated water. This is still not likely to be at Port Samson, where limited depth is available, but rather in its vicinity, possibly, as suggested, at Cape Lambert. The effect of such developments on Roebourne is hard to foresee, because, if conditions in the new town are pleasant enough, a drain of population may result, and more and more services may be transferred. Roebourne, with only one hotel, was unable to cope with the rush of newcomers who were seeking employment on the Hamersley ore project, even though most of the recruiting was done in Perth.

Port Hedland, even in its early years, had the advantage of being the administrative center and the port as well. It is also the only port for the vast Pilbara Gold Field, which extends southeastward to include the important mineralized areas around Marble Bar, as far as the Tropic, and including also most of the sedimentary Canning Basin to the east. Outside the township, the whole Port Hedland Shire is divided among 14 sheep raising stations, which have 1.5 million hectares of grazing land on which in 1965 lived 135,000 sheep, producing 430 tons of merino wool during the year. This wool, and much of the similar amount produced in the Marble Bar and Nullagine Shires further inland, is shipped through Port Hedland.

The old Port Hedland harbor could be approached only at high water and by ships of not more than 5,000 tons, because of extensive sand banks nearly exposed at low tide. The range of spring tides is about 6 meters. The entrance to the harbor is narrow, but the port is well sheltered. It is well known in the scientific world because it has the only tidal gauge on the northwestern coast of Australia.

With its many limitations the old port shipped large amounts of minerals. Until World War II it was the terminus of the only railway line in the northwest, the narrow gauge railway to Marble Bar. Traffic was never intense, but there had been a steady output of gold, occasionally of tin, and supplies were railed in. After the war the rails were lifted and sent south, and all traffic was by road.

The only localities each producing more than 100 grams of gold in 1965 were, in order of quantity, Bamboo Creek, Pilgangoora, Lalla Rookh and Marble Bar; the district's total was 15,547 grams. During 1966 production at Bamboo Creek increased, while the modest output of the other centers remained almost unchanged; the district's total rose to 28,429 grams. Many years ago Marble Bar was the main goldmining center of the hinterland; its total gold output to the end of 1966 amounted to 57,273 kilograms, or 55 percent of the Marble Bar Mining District total and 37 percent of the entire region's total.

Tin is found at many localities, of which Moolyella, 16 kilometers from Marble Bar, is the best known. With the end of asbestos mining, tin is taking third place in value among nonprecious minerals. The main producer is Pilbara Tin, which has installed a 2 million dollar gravity plant at Moolyella. A survey of water resources has revealed water in the deeply weathered granitic rocks (Allen, 1966). The water lies between 4 and 12 meters below the surface, under some pressure beneath the dense clay formed by the deep weathering of the rock. It has a temperature of about 31°C. Of 34 wells, 13 showed a slow seepage, or at most gave a small supply, 17 yielded up to 500 liters per hour, and only 5 yielded more. As to depth, only 4 wells reached beyond 30 meters below the surface. The salinity of the water does not exceed 770 parts per million and allows domestic use (Allen, 1966).

Another tin mine is at Cooglegong, 65 kilometers southwest of Marble Bar, not far from the Shaw River and from an existing road; the investment here has been $850,000 (Kerr, 1967). The third mine in order of importance is at Hillside, some 30 kilometers south of Cooglegong. Special mention must be made of the production of tin by the shallow dish (*yandy*) method by groups of Aborigines who had left the sheep stations after 1946. Their cooperative organizations were first known as the Group, from which was formed the Northern Development and Mining Corporation, active from 1951 to 1954. Then, in 1955, the Pindan Company was formed, from which a dissident group formed Nomads Limited in 1959 (Kerr, 1967). The struggle for social and economic independence was won only because of the income provided by these small mining ventures (Stuart, 1959), in which some 200 Aborigines are still engaged today. The total output of tin in the Pilbara Gold Field, including the major mine plants, was 650 tons in 1965 and 569 tons in 1966.

The real strain on Port Hedland's port facilities came with the increased production of manganese at Woodie Woodie, in the Gregory Range, nearly 400 kilometers away, and at Ripon Hills, some 70 kilometers to the west, not far from the Nullagine River. The ore is crushed near the mines, and brought to Port Hedland by truck and trailer combines with a capacity of 90 to 95 tons. Since 1953 production has exceeded 40,000 tons per year, with the exception of the years 1955, 1963 and 1964. The ore is stockpiled at the port and taken to the ship by container trucks; each container is lifted and emptied in one operation. This method allows a loading rate of 1,500 to 2,000 tons per day (Kerr, 1967). With the dredging of the harbor for the iron ore shipments, larger ships can enter, and bigger shipments of manganese ore have also been possible; the 1965 total reached 68,000 tons, and the 1966 total 96,700 tons, all exported to Japan. The value of the 1966 production was 3 million dollars.

With the establishment of large-scale iron-ore exports, the port had to be dredged to take ships up to 45,000 tons, almost exactly ten times the maximum allowed by the old port. The general harbor zone has two government wharves, one of which is to be made available for salt exports. Dredging has been carried out at a daily rate of up to 30,000 cubic meters, for a total of 9 million cubic meters. Some mangrove swamps have been cleared and filled. The approach channel is 12 meters deep, and the berths up to 27 meters. At Nelson Point, in the late 1960s, ships of up to 70,000 tons had docked, but 100,000 tons is the expected limit. Finucane Island, 800 meters away across the harbor, which it partly shelters, has been linked to the mainland by the causeway supporting the Mount Goldsworthy railway; ships use a berth on the southeastern side of the island, part of a 215-meter wharf. Loading capacity is 2,700 tons per hour. A good depth of water has been made available, including a large turning circle in the sheltered area.

In 16 months, 100 ore carriers have called and loaded. It is expected that by 1975 the port will handle 500 ships and 30 million tons per year, more than twice the cargo handled at Sydney. This spectacular development has been brought about by the large-scale exploitation of recently discovered deposits of iron ore of gigantic size. Port Hedland finds itself the terminal of the Mount Goldsworthy and Mount Newman iron-ore projects and the site of the large Leslie salt works. The consequent improvements in the harbor have led to a speeding up of manganese exports.

An agreement concluded in 1962 between the Western Australian government and a group of associated mining companies (Cyprus Mines, U.S.; Utah Construction and Mining, U.S.; Consolidated Goldfields, Australia) covers an initial period of 21 years and provides for the mining and export of up to 15 million tons of iron ore from the Mount Goldsworthy deposits, situated about 115 kilometers east of Port Hedland. The rate of export should not exceed one million tons per year. The known reserves are some 30 million tons. The companies have established and equipped the mine; ultimately there will be 18 benches reaching down 230 meters below the top and 105 meters below sea level. After the signing of contracts with Japanese steel mills, large-scale mining operations began in 1966. A small town for 300 employees and their families was laid out, with housing, local roads, a school and amenities. Water comes from the bed of the DeGrey River, 13 kilometers away. The 115-kilometer standard-gauge railway was built from the mine to the wharf on Finucane Island. It crosses the DeGrey River on a bridge, a few major creeks on low-level concrete crossing (submerged during floods from tropical cyclones), many minor creeks on culverts, and the muddy West Creek salt flats at Port Hedland on a 6-kilometer causeway to Finucane Island, side by side with the road. Three or four trains a day use the line; a train of about 50 hopper cars carries some 3,500 tons of ore. The total outlay is likely to exceed 24 million dollars. Originally the terminal was to be at Depuch Island, but it was declared a reserve in order to protect its fine Aboriginal rock engravings (Ride and others, 1964). So Finucane Island was selected instead. The Mount Goldsworthy ore shipped from June to December 1966 weighed 862,000 tons and was worth 7.5 million dollars; its iron content was 65.15 percent.

Construction work on the giant Mount Newman iron-ore project began in May 1967. The total cost may reach 200 million dollars, far more than has been spent on any other mining venture in Australia. Hematite was first found at Mount Newman, on the eastern end of the Ophthalmia Ranges, in 1956; the following year a deposit estimated at 1,000 million tons of high grade (64.6 percent) hematite was discovered on Mount Whaleback, also part of the Mount Newman system. The ore is so pure that it can be fed directly into blast furnaces. A first survey of mining sites, rail routes, and open-sea loading sites was made by a United States firm in 1963; an Australian firm became a partner in 1964. A preliminary agreement to supply Japan with 100 million tons of ore over 20 years was signed at the beginning of 1965. During 1966 the financial partnership underwent several changes, finally resulting in a 60 percent Australian ownership, of which half is controlled by Broken Hill Proprietary and half by Colonial Sugar Refining. The remaining interests are United States, Japanese, and British. In October 1966 the agreement with the Japanese steel mills was modified to allow for the delivery of 100 million tons of ore within 15 years. In addition, the Broken Hill Proprietary Limited agreed to take 70 million tons for its own use. It is likely that in the future some shipments will go to Europe. The company holds three iron-ore reserves totaling 1,910 square kilometers. The main reserve is the Mount Newman one, which includes the deposits being exploited on Mount Whaleback. To the northwest is the Weeli Wolli reserve, at the eastern end of the Hamersley Range. Further south is the Pamelia Range reserve.

The open-cut mine at Mount Whaleback was to begin with an output of 10,000 tons per day, rising in five years to 40,000 tons per day. A proper townsite is being established for 1,300 persons. The ore will be crushed and screened at a rate of 4,000 tons per hour. Crushed ore will be heaped in a 30,000 ton pile, under which 5 chutes drop the ore on a train waiting in a loading tunnel 100 meters long. Ore trains will usually consist of three 3,000-horsepower diesel locomotives hauling up to 135 hoppers carrying 12,000 tons of ore. Loading can be done in one hour, and the trains can go to Port Hedland, unload, and return in a day. The railway line, 426 kilometers long, is the longest private railway in Australia.

The largest salt-producing company of the United States, Leslie Salt, is establishing eight 800-hectare evaporating ponds some 50 kilometers east of Port Hedland. The brine will be pumped 29 kilometers to a crystallizing pond near the harbor. Exports were to reach 475,000 tons per year by 1969, rising to one million tons when full capacity is reached. A further increase to 2 million tons per year might be possible by 1975.

Port Hedland, as a result of these various activities, increased in population from 1,120 in 1961 to 1,785 in 1966, and neared 4,000 by the end of 1967. It is estimated that it will exceed 5,000 by the early 1970s, and probably 12,000 by 1980. The water supply comes from the Turner River, which flows only occasionally. Surface flow reached its mouth in 1961 and 1966; in 1965 it came to about 20 kilometers from it. Water is stored in the sands of the river bed, where it is tapped by wells sunk by the Public Works Department. The maximum supply is 8,000 hectoliters per day, and the recharge is slow and irregular, depending on flooding by the river, usually only after tropical cyclones. After the 1966 cyclone the level of water in the wells rose by amounts varying between 50 centimeters and over 3 meters. Pumping, if there is no flood, reduces the level of the groundwater by 5 or 6 centimeters per month, so that even a full recharge of the bed would not last much more than 3 years (Farbridge, 1967). The old 15-centimeter pipeline has been replaced with a 375-millimeter asbestos pipeline.

The Yule River, 68 kilometers west of the town, has been investigated in order to find out whether the 1978 needs may be met, and it is estimated that it can yield nearly 30,000 hectoliters per day without any depletion of its groundwater storage. There are three separate aquifer levels, and the area underlain by good quality water is nearly 100 square kilometers. The Yule River flows only after cyclonic rains; between 1945 and 1966 it is known to have run on at least six occasions, so that

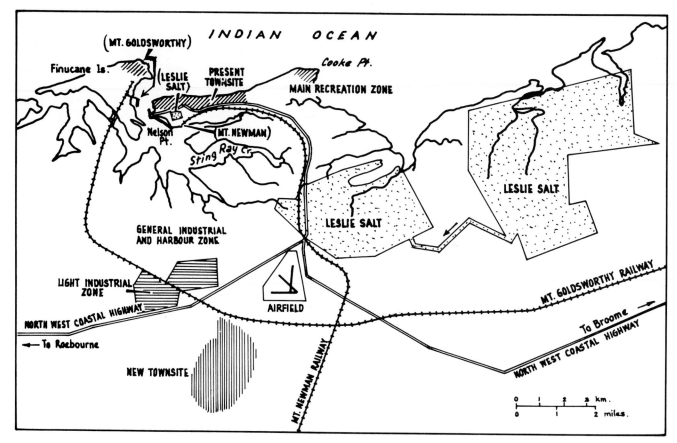

Fig. 24-6. The Port Hedland area.

there are periods of several years between recharges (Whincup, 1967). The cost of deriving water from Yule River wells is estimated at 1.8 million dollars; of this, the Mount Newman company will pay 1.3 million dollars.

By the end of 1967 expansion of harbor activity at Port Hedland had generated so much noise and dust from ore handling, that the establishment of a new townsite was proposed, this to be located closer to the airfield, just south of the highway and the Goldsworthy railway.

During 1964–65 a total of 129 ships displacing 200,000 tons left Port Hedland; 22 went overseas, 55 to the eastern states, 52 to other West Australian ports. During 1967–68 there came 300 ships unloading 283,000 tons of cargo, and loading 3,943,000 tons, mostly iron and manganese ores. The number of passengers using the airport rose from 3,528 in 1955–56 to 16,785 in 1966, and the weight of air cargo from 268 tons in 1955–56 to 552 tons in 1966. Port Hedland is one of the busiest bases for the Royal Flying Doctor Service; in 1958–59 it handled over 14,000 telegrams and was in contact with 61 outposts; 157 flights were performed over a distance of 31,000 kilometers, carrying nearly 100 patients, half of whom were Aborigines. Nearly 400 medical consultations were given over the air. The radio base is also used for educational broadcasts and correspondence schooling. A national broadcasting station began transmitting in 1968.

The recent discovery of offshore deposits of natural gas has led to a complete reappraisal of the future plans for the industrial development of the region. The deposits are being developed by Burmah Oil.

The eastern margin of this subregion is quite undeveloped; its iron ore (from Mount Goldsworthy) is railed to Port Hedland, the coastal highway has to cross the dreaded Pardoo Sands, and the only wealth comes from the few sheep stations. However, deposits of iron and manganese 150 kilometers east of Port Hedland are being investigated by the Sentinel Mining Company, owned by the Ludwig Organization of the United States; a possible site for a port may be at Cape Keraudren, 130 kilometers northeast of Port Hedland. Should this development take place, and if the manganese deposits at Woodie Woodie and in other eastern parts of the hills prove large enough, there is the possibility of new rail links being established in this area, independently of Port Hedland. It is certain, however, that only the most outstanding mineral deposits can justify such an expense.

The Eighty Mile Beach (Subregion V)

In subregion V winter rains rarely occur, and the monsoonal thunderstorms bring occasional showers during the summer. Tropical cyclones may bring erratic but extremely heavy rains in the autumn; they caused damage or loss of life in 1887, 1890, 1908, 1910, 1917, and 1935. The poor pastures and higher temperatures make the raising of sheep too difficult, and the few northern stations all raise cattle. Economic life is restricted to a narrow coastal strip, although the telegraph line enters this subregion some 80 kilometers away from the coast, to which it draws gradually closer as it proceeds north.

The main cattle station is Anna Plains, owned by Linkletter Enterprises from the United States. It occupies over 300,000 hectares and holds about 20,000 cattle.

There is a faint possibility that the Canning Basin may contain oil; exploratory drilling has been carried out at various places, the most remote one being Sahara Well, over 600 kilometers east of Port Hedland.

The Southern Littoral

Little development, apart from the gradual eastward expansion of sheep and cattle raising, has occurred in the southern littoral. The Eucla artesian basin occupies 140,600 square kilometers in Western Australia and 49,200 in South Australia. Over 90 percent of its area the water contains more than 14,000 parts of salt per million and is therefore unusable (Australian Water Resources Council, 1965). Toward the southern margin the water contains 3,000 to 7,000 parts of salt per million and can be used for watering sheep. The depth of the wells varies from 100 to nearly 700 meters. The only true artesian well is at Madura, W.A., with an initial flow of about 24,000 liters per day. Other wells (for example, at Cook, S.A.), are subartesian; the water has been pumped at rates between 300 and 10,000 hectoliters per day. The limestone which overlies the sand and gravel artesian aquifers holds a fair amount of groundwater, far too saline for economic use, except toward the north and west, where it is used for pastoral and railway purposes. The water table lies at some 15 meters below the surface in the south and at more than 60 meters below the surface in the north.

The Eyre Highway which links Western and South Australia runs along the southern coast. It is gradually being paved; the Western Australian section was to have been completed in 1969. Some hundreds of cars per week use it, as do many heavy trucks and semitrailers. An interstate tourist bus service is also available.

There has been slightly more rain than usual in late winter during recent years, and this has favored the growth of grass to such an extent that pastoralists have greatly increased the number of sheep, now numbering some 100,000, in the belief that the carrying capacity of the land may be greater than was known in the past. A possible return to the rainfall regime of earlier years may cause losses. Another consequence of the changed rainfall pattern is the extreme danger from fires lit near the highway by camping parties.

Apart from the pastoral stations, the only permanent settlements in the coastal belt are the few points offering services to travelers. The railway is too far inland to affect the littoral region.

Bibliography and References

ALLEN, A. D.
: 1966 Groundwater from granitic rocks, Moolyella Creek area, near Marble Bar, North-West Division. In Western Australia, Geological Survey, Annual Report 1965, p. 24–26. (Extract from the Report of the Department of Mines) By Authority: Alex B. Davies, Government Printer, Perth. 88 p.

ANONYMOUS
: 1962 Plan for the Pilbara. Journal of Agriculture of Western Australia, Ser. 4, 3(12):924–934.

AUSTRALIA, BUREAU OF CENSUS AND STATISTICS, WESTERN AUSTRALIAN OFFICE
: 1964 Official Year Book of Western Australia, 4. Perth.
: 1967a Official Year Book of Western Australia, 5. Perth.
: 1967b Statistical Register of Western Australia for 1964–65. Perth.

AUSTRALIAN WATER RESOURCES COUNCIL
: 1965 Review of Australia's water resources (stream flow and underground resources), 1963. Department of National Development, Canberra. 107 p.

BOWEN, B. K. (ED)
: 1964 The Shark Bay fishery on snapper *Chrysophrys unicolor*. Western Australia Fisheries Department, Report 1.

COLEBATCH, SIR H. P. (ED)
: 1929 A story of a hundred years: Western Australia, 1829–1929. F. W. Simpson, Gov't. Printer, Perth. 476 p.

CONDON, M. A.
: 1954 Progress report on the stratigraphy and structure of the Carnarvon Basin, Western Australia. Australia, Bureau of Mineral Resources, Geology and Geophysics, Report 15.

FARBRIDGE, R. A.
: 1967 Port Hedland town water supply, Turner River exploratory drilling. In Western Australia, Geological Survey, Annual Report 1966, p. 21–27. (Extract from the Report of the Department of Mines) By Authority: Alex B. Davies, Government Printer, Perth. 84 p.

FITZGERALD, K.
: 1955 Buffel grass. Journal of Agriculture of Western Australia, ser. 3, 4:83–90.

FRASER, A. J. (ED)
: 1962 The results of an expedition to Bernier and Dorre Islands, Shark Bay, Western Australia, in July, 1959. Western Australia Fisheries Department, Fauna Bulletin 2.

GENTILLI, J.
: 1948 Bioclimatic controls in Western Australia. Western Australian Naturalist 1:81–84, 104–107, 120–126.
: 1949 Foundations of Australian bird geography. Emu 49:85–129.
: 1951 Bioclimatic changes in Western Australia. Western Australian Naturalist 2:175–184.

GENTILLI, J. (cont'd)
  1952 Seasonal river regimes in Australia. *In* International Geographical Union, 8th General Assembly and 17th International Congress, Washington, D.C., 1952, Proceedings, p. 416–421. United States National Committee of the International Geographical Union, National Academy of Sciences/National Research Council, Washington, D.C. 776 p.
  1956 Tropical cyclones as bioclimatic activators. Western Australian Naturalist 5:82–86, 107–117, 131–138.

HOLMES, J. M.
  1963 Australia's open North, a study of northern Australia's bearing on the urgency of the times. Angus and Robertson, Sydney. 505 p.

KERR, A.
  1962 Northwestern Australia. [Written] in association with M. Lightowler and B. Willmott. Perth 440 p.
  1967 Australia's North-West. University of Western Australia Press, Nedlands, Perth. 437 p.

LAWSON, J. A.
  1954 Peanut growing on the levee soils of the Gascoyne River. Journal of Agriculture of Western Australia, ser. 3, 3(2):187–192.
  1958 Producing lucerne hay for north-west pastoral areas. Journal of Agriculture of Western Australia ser. 3, 7(2):137–149.
  1961 Banana cultivation at Carnarvon. Journal of Agriculture of Western Australia, ser. 4, 2(4):293–316.

MEADLY, G. R. W.
  1956 Weeds of Western Australia: Mesquite (*Prosopis juliflora* D.C.). Journal of Agriculture of Western Australia, ser. 3, 5(1):65–74.
  1962 Weeds of Western Australia: Mesquite (*Prosopis juliflora* D. C.). Journal of Agriculture of Western Australia, ser. 4, 3(9):728–739.

MEIGS, P.
  1966 Geography of coastal deserts. Unesco, Paris. Arid Zone Research 28. 140 p.

MILLER, ROBIN
  1971 Flying nurse. Rigby, Adelaide. 220 p.

NUNN, W. M.
  1956a Pastoral research. Journal of Agriculture of Western Australia, ser. 3, 5:357–364.
  1956b Irrigation projects on north-west stations. Journal of Agriculture of Western Australia, ser. 3, 5(6):709–713.

  1958 Solving pastoral problems. Journal of Agriculture of Western Australia, ser. 3, 7(6):593–596.
  1960 Water usage trials with bananas on the Gascoyne. Journal of Agriculture of Western Australia, ser. 4, 1(10):869–871.

RIDE, W. D. L., AND OTHERS
  1964 Report on the aboriginal engravings and flora and fauna of Depuch Island, Western Australia. Western Australian Museum, Special Publication 2. 87 p.

SERVENTY, D. L., AND A. J. MARSHALL
  1964 A natural history reconnaissance of Barrow and Montebello Islands, 1958. C.S.I.R.O., Australia, Division of Wildlife Research, Technical Paper 6. 23 p.

STUART, D.
  1959 Yandy. Georgian House, Melbourne. 156 p.

SUIJDENDORP, H.
  1955 Changes in pastoral vegetation can provide a guide to management. Journal of Agriculture of Western Australia, ser. 3, 4(6):683–687.
  1962 Lambing trials at Abydos Research Station. Journal of Agriculture of Western Australia, ser. 4, 3(12):936–938.

TEAKLE, L. J. H., AND B. L. SOUTHERN
  1935 An investigation of the terrace soils of the Gascoyne River at Carnarvon. Western Australia, Department of Agriculture, Leaflet 451.

WALKER, K.
  1967 The dawn is at hand; Poems. Jacaranda Press, Brisbane. 49 p.

WARD, L. K.
  1951? Underground water in Australia. Tait Publishing Co., Melbourne. 75 p.

WESTERN AUSTRALIA, DEPARTMENT OF FISHERIES AND FAUNA
  1966 The Shark Bay prawning industry.
  1966–67 Tuna survey. Bulletin 1–10.

WESTERN AUSTRALIA, DEPARTMENT OF MINES
  1966 Annual report 1965. Perth.
  1967 Annual report 1966. Perth.

WHINCUP, P.
  1967 Port Hedland town water supply, hydrogeological investigation of the Yule River area. *In* Western Australia, Geological Survey, Annual Report 1966, p. 13–18. (Extract from the Report of the Department of Mines) By Authority: Alex B. Davies, Government Printer, Perth. 84 p.

INDEX

Abu Dhabi, 55
African coastal deserts, 68, 69, 72. *See also* Namib; Sahara
agricultural development, 26, 30; Australia, 194–195; Brazil, 153; Chile, 129–130; Namib, 183–184, 195; Peru, x–xi; Sahara, 162–165, 167–169
agricultural markets, 30, 51, 130; Australia, 195
altiplano, 113, 114, 119, 120, 123, 126, 130
anticyclones: Australia, 89; South Pacific, 119–120, 150
Atacama, 123–135
atmospheric humidity: Chile, 117–118; Peru, 98, 102, 104
Australian coastal deserts, comparative climatic data, 67, 68; northwestern littoral, 189–203; southern littoral, 203

Brazil, northeastern, 153–154

Caribbean, 64–66, 75–90
Chile
—climate, 111–112, 115–123, 147; drainage, 112–113; energy needs, 125, 132, 134; fisheries, 126, 130–131, 137–146; geography, 123–135; geomorphology, littoral, 111–112, 147–151; minerals, 111–114, 128–129, 131, 135, 137; vegetation, 113–114; water resources, 132–134
climate of coastal deserts, 3–6, 67–106; Australia, 189–190, 193, 195, 196, 202; Chile, 115–120; Israel, 172; Namib, 178; Sahara, 159–165
—mapping, 7–12
climate farming, 30
climatic analogs, coastal deserts, 67–72
climatic anomalies, 69–72, 75–90, 91, 96, 120, 132
climatic change, Chile, 147–151
climatic-vegetal relationships: Chile, 113–114; eastern hemisphere, 157–158
clouds: Chile, 117, 119; Peru, 98, 102
"coast-wise subsidence," 69
competing uses (water), 38–39, 41
condensation, Namib, 179
cooperatives, 61
cultural change, 38, 43
cyclones, Australia, 189–190, 194, 195, 196, 201, 202

demography, 28, 33–36; Sahara, 167–169
desalination, ix, 3, 26–27, 42, 45–53, 55–61; Abu Dhabi, 55, 60–61; Australia, 197, 199; Chile, 132, 133, 135; Israel, 44–46, 48, 50–51, 173; Namib, 179; Puerto Peñasco, 3, 55–60, 133
—agricultural applications, 26, 42, 51–53, 55–61; costs, ix, 27, 30, 46–53, 55, 173
desertification, Chile, 113
diamond-polishing industry, Israel, 175
diamonds, Namib, 182
divergence, 69–71, 79–80, 83, 85, 87–88, 90
drainage, 28, 40; Australia, 190; Chile, 112–113, 123
—soils, xi, 158

economic aspects, development, 28, 29; Australia, 189–203; Chile, 123–135; Israel, 175; Sahara, 167–169
—desalination, 46–52
educational programs, 31–32
Eighty Mile Beach, Australia, 189–190, 202–203
Eilat, Israel, 171–175
energy, 46–49; Chile, 125, 132, 134; Puerto Peñasco, 55–56
environmental aesthetics, 31, 175
environmental degradation, ix, 28, 29, 31, 38, 39; Chile, 113–114
environmental engineering, x, 25, 28, 29, 31, 32

field crops, x, 158; Brazil, 153; Sahara, 163
fisheries, ix, 25, 29; Australia, 190, 193–194, 195, 196, 198; Brazil, 153, 154; Chile, 126, 130–131, 137–146; Israel, 173, 174; Namib, 180, 181, 182, 185
floods, 25, 27, 41, 115, 149
fog deserts, 6; Chile, 117, 132–133; 147; Namib, 178
food supply, ix, 26, 55–61
fruit crops, x, 30, 158; Australia, 194–195; Brazil, 153–154; Sahara, 162

geomorphology, Australia, 197; Chile, 111–112, 114, 115; Israel, 172; Namib, 178–179; Peru, 109–110; Sahara, 159–165

geothermal power, Chile, 134
grazing, 42; Australia, 194, 196, 198, 199; Brazil, 154; Namib, 183–184, 185–186
greenhouse agriculture, 3, 51–61, 133
groundwater, Chile, 112, 114; eastern hemisphere, 157–158; Peru, 109–110; Sahara, 161–165
guano, Namib, 183
Gulf of California, Puerto Peñasco, 3, 55–60, 133

Humboldt current, 111, 114, 144, 147, 148, 151
humidity: Chile, 112, 116–118, 120; Namib, 178; Peru, 98
hydrogeology, Peru, 109

industrial development, x, 25–32; Israel, 173, 175. *See also* fisheries; mineral resources development, petroleum resources
—nitrate deposits, Chile, 111, 112, 113,114, 125–126; oil, Australia, 196–197, 203; salt, Australia, 194, 195, 196, 198, 201; salt, Brazil, 153; wax, carnauba, Brazil, 154
insecticides, restrictions, x
inversion: Chile, 111, 117–119; Peru, 92–98, 100–102, 104
Iquique, Chile, fish-meal industry, 137–146
irrigation agriculture, x–xi, 26–28, 30, 32, 40, 42, 50–53, 55, 58–61, 129, 158, 162, 172
—field crops, x, 130, 162; flowers, 30; orchards, x, 30, 130, 162; vegetables, 30, 51, 55, 58–60, 130, 162
irrigation practices, 28, 32, 40, 50
islands, 63–66
Israel, desalination plants, 45–46, 48, 50–51, 173; Eilat, 171–175

Kuwait, population, 33–34, 36

Land use planning, 29, 31; Sahara, 159–165
Livestock, Australia, 194, 196, 198, 199, 203; Brazil, 154; Chile, 130

Marine terraces, 149
Mexico
—population, 34–36; desalination plant, Puerto Peñasco, 3, 55–60, 133
mineral resources development, ix, 27, 28, 29; Australia, 190, 197–202; Chile, 128–129, 131, 135, 137; Israel, 174; Namib, 182–183; Sahara, 167
multiple-purpose projects, x, 31

Namib, 177–186
—climate, 178; fisheries, 180, 181, 182; landforms, 178–179; mining, 182–183; natural resources, 182, 186; ports, 180–181; settlements, 179–182; tourism, 184–185; transportation, 184
El Niño phenomenon, 25, 106, 131, 144
nomadism, 38; Namib, 183; Sahara, 162–165, 167
nuclear power plants, 46–50

oases: Chile, 129; Sahara, 162–165, 167–169
orography, 25, 76, 83, 87, 103, 105, 119, 120

Pakistan, population, 33, 35
Pampa del Tamarugal, 118–120, 123, 126, 132, 137
Peru, x–xi

—coastal climate, 91–106; littoral geomorphology, 109–110; Mallares Farm, x–xi; population, 33–35; Rio Chira, x–xi
petroleum refinery, Israel, 175
petroleum resources, ix; Sahara, 167
piedmont plains, Chile, 111–113, 115, 118–120; Peru, 109
plant growth, 59
Pleistocene Epoch, Chile, 148, 149, 150
Pliocene Epoch, Chile, 147–148
political aspects, 40, 41, 46, 52
population growth in coastal deserts, ix, 26, 27, 33–36, 39, 55; Australia, 190, 193, 198–203; Chile, 137, 142; Namib, 181; Sahara, 167
ports: Australia, 191, 195, 197, 199–202; Chile, 126, 131–132, 137, 139, 143, 145; Israel, 175; Namib, 180–182, 185
precipitation, 27, 67–71; Australia, 190; Caribbean, 75–78, 87–88; Chile, 118, 120, 149; Peru, 98, 103
Prosopis tamarugo, Chile, 113, 115, 130
Puerto Peñasco, Mexico, desalination plant, 3, 55–60, 133
pumping costs, 50–51, 173

Quaternary Period, Chile, 147–151
quebradas, Chile, 113, 115, 148

radiosonde records, Peru, 91, 94–95
reclamation of coastal deserts, eastern hemisphere, 157–158
relict forests, Chile, 148–150
remote sensing, 13–24
—Apollo and Gemini flights, 15–18; artificial satellites, 13–24
resort areas, 29–30, 31; Baja California, 4–5; Eilat, Israel, 174–175; Mediterranean, ix, 4; Namib, 185, 186
river systems: Australia, 190, 193, 197, 201; Chile, 123, 132; Peru, x–xi, 109

Sahara, northern littoral, 4
—climate, 159–162; geomorphology, 159–162; land use, 162–165; oases, 167–169
salinity, 40
salt flats: Chile, 112–113, 115, 123; Sahara, 162
Salton Sea, 17–20
sand dunes, 157–158, 181; Chile, 149; Sahara, 159
Sechura Desert, Peru, 109
sedentarization, 38
semiarid lands, development, 30–31; Saharan coastal steppe, 159
settlements, 28; Australia, 194–203; Caribbean, 63–66; Chile, 129–130; Namib, 179–182; Sahara, 167–168
social aspects, 37–38, 39, 63–66; Chile, 137–146; Sahara, 167–169
soils of coastal deserts, 27, 179; Australia, 190; Chile, 112, 149
—sand dunes, 157–158; solonchaks, 157
solar energy: Chile, 134, 135; Soviet Union, 132
solonchaks, 157
South American coastal deserts, 25. *See also,* Chile; Peru
—climate, 67–106
South West Africa, 177–186
subsidence (air), 68–71, 81, 85–88, 92, 95–96, 119

Index

temperature: Chile, 116, 118; Peru, 99, 104
thermo-tidal winds, 71–72
trade winds, 83, 92–95
—inversion, 92, 95–97, 104–106
transportation, ix; Australia, 191, 193, 195, 196, 198, 200, 202, 203; Israel, 175; Namib, 184
"trench farming," USSR, 158
turbulence, thermal, 69, 83

urbanization of coastal deserts, ix, 28, 29, 31, 33–36, 38; Chile, 131–132; Namib, 181–182

vegetable crops, 30, 51, 55, 58–60, 158
vegetation, Chile, 113–114; Namib, 179; Sahara, 159, 162

water allocation (policy), 30, 37
water budgeting, 40
water costs, 30, 40–41, 49–53
water distribution (applied), 50
—delivery costs, 51
water management (applied), 38, 43
water resources development, 38, 41; Chile, 132–135
water shortage, ix, 27, 30, 37, 40
water storage, ix, 50
water supply, 37, 39; Australia, 194, 197, 198, 200, 201; Israel, 173; Namib, 179, 181
water use, 38, 39
—agricultural use, 26, 30, 32; Peru, x–xi; Sahara, 162–165, 168
wildlife, Namib, 185, 186
—extinct, Chile, 148
winds, 71–72; South America, 78–83, 86. *See also* thermo-tidal winds; trade winds